Human
Biology

Human Biology

Third Edition

An introduction to human evolution, variation, growth, and adaptability

G.A. HARRISON

Department of Biological Anthropology, University of Oxford

J.M. TANNER

Department of Growth and Development, Institute of Child Health, University of London

D.R. PILBEAM

Department of Anthropology, Harvard University

P.T. BAKER

Department of Anthropology, Pennsylvania State University

Oxford   New York   Tokyo
OXFORD UNIVERSITY PRESS
1988

Oxford University Press, Walton Street, Oxford OX2 6DP

Oxford New York Toronto
Delhi Bombay Calcutta Madras Karachi
Petaling Jaya Singapore Hong Kong Tokyo
Nairobi Dar es Salaam Cape Town
Melbourne Auckland

and associated companies in
Beirut Berlin Ibadan Nicosia

Oxford is a trade mark of Oxford University Press

British Library Cataloging in Publication Data
Human Biology : an introduction to human evolution, variation, growth and ecology. — 3rd ed.
1. Human physiology
I. Harrison, G. A. (Geoffrey Ainsworth)
612   QP34.5
ISBN 0-19-854144-9
ISBN 0-19-854143-0 Pbk

Library of Congress Cataloging in Publication Data
Human biology.
Includes bibliographies and indexes.
1. Physical anthropology. 2. Human biology.
I. Harrison, G. A. (Geoffrey Ainsworth), 1927-
[DNLM: 1. Evolution. 2. Genetics, Medical. 3. Physiology.
4. Variation (Genetics) GN 281 H916]
GN60.H83 1987 573 87-7716
ISBN 0-19-854144-9
ISBN 0-19-854143-0 (pbk.)

Set by
Burns and Smith, Derby

Printed and bound in Great Britain by
Biddles Ltd, Guildford and King's Lynn

Foreword to the first edition P. B. MEDAWAR

What is 'human biology'? Is it just an attempt to market old and familiar goods under a new name or does it indeed stand for a new conception of the natural history of mankind?

'Conception' is, I think, the right word; for human biology is not so much a discipline as a certain attitude of mind towards the most interesting and important of animals. Human biology portrays mankind on the canvas that serves also for other living things. It is about men rather than man; about their origin, evolution, and geographical deployment; about the growth of human populations and their structure in space and time; about human development and all that it entails of change of size and shape. Human biology deals with human heredity, the human genetical system, and the nature and import of the inborn differences between individuals; with human ecology and physiology, and with the devices by which men have met the challenges of enemies and of hostile environments. Human biology deals also with human behaviour — not with its wayward variations from one individual to another, but rather with the history and significance of, for example, family life; or of love, play, showing off, and real or sham aggression. Finally, and most important — because most distinctively human — it must expound and explain the nature, origin, and development of communication between human beings and the non-genetical system of heredity founded upon it.

Man (if I may relapse into universals) is a very proper study for biologists, particularly for beginners in biology. Many of the so-called 'principles' of biology can be taught as well upon a human text as upon any other, and some can be taught better. Man is the great amateur among animals. Where other animals have specialized in ways that commit them to one or another particular and restricted way of life, human beings have retained their amateur status. As nearly as any animals can be, they are biologically uncommitted, disengaged. As mammals go we are quite simple creatures, without any very noteworthy anatomical (or genetical or developmental) singularity not already foreshadowed by lower primates. Binocular vision and the crossing-over of the optic tracts, the enlargement of the cerebral cortex, the liberation of the forelimbs for manipulation, the prolongation of childhood — these are not innovations of mankind, but exaggerations of tendencies already to be seen in apes. As it happens human beings, considered as a biological enterprise, succeeded — but only just, and only quite recently. People forget or do not realize how precarious was the hold of human beings upon the earth for their first many tens of thousands of years. One way or another the biologist has much to learn from human biology, and chiefly this lesson: that the distinction he has been wont to draw between Nature, on the one hand, remote and wild, and, on the other hand, Man and his works, is one that damages his understanding of both.

Human biology is also a very proper study for medical students and for those who practise medicine. The 'biological principles' bandied about by an earlier or older generation of physicians are, most of them, nonsense — among them the deep-seated and all but ineradicable belief that natural dispositions and adaptations are well-nigh perfect, and that sickness and other disabilities are part of a long-drawn-out expiation for leaving nature and leading unnatural lives. A case can be made for thinking out medical education anew and rebuilding it upon a foundation of human biology. Certainly medical education has defects, of both a medical and an educational character, that cannot be made good by mere juggling with syllabuses. How much of what is to be read in the following pages can be found in a medical syllabus today? A good deal, somewhere or other, as it happens, but not where it ought to be, at the *beginning* of the medical course, where plenty of room could be found for it if only the dogfish and the earthworm could be given their *congé* after so many years of undistinguished service to the medical profession. As matters stand, human genetics makes its appearance, if at all, as a supernumerary course fitted somewhere into the clinical years; demography is mixed up with sanitation, and human ecology is treated as something which, though it may help us to understand the medical predicaments of foreigners, is barely relevant to our own cosy domestic medical scene.

For all these various reasons I believe that this excellent text, by four authors, who are pioneers of human biology and who have made distinguished contributions to the subjects they expound, will have an important influence on the development of all the individual sciences that deal with mankind.

Preface to the third edition

The aims of the present edition of *Human biology* remain those of the original 1964 edition: to synthesize, at an introductory level, present knowledge of the biological organization of past and present human populations. The success of the original venture led to a second, updated edition in 1977; since this was prepared, however, we have suffered the loss of two of the original authors, J. S. Weiner and N. A. Barnicot. Their deaths have been a profound loss to the whole discipline of human biology and the two remaining authors initially feared that without them it would scarcely be possible to continue this book. However, there is still considerable demand for the work and it does appear to occupy a unique place in the development of human biology as an academic discipline. When, therefore, the two new authors willingly accepted an invitation to help update the book, we set about preparing a third edition.

The structure of this edition very much follows that of its two predecessors except that the sections on human genetics and human variation have been combined. Developments in the field and change of authorship did however necessitate considerable reorganization and rewriting. The sections on Human evolution and Human adaptability are essentially new, and those on Human genetics and variation and Human growth and constitution greatly modified (though many items originally written by N. A. Barnicot on variation are preserved). Human adaptability has been re-titled (from Human ecology) because it concentrates more on the substantial progress which has been made in recent times on how our species relates to its natural and cultural environments and particularly on the adaptation processes. A serious attempt has been made to keep the book to around its former size, so that including new material has meant the removal of some old sections. Thus, for example, former chapters dealing with familial genetics have been greatly compressed in the belief that those using this book will already be familiar with basic genetic principles.

We very much hope that the book will continue to be valued by our friends, colleagues, and students worldwide. We wish again to record our thanks to the staff of Oxford University Press for their continual help and encouragements. Acknowledgement is also due to the various authors and publishers (mentioned individually in the text) who kindly allowed reproduction of figures.

G.A.H.   J.M.T.

D.R.P.   P.J.B.

1987

Contents

Part II · Human genetics and variation
G. A. HARRISON

Part III · Human growth and constitution

J. M. TANNER

Part IV · Human adaptability

P. T. BAKER

Part I

Human evolution

D.R. PILBEAM

1 Studying human evolution

Introduction

How did human beings, *Homo sapiens*, come to be so different from their animal relatives? This question is clearly composite. To answer it we need first, a list of differences between humans and non-humans which sets the agenda for any scientific study of human origins (and this list is explained fully in subsequent sections of this book). Secondly, we would like as far as possible to identify and describe the extinct species that are relevant to understanding the evolution of *H. sapiens*. Thirdly, these species must be linked in ancestor–descendant sequence. Finally, each species in which we are interested must be 'fleshed out' or reconstructed, as much as possible as though it were alive and functioning in the world today.

When scientists try to explain the course of human evolution rather than simply describe, they are faced with a problem. Humans are unique. Only the single human species, *H. sapiens*, survives out of all the hominids that lived in the past (hominids are human ancestors or close relatives that existed since the divergence from apes). However, a unique result implies unique events and unique events suggest unique causes. Unique causes by their very nature cannot be repeated and the only way scientists can explain such unique events is to break down the problem into smaller components for which understandable linkages of cause and effect can be invoked. These components should be more general; for example, they may be seen to have operated in the evolutionary histories of other animals, or animal groups may be found in which similar causes and effects are operating at present. Hence, human evolution must always be studied with a broader, comparative perspective.

It is worth remembering that if we were not studying ourselves, we probably would not bother too much with hominids. A visiting Martian zoologist would be unimpressed by a mammal group which had only one surviving member. If she observed earth only 10 000 years ago (a mere eyeblink in the millions of years of human evolution), she would have seen humans as hunters and gatherers who were then a relatively minor part of the total animal population, and who still had little impact on their surroundings and environment.

Yet we humans are understandably fascinated by our history, and we greet every new fossil or other relevant discovery with interest and enthusiasm. We know much more than we did even a decade ago about the events of human evolution. The fossil record grows rapidly, for hominids as well as for the plants and animals the hominids rubbed shoulders with. Our ability to date the past is now at a very sophisticated level. Ways of analysing the fossil record to extract interesting and important information have improved because of two factors: better ways of extracting biologically useful information from fossils, to a considerable extent due to the expansion of 'technological' analyses of various kinds; and a better understanding of living biological systems. Comprehending the present is an essential step in the process of understanding the past, because we need to understand the links between ecology, behaviour, function, and morphology in order to make sense of the information we have about the past.

Important human attributes

Exactly which human attributes are important out of thousands that can potentially be listed has been debated for hundreds of years. It has generally been assumed that some subset of the differences between humans and non-humans accounted for our success — indeed, did so almost self-evidently — and that identifying the differences essentially explained their origin. Both assumptions are false. Human evolution cannot be considered a self-evident success story; and the biological role of features generally changes in evolutionary time.

TABLE 1.1.
Characteristic human features

1. Usually non-forested habitats
2. Relatively very large brain
3. Slow maturation; female ovulation concealed, no obvious oestrus
4. Omnivorous; food being hunted, gathered, transported, shared, prepared, stored
5. Bipedal; central place foraging
6. Complex cultural behaviour, including language
7. Social organization built around marriage; prolonged infant care
8. Tool-making, technological skill, dependence on equipment

However, palaeoanthropologists are now generally agreed on which important features of humans need explanation (Table 1.1). First, our possession of an enormous brain — at 1400 cm^3 three times the brain volume of apes similar in body size to us (and the apes themselves already have relatively large brains). Secondly, human behaviour is controlled and patterned in a unique way, largely by its use of symbols (culture). Symbolic capabilities enable human behaviour to be versatile and flexible. Modern humans can adapt behaviourally to changing circumstances at rates impossible for non-humans. The most obvious examples of our symbolic abilities are spoken and written languages which are peculiarly human attributes. Much evidence links both size and internal organization of our brains to these language abilities. The volume of information stored and transmitted by human languages, including computerized and other non-verbal ones, is extraordinary and growing exponentially.

The third feature is the technological expertise of modern humans. We are totally dependent for survival on our technological abilities and tools, be they digging sticks and stone hammers, or microchips and laser probes. Our current range of toolmaking skill is clearly a reflection of our large and complex brains, and of the sophisticated links between eyes, brain, and our skilful hands. Fourthly, our range of foods is wide and includes animals as well as the plants which are more typical in non-human primate diets. There is good evidence that until about 10 000 years ago all groups of *H. sapiens* obtained their food by hunting game and gathering wild plants ('food gathering'). Language and relatively simple tools were integral parts of this behaviour pattern, from its planning through to processing, storage, and distribution of the food. One critical aspect of human behaviour is sharing, and the sharing of food within and especially beyond the confines of the family has been particularly important.

Fifthly, humans use space in an interestingly different way from their closest relatives, the apes. They forage from a central place, be it a hunting campsite in the African bush, a Moorish castle, a fancy New York apartment, or a modest dwelling house: dispersing from and returning to a fixed place after the day's tasks. The home base is a place for sharing, cooking, and processing food, for toolmaking, often for sex, and for a whole range of social activities. Humans generally disperse to find food (or paid work with which to provide it), but aggregate to prepare and share it.

Sixthly, human social groups are organized in quite specific ways, in families. Family structures are highly variable, from the monogamous nuclear family to polygamous extended families. Whether they include one or several wives or husbands, families are formally composed following the culturally determined rules governing marriages, and perhaps the best way of describing that which is uniquely human about family structure is to use the term 'marriage' rather than one such as 'pair bonding'. Marriage is a child-raising system, involving a complex of economic, reproductive, and other factors. Modern human marriage systems are highly variable. Marriage is as characteristic of humans as the possession of a large brain or the use of a language, but it is only possible because language permits us the conceptual and symbolic ability to arrange, promote, and maintain it. Relationships between individuals, and also between social groups, are a complicated amalgam of the nice and the nasty, the co-operative and the competitive, the gracious and the aggressive. Modern societies are complex and highly structured; power and wealth are seldom, if ever, evenly distributed, and their transmission is governed by rules only made possible by language.

There are a few basically anatomical features that distinguish *H. sapiens*. Humans are upright, walking habitually on two limbs and not on four. This bipedalism freed the hands of early hominids. Limb proportions, joint structure, the anatomy of the hands, feet, pelvis, and spine all bespeak the pervasive nature of bipedal adaptations. We have an odd distribution of hair on our bodies. We sweat abundantly, and can survive heat only if we drink water at regular intervals. Despite the diversity of human cultural behaviour, modern human groups are genetically very similar to one another. Perhaps the most important characteristic of human biology is that we have evolved a genetic structure which permits an apparently endless variety of cultural behaviours. This genetic structure has existed at least since modern humans appeared in the fossil record, and that is for at least 40 000 years.

How the past has been used

Why do we study the past, be it history or prehistory? For some students of human evolution or of history the answer would run something like this. Only by understanding the past can we truly understand the present and have some notion of the way the future might be. We adopt such explanations when we use the apparent aggressiveness of our ape-like ancestors to account for community violence, yet stress the co-operative facets of ape behaviour to explain our more agreeable features. In fact, reconstructions of the past have often been projections of the present, consciously or unconsciously reflecting the times in which they were made and the cultures from which they arose. The story of human evolution, as generally told, bears strong resemblances to a traditional hero or success story. Humanity had humble beginnings, we were tested in various ways, and after enduring trials and privations, we triumphed. We are the star, and also the author, director, and producer of this production.

Stories of human emergence tend to be put together in similar ways, and a few of these common features are worth reviewing here. First, a particular human characteristic is picked out as being our ultimate characteristic. The same characteristic is then isolated in our remote ancestor as the prime mover, if not the sole cause, of our progress. For example, at the turn of the century, the large human brain was the quintessence of our humanity, and the cause of our initial separation from the apes. Evidence now at hand suggests that this is not so. Prime movers, like fashions, change and generally tell us more about the times and prejudices of particular palaeoanthropologists than they do about the causes and events of human evolution.

A further problem with nearly all attempts at reconstructing our beginnings is that they assume, tacitly if not explicitly, that the function of a feature does not change through time. However, this is rarely the case in evolution. Human fingers that are nimble and talented enough to play the violin obviously did not originally evolve to do so; they were perhaps selected so that their owners could climb trees, pick and peel fruit, make simple tools, and groom fellow primates. These were the selection pressures which moulded the structures which can quite fortuitously play the musician's violin and wield the surgeon's scalpel. Reconstructing the past with these misleading assumptions inevitably makes it seem more modern and more like the present than it actually was. Many accounts of human evolution also smack of inevitability. Of course, earlier hominids evolved into modern humans. What other goal could there have been than ourselves? Any prehuman with just a few of our traits would inexorably evolve into us — all roads lead, not to Rome, but to humanity. How often have you wondered which species will become like us and take our place if we were to become extinct? Buried in those musings is the assumption that there is something inevitable and quintessential about becoming human.

If you approach the past with the perspective that it is merely a way station to the present; if the past is studied solely as though it were an eye in which the present is a gleam, then almost inevitably the past comes to resemble the present. However, if what happened was not inevitable and if we accept that the past should be studied for its own sake and on its own terms, then we could begin to tell a story which sees each step as uniquely important; a story in which each step does not lead inexorably to the next. *Homo erectus* did not know it would evolve into *H. sapiens*. *Homo erectus* was in a fundamentally important sense living for itself, in its 'own' present. Most of us now agree that there was no inevitability to becoming human, and that it is essential to approach the past as though one were studying a kind of present, but a present that was different from ours.

How the past is reconstructed

Given all these caveats how should the student of human evolution proceed? Imagine that a time machine has transported you two million years back into the past and that you are sitting with a group of extinct hominids. Which questions would you be asking? What would be on your 'check list' of data to collect? Surely, they would be the same questions you would ask if you were studying chimps or humans or hunting dogs in their natural habitat today (Table 1.2).

You would probably start by asking about the ecology of hominids. Which environments and habitats do they favour, and what is their geographical distribution? Then you would ask questions about the animals themselves. How big are the males and females? How long are their pregnancies? At what age do they first reproduce? How long do they live? How do they move about? How much time is spent on the ground, and how much in the trees? How far do individuals range? Which foods do they eat? How do they procure and process food, and do they store or share it? How do the sexes behave among themselves and to each other? Are groups made up of related females, or related males? Are relationships among males generally tolerant or are they aggressive? Are individual males and females bonded in some way? How do social groups behave toward each other? Are tools used, and for what purposes? Are they simple modifications of natural objects, or complexly manufactured implements? How and to what extent do individuals communicate? Are there ways of storing and transmitting information?

TABLE 1.2
List of critical behavioural–ecological attributes

1.	Ecological variables
2.	Body size and dimorphism
3.	Brain size
4.	Life-history patterns
5.	Diet and subsistence behaviour
6.	Positional and ranging behaviour
7.	Male–male relationships
8.	Female–female relationships
9.	Male–female relationships
10.	Adult–infant relationships
11.	Intergroup relationships
12.	Communicative behaviour
13.	Technological behaviour

These are some of the questions we would like to answer, but of course, for fossil species, many of them can only be answered in the most general terms, and some cannot be answered at all. We must be honest about which questions we simply do not have enough evidence to answer, but we should use the attribute list as a goal towards which our ingenuity leads us.

While we are attempting to reconstruct fossil species, and trying to picture them as though they were alive today, we are at the same time trying to say something about their evolutionary relationships. We want to know which species is ancestral to which and to assess this we need to make judgements about the relative importance of the similarities between fossil species. Similarities are only useful for determining relationships if they are special in certain ways. They have to be newly evolved features that some species share, and others lack. There is, of course, a general relationship between geological age and degree of primitiveness; the older a species, the more primitive it is likely to be, but this is only a general relationship. Ultimately, it is the analysis of anatomical features that tells us how two species are related.

This then is the agenda: a list of human differences to explain; the ability to link extinct species in such a way as to outline the evolutionary road which led to modern humans; and an approach to the past which strives to make its reconstruction as dynamic as possible.

2   Patterns of evolution

Species and niches

Life began more than 3000 million years ago on an earth which came into existence at least 1000 million years earlier. Life is obviously not a continuum; living matter is subdivided into packages known as species. Species are natural populations of individuals which are genetically similar enough to be capable (or potentially capable) of interbreeding with the production of fully fertile offspring. At least 2 million species survive today, but that is less than 0.1 per cent of all species that have existed since the dawn of life. The living world is abundantly populated with species and these species are highly diverse. Each is finely tuned to its environment and has morphological, physiological, or behavioural features which adapt it to its niche. This pattern exists because over immense periods of time species have split, evolved, and diversified to produce the wonderfully adapted world we observe. Evolutionary theory attempts to explain how these patterns came about; Charles Darwin is the person who first proposed a plausible mechanistic and causal explanation for how evolution had occurred. He adduced a great mass of supporting information from a variety of sources: comparative anatomy, embryology, the fossil record, and the distribution and diversity of living species.

Each species is part of a complex network or ecosystem made up of many plant and animal species — the community — and all communities have structure. Ecosystems can be visualized as comprising several levels: a primary level of plants which convert solar energy and nutrients into living cells; a secondary level of herbivores which consume plants; and a tertiary set of carnivores which eat herbivores. These levels are also called trophic levels. Energy flows through the system, but only about 10 per cent transfers from one trophic level to the next.

At its most basic level the structure of communities reflects each species' ecological niche. No two species have the same niche, and even very similar species will differ in at least one important feature of niche structure. The concept of niche is complex and perhaps best considered metaphorically. Imagine species as filling roles in a community, defined or recognized by criteria such as habitat preference (arboreal or terrestrial), activity (nocturnal, diurnal), preferred food, climate, and so forth. This role is the species' niche, reflecting its specialized relationships to food, physical factors, competitors, and so forth. More formally, it can be conceived as a multidimensional hypervolume, with the above criteria as axes, in which a population does (realized niche) or theoretically could (potential niche) exist. There is some debate over the extent to which competition between species for critical resources plays a role in shaping community structure by influencing niche size. At least at the carnivore level this is probably quite important.

Darwin and another naturalist, Alfred Russell Wallace, are perhaps best known for the theory of natural selection, which explains how population variation at a single time gets translated into variation — change — through time. They focused attention on the natural variation of individuals within populations and on the uniqueness of individuals. Differences between individuals became the most important and basic of biological facts. Individuals are genetically unique, and this uniqueness is understood now down to the level of DNA. Individuals struggle for survival, to find food and mates (if they reproduce sexually), and to avoid predators. This struggle is integral to Darwin's and Wallace's evolutionary mechanism: natural selection.

Natural selection can be summarized briefly as follows. Individuals within populations differ genetically. Offspring resemble parents genetically more than they resemble non-related individuals. Some individuals survive and reproduce, others don't; some lineages survive, others die out without issue. Some of the individuals or lineages that survive do so because they are more effective in gaining access to critical resources. As a consequence of this differential survival, the genetic structure of the population will change over time, and individuals in the population will be better adapted. Genetically-controlled morphological, physiological, and behavioural characters which promote successful survival will thus be selected for.

Genetic variability between individuals is maintained by several different mechanisms, including recombination and mutation. These mechanisms operate randomly relative to any trajectory of evolutionary change. The editing and shaping hand of natural selection is necessary to generate change in which adaptations appear.

Darwin noted what others before him had: many characters of organisms are very well fitted or designed to promote successful survival in a particular environment. Such characteristics are adaptations. In colloquial language we often use the term to describe a physiological response, like an athlete's lowered pulse rate, and both pre-Darwinians and 'creationists' believe that adaptations are evidence of God's design. Evolutionary biologists use adaptation in a restricted sense: adaptations are characters shaped by natural selection to play a particular role or fulfil a particular function.

Gibbon hands with their long and curved fingers and metacarpals are adapted for aboreal swinging, climbing, and brachiating; human feet have longitudinal and transverse arches, with big toe parallel to the others, as adaptions to bipedalism; the composition of the milk of various animal species, especially protein and fat content, is correlated with their neonatal growth rates. Adaptations are inherited characters, designed or shaped by selection to maintain or improve survival and reproduction. ('Adaptation' can also apply to the process of becoming adapted.) What an adaptation does is its function.

However, it seems probable that not all characters are adaptations or are adapted at a particular time. 'Adaptation' implies the notion of shaping or design, by natural selection, for a particular current function. Since this shaping happened in the past, the implication is that the use of the character hasn't changed over time, and that its current utility is the same as the original utility as shaped by natural selection. In turn, this implies that the historical genesis is well known. However, there are situations for which we can quite plausibly say that utility has changed, and that current use differs from original use. Thus, we can use our hands now to perform many activities because we are bipedal, but few if any of these activities would have been involved in the original shift to hominid bipedality. A useful character not built by natural selection for its current role is an exaptation. This is a useful concept and enables us to label and so to think about a phenomenon which must be frequent in evolution, the change in a character's role through time. It has been suggested that, as the operation of an adaptation is referred to as its function, the operation of an exaptation should be called its effect.

The term 'preadaptation' is often used to describe a situation that is frequent in evolution: a character evolved for use in one situation subsequently comes to be used in another different situation. Exaptation is a better term to use in describing the later utility, and if the feature has to be mentioned in its original condition with reference to subsequent usage, 'preaptation' would be a better term than preadaptation.

In addition to exaptations, some characters may have no functions or effects or never have had, and these can be termed non-aptations. A number of mechanisms are well understood which would allow gene frequencies to change without the action of natural selection. These include random fixation of traits because of small breeding group size, developmental side-effects of genes selected for other reasons, the accumulation of mutations because the DNA is unexpressed or because no amino acid change results, and the accumulation of 'neutral' amino acid changes which apparently do not affect protein function. Not all evolutionary change — changes in genes and shifts in gene frequencies — is adaptive. Some is maladaptive. Change is constrained by history, by the past genetic and phenotypic structures of the species, and at least in the case of morphology and behaviour many characters are compromises. A particular anatomical feature might function in a variety of behaviours, and for none of them would it be 'optimally' designed. This sometimes makes the description and explanation of evolutionary change difficult. However, it also makes it challenging.

Macroevolution

Patterns of evolutionary change above the species level are usually termed macroevolutionary in contrast to the finer-scale microevolutionary patterns associated with selection and so-called non-Darwinian mechanisms of genetic change. The generation of new species occurs in two ways: by non-abrupt change, perhaps of varying rate, within a lineage (a succession of species populations through time) so that, given sufficient time, enough change accumulates to warrant a new species name (phyletic evolution, anagenesis, progressive evolution, lineage evolution). Or lineages may split and produce two new species (cladogenesis, splitting, speciation). A lineage can also exhibit no change (stasis, equilibrium), and it can also terminate and become extinct, which is by far the commonest fate of lineages.

The way in which new species are produced is a hotly debated topic. Does it occur mostly by the gradual accumulation of adaptive genetic changes within a lineage? Or do most new species arise by splitting, either with populations separated geographically (allopatric) or within the same area (sympatric)? In this second case, does genetic change arise in small adaptive increments, or through 'macromutations' — genetic changes which produce major anatomical, physiological, or behavioural changes? The brief answer is that we do not know, although plausible models exist for these several and rather different modes. Instances where speciation can actually be documented in the fossil record are extremely rare. Discussions of the origin of particular hominid species, for example, are actually statements about ancestors and their inferred descendants (one of which may be hypothetical) and about the 'gap' between them, rather than about how the gap was bridged.

However obscure the production of new species is, it is clear that evolutionary rates do vary, and are sometimes measurable in the sense that a quantifiable morphological feature can be monitored in a lineage over hundreds of thousands or even millions of years. This leads to another controversial area in evolutionary biology: patterns of lineage evolution. There are two principal and different proposed patterns: punctuated equilibrium and phyletic gradualism (Fig. 2.1).

Punctuated equilibrium describes a pattern in which lineage evolution is minimal to zero, and in which the bulk of evolutionary change is therefore concentrated in speciation events. New species are generally believed to arise rapidly in small, peripheral, isolated populations. Morphological gaps in the fossil record between species are in a sense real, because the probability of sampling lineages in transition is extremely low. In phyletic gradualism most lineages change, at varying rates, and new species form both by lineage change and by lineage splitting. Evolutionary change is partitioned between that occurring within lineages and that occurring when lineages divide. Gaps in the fossil record are then likely to be due to the patchy nature of that record.

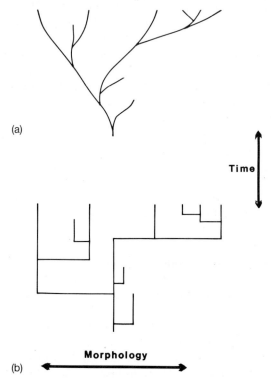

(a)

Time

Morphology

(b)

FIG. 2.1.   Comparison of (a) phyletic gradualism and (b) punctuated equilibrium patterns of evolutionary change. (J. Sept.)

As with modes of speciation, these patterns are hard to demonstrate empirically, and they are in important senses models. Some groups are poorly enough represented that it is unclear which pattern is operating; the primates are, for the most part, such a group. Exactly which pattern applies has important implications for us in trying to explain the changes we see in the fossil record.

When species reach environments or ecosystems which are relatively 'empty' they often undergo rapid diversification into many new species which are able to capitalize on some initial adaptation and, by adding new adaptations, are able to expand into many new niches. A famous example of this are 'Darwin's finches' on the Galapagos Islands. A single finch species initially colonized the islands from the mainland. Niches that were filled on the mainland by competitors were empty on the islands, and new finch species evolved which often filled rather unfinch-like niches: warbler-like forms, cactus-feeders, tree-finches which included a woodpecker-like species. Similarly, the first New World monkey (Ceboidea) to reach South America found environments ideal for arboreal primates, and relatively unexploited. Over the past 35–45 million years the ceboids have diversified in South America and evolved forms ranging widely in body size, positional behaviour, dietary preferences, and social behaviour. Some species came to resemble certain monkeys (cercopithecoids) or apes (hominoids) of the Old World anthropoid primates, the Catarrhini.

These patterns of diversification and exploitation of a basic adaptive pattern are examples of adaptive radiations. Adaptive radiations can be of varying magnitude, from small, recent, and local, to very diverse and complex. It is often the case that radiations in two different areas, starting with different though rather similar species, can end up producing patterns that are in certain ways surprisingly similar. The structures of the Miocene (24–5 Ma) faunas of South America and North America were totally different taxonomically (at the level of individual species), but several South–North pairs showed varying degrees of similarity in feeding habits (inferred from teeth) or positional behaviour (inferred from limbs). Such examples can be multiplied. The New and Old World anthropoid primate radiations contain pairs of species that are — though different overall — similar in certain interesting features. *Ateles*, the South American spider monkey, and *Hylobates*, the East Asian gibbon, are small arboreal creatures which often use their arms to feed and move below branches. Features of the arm and thorax — for example, the shape and articulation of the humeral head, or the scapula — are similar in the two, despite the fact that their distant common ancestor was quite different.

Adaptive radiations occur at various scales, and do so when sufficient ecological 'space' exists for species to radiate into. It now seems that opportunities for radiations are created in two broad ways. Species can enter new areas, as geographical barriers disappear or as a newly adapted species capable of immigrating evolves. Or species can become extinct, to be subsequently replaced by a radiation into 'vacated' ecological space. Such extinctions can occur for a variety of reasons, prime among them being climatic change leading to habitat change. Climates can alter for reasons intrinsic or extrinsic to the earth: intrinsic factors would involve long- or short-term fluctuations in temperature, for example. Recently, it has been suggested that extrinsic factors — impacts from asteroids or meteor showers — can produce sudden and massive climatic deterioration causing major extinctions.

Individual extinctions will have particular causes, though these will rarely be known. The general assumption that long-term faunal evolution is shaped by competition and reflects the rise of superior adaptations may not be true, or may be true only partly. Other essentially random factors may impose a shape on evolution which is in important senses non-deterministic.

Homology and homoplasy

Characteristic similarities between distantly related species in which the common ancestor differed in the character are described as convergences or parallelisms (depending on how distantly or closely related, respectively, the species or groups under consideration are). The single term homoplasy is useful in describing similarity not due to inheritance from a common ancestor possessing the trait. Homology is similarity due to inheritance from an ancestor with the trait. Homologies reflect the genealogical histories of lineages and when properly identified allow us to organize species into evolutionary 'trees' or 'bushes'.

The concepts of homoplasy and homology are important. The study of convergences makes it possible to identify important functions. If two distantly related species evolve to resemble each other in some character, we may be able to infer something about common function. For example, thick enamel is rare among primates, but does occur in several distantly related species, including early hominids, which otherwise differ markedly. A search for some common behavioural feature related to diet might give us clues to the adaptive function of this trait, and this might in turn help to understand the function of thick enamel in the now extinct early hominid, *Australopithecus*.

Not all homologies are, of course, similarities: horse hooves and bat wings are homologous, derived from the five-fingered hands of distant shrew-like ancestors. In an evolutionary and developmental sense, the 'source' component parts have been modified and rearranged to produce radically different patterns. This conservation is frequently found in evolution. Truly 'new' structures are rare, and new configurations are produced by tinkering with old structures.

Thus, the search for homology is important in explaining some evolutionary patterns, particularly the evolution of novelties. For example, it now seems likely that the human brain, with its language-generating powers, is wired in a way that fundamentally resembles that of the non-linguistic macaque monkey. The connections between different parts of the system are similar. However, the relative proportions and the functions of particular parts have changed, making the outputs of the systems very different. As with bat wings and with horse hooves, evolutionary novelties can be viewed as resulting from relatively small adjustments and rearrangements of a basically conservative overall plan.

Phylogenetic reconstruction

How are evolutionary sequences (trees, bushes) actually put together by the practising evolutionary biologist? There are several basic steps. First comes the identification and description of species which involves sorting individual fossil specimens, often broken or fragmentary, into groups that correspond as far as possible to real species. Morphological patterns and ranges in living species are standards for the arrangement of fossils. Next is the recognition of relationships among species: which are closely related sharing a more recent common ancestor; which are more distantly related, with a more ancient ancestor. These relationships need to be inferred ultimately for living species and for fossil species together, although it might often be easier in particular cases to start with living forms. The determination of relationships depends on the interpretation of similarities and differences: morphological, physiological, genetical, and behavioural, with the aim of ultimately deciphering the genetic changes that occurred during the evolution of the group under study.

Once relationships are understood they can be described in a cladogram, a diagram which shows degrees of relationship via a hierarchic branching network. Then geological time is added, and the network is transformed into a phylogeny, a pattern of species linked as ancestors, descendants, and collaterals. Finally, comes interpretation and explanation of why particular adaptations appeared and evolved. This is often called a scenario.

Let us look more closely at some of these steps. Even with relatively complete fossil material there are still many problems in grouping individual specimens into species. The only criterion available for recognizing species in the fossil record is morphological similarity, and while the interbreeding criterion is, in practice, rarely used for living species, it is available in principle. While members of living species are indeed phenotypically similar, one can never be absolutely sure that a fossil 'species', defined morphologically, does not include several very similar species, or that two fossil 'species' might not be males and females of a single dimorphic species.

A variety of statistical and morphological procedures have been developed to assist in making decisions of this kind. Ultimately, the procedure depends on setting up criteria for species recognition, based on known patterns of variability in the living world, and applying these criteria to extinct forms. In turn, this rests on the uniformitarian assumption that physical, chemical, and physiological principles are invariant in time, even though past configurations might differ in many ways from extant ones. As we shall see there are a number of practical problems with the application of this very reasonable assumption.

Once we have done our basic sorting into species we can begin to look at relationships by examining patterns of similarity and difference between species. To make comparisons we need to think of individuals and species as being composed of characters: anatomical (to the level of DNA), physiological, or behavioural features of an organism. Preferably, characters are inherited and simple to define, recognize, and analyse functionally, developmentally, and genetically. (We would ideally like to be able to track genetic change through time, but we can only monitor it in fossils by using morphology, at best an approximate mirror.)

There are several approaches to the inference of relationships. One used widely by palaeontologists involves looking at samples that are closely spaced stratigraphically, and therefore (normally) temporally, and linking groups together by assessing overall similarity. Overall similarity is often termed phenetic similarity and this procedure has been called 'stratophenetic'. In general, there will be correspondence between degree of relatedness and some kind of general similarity, but not always. An alternative procedure (actually palaeontologists frequently use both approaches) is one of phylogenetic analysis, or cladism (Fig. 2.2). The aim of this approach is to distinguish between various kinds of similarity. A couple of examples will be helpful.

Suppose we have three species: a bird, a bat, and a monkey. We wish to know which two are most closely related (form a 'monophyletic group') and which one is most distantly related (is the 'outgroup'). This is the three taxon problem (a taxon, plural taxa, is any unit in the classificatory hierarchy) worked with species from three very different groups. We know, in fact, that birds are the outgroup; bats and monkeys are both placental mammals and are more closely related. Birds and bats have wings, but their wings are not homologous. Both developmental and palaeontological information tells us clearly that the similarity is a homoplasy. Bats differ from monkeys; bats have wings, monkeys hands and feet, and these differences contribute no information to understanding their relationships. Both have fur, which birds lack, and which we infer was present in the bat–monkey ancestor. It is a homology, and this similarity is the critical one in linking monkey and bat in the monophyletic group.

However, not all homologous similarities tell us about relationships. Take another three taxon example, this time with opossum, horse, and monkey. The opossum is a marsupial while the other two are placentals. Opossum and monkey have five-fingered hands, while the horse's forefoot has only one. The five-fingered condition is clearly homologous yet it does not give us correct information about the monophyletic group (horse plus monkey). It is a feature present in the ancient ancestor of all three taxa, retained in the later horse–monkey ancestor, but subsequently modified uniquely in the horse lineage. To uncover the horse–monkey link we need a homologous character present in their ancestor only, and not in the ancestor of all three species. Uterine morphology or reproductive behaviour would be two such characters.

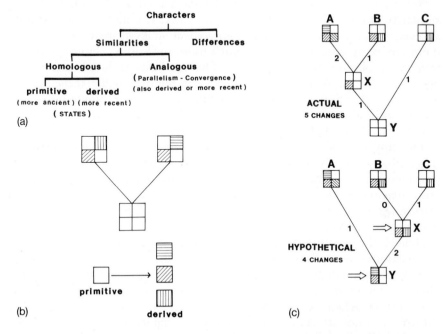

FIG. 2.2. Phylogenetic (cladistic) analysis. Similarities can be analogous (homoplasic), homologous primitive, or homologous derived (a). Primitive and derived character states are shown hypothetically in (b). In (c), top diagram shows an evolutionary sequence in which ancestors X and Y are known. A total of five character state changes occurs: A and B are 50 per cent similar, 25 per cent in shared primitive (open square) and 25 per cent in shared derived (cross hatch). Lower diagram of (c) shows situation when X and Y are unknown and then reconstructed hypothetically. B and C are grouped together because they are 75 per cent similar; a total of only four character changes is inferred. However, the similarity is 50 per cent shared primitive and 25 per cent homoplasic. Grouping using phenetic or overall similarity gives an incorrect solution, despite fewer state changes. (After Professor R. D. Martin.)

Before we discuss how to partition similarities into those which are useful in phylogenetic reconstruction and those which are not, we need a little more terminology. A characteristic occurring earlier in evolution is termed 'primitive'; one appearing later is 'derived'. Within a group of species, characters resembling the common ancestor more closely are relatively primitive while others are more or less derived in proportion to their departure from the ancestral condition. Suppose we treated overall morphology of the pelvis as a character (it is actually better analysed as several), we could describe the pelvis in hominoids as exhibiting two character states: deep as in apes, broad as in hominids. Using our definitions, the deep form of the ape pelvis is primitive, the hominid shape derived.

Return now to our examples, and consider kinds of similarity. The similarity of bird and bat wings is a homoplasy. It is an obvious example, as closer morphological or palaeontological analysis reveals, although there are many others which are much more misleading. Homoplasy is a frequent evolutionary phenomenon. The similarity in hand structure of opossum and monkey is a homology, but relative to the three species under consideration the feature is primitive. Neither homoplasies nor shared primitive features help in deciding which two taxa are a monophyletic pair.

The only kind of useful similarity is that due to the sharing of relatively (it's always relative) derived character states. For example, the reproductive similarities of horses and monkeys, relative to the marsupial opossum, are derived (hence 'shared-derived'). The other kind of derived feature we noted, the uniquely derived horse's hoof, is of course useless in assessing relationships.

Hence, the critique of the use of overall similarity is that phenetic comparisons use similarity due to homoplasy and shared-primitive states, as well as the phylogenetically useful shared-derived conditions, and that phenetic comparisons can only give good estimates of phylogenetic relationships when evolutionary rates of all characters are reasonably similar. If rates of change are widely discrepant between characters, overall similarity will give misleading estimates of true relationships. An example, discussed at greater length later, is that of the relationships of chimpanzees, gorillas, and humans. It now seems probable that, despite the great overall similarity of the two African apes, chimps and humans form the monophyletic group.

Hence, we (ideally) follow these steps in determining phylogenetic relationships. Individuals are assigned to species and an approximate sorting of species is done using overall similarity, generally taking note of geological age. However, we are always very cautious about drawing firm conclusions at this stage. A careful character analysis then follows; if we are dealing with fossils this is exclusively a matter of morphological characters. We define reasonable characters, reasonable in the sense that we believe they have some genetical, developmental, and functional integrity, and then assign character states. Deciding which state is primitive and which is derived is called determining the polarity of character states. It involves making an hypothesis, based on variable amounts and varying kinds of data, and it is one of the most important steps in the analytical process.

Once these decisions have been made, the cladogram, summarizing the degrees of relatedness, is put together. Ancestry is inferred by the sharing of derived homologies. In the best of all possible worlds, all characters analysed will lead to the same cladogram. Unfortunately, this rarely happens, and in most cases where several characters are analysed, several possible cladograms can be inferred. Such incongruencies arise either because homoplasies are missed or because polarities are incorrectly inferred. Unfortunately, there are no ways of making unambiguous decisions; the inferences remain hypotheses with varying degrees of probability.

There are disagreements among biologists about how species ought to be sorted into higher categories, such as genera, families, suborders, and so forth, and many cladists argue that classification should strictly follow branching order. This might be a generally acceptable procedure, but there will always be particular problems. For example, if indeed chimps and humans are closest relatives, a strict cladistic approach requires grouping them together, with gorillas in a separate group.

From cladograms we proceed to phylogenies, and do so by adding information about dating and geographical distribution of fossil forms. Hypotheses are made about ancestor–descendant relationships of species. We would also like to minimize unsampled periods of geological time, to keep morphological gaps between putative ancestors and descendants to biologically reasonable proportions, and indeed to use all conceivable information. Several different phylogenies can usually be generated from the same cladogram; each one in turn is an hypothesis, the probability of which can be raised or lowered as new information appears or as earlier data are reanalysed. As we shall see in our later discussions of fossil material, alternative phylogenies can often be difficult to rank. And the fleshing out of phylogenies into explanatory and biologically dynamic scenarios, although the most interesting part of the entire exercise, can become frustratingly complicated.

Most phylogenetic analyses of living species have been based on the comparative study of morphological and physiological characters, while solely morphological characters have to be used for fossils. In the past 25 years another set of data has come to play an extremely important role in the analysis of living species relationships, that of molecular and genetical information from chromosomes, proteins, and DNA. The first work of this kind was conducted early this century, but was only fully developed about 25 years ago when comparative immunological studies gave indirect estimates of amino acid sequence differences in homologous proteins between pairs of living species, and these were used to infer evolutionary relationships. This early work on primates revealed a number of interesting patterns, including the observation that the African apes, chimpanzees (*Pan*) and gorillas (*Gorilla*), were similar to humans while the Asian apes, orangutans (*Pongo*), gibbons, and siamangs (*Hylobates*), were more different. These results were confirmed by careful quantitative assays of a single serum protein, albumin. If it is assumed that measures of phenetic similarity in antigenic reactivity are proportional to degrees of phylogenetic relatedness, this produced the surprising conclusion that humans and African apes shared a common ancestor long after the ancestor of all hominoids, and that hominids were in a genetic sense odd African apes.

One response to these phenetic patterns was to argue that humans and African apes were conservative while the orangutan, for example, had evolved more rapidly and was hence more different. This alternative explanation can be evaluated in the following elegant way. Make humans, chimps, gorillas, and orangutans members of a monophyletic group, and pick an outgroup, for example, one of the Old World monkeys. If the orangutan has evolved significantly faster (= more) than African apes and humans, it should be more different from monkey in albumin structure than they are. It is not (Fig. 2.3). This demonstrates that the similarity of

ALBUMIN DISTANCES   (Cronin 1975)

	Go	Hu	Ch	Or	Gi	OWM
Gorilla	—	4.5	8.0	10.5	11.0	32.0
Human		—	5.5	11.5	11.5	40.0
Chimpanzee			—	9.5	13.5	37.5
Orangutan				—	11.0	40.0
Gibbon					—	38.5
OW monkey						—

Column average		6		10.5	11.8	37.6
Column range			3.5	2.0	2.5	8.0
% Ra /Av			58	19	21	21

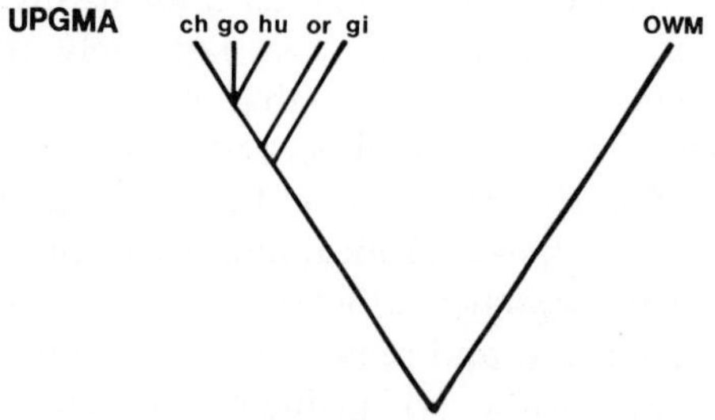

FIG. 2.3.   Matrix of albumin distances in catarrhine primates from Cronin (1975). The distances are quantitative measures of degrees of antigenic similarity between pairs of species. The branching sequence is derived from the matrix by the unweighted pair group method of analysis (UPGMA). (J. Sept).

humans, chimps, and gorillas is not due to conservative slowness, but to the fact that they are more closely related.

This test is called the 'rate test', and the conclusions drawn from it are profound indeed. Relative to the outgroup monkey, the difference in albumin structure — measured indirectly by immunological reactivity — of all the hominoids is more or less the same. Indeed the result can be generalized. It is usually the case that, for any monophyletic grouping of the species, the phenetic similarities (sometimes called distances) of each species to an outgroup species are approximately equal. This implies that roughly equal amounts of change in albumin accumulate along different lineages from a common ancestor.

It has been further claimed that the change accumulates at a constant rate. Minor variations in distance measures are said to be due to measurement error and to minor fluctuations in evolutionary rate, which would tend to average out over long periods of time. Rates of change of albumin have been estimated for mammals, using a few fossil-based branching points for calibration and assuming that albumin changes in a linear fashion. These rates were then used to estimate other branch points, in particular that for hominids and African apes (Fig. 2.4). This falls at what was a surprisingly young age, 4 or 5 million years ago.

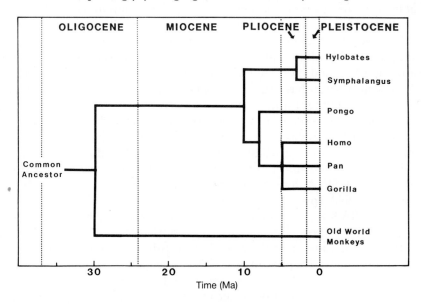

FIG. 2.4.   Evolutionary tree of catarrhines derived from albumin comparisons assuming a hominoid–cercopithecoid split at 30 Ma. (J. Sept.)

How could such a system work? Only if amino acid replacements were, for the most part, functionally and thus selectively neutral. This explanation was preferred because, a little earlier, it had been suggested that many point mutations (nucleotide changes in DNA) were selectively neutral and would thus tick away unseen by the editorial forces of selection, and therefore provide a potential 'molecular clock'. This proposal, although by no means universally applauded, has now been generally accepted, at least as far as its main tenets are concerned.

However, cladists have had more fundamental criticisms. Phenetic comparisons look at much more than shared-derived similarities and so, in principle, either ought not to work or at the very least must be treated with great care. In reply can be noted the consistency of the rate tests, and how difficult these are then to explain away.

Cladistic analyses are possible using characters derived from cytogenetic comparisons of chromosome structure, amino acid sequences of several proteins, nucleotide sequences of mitochonidrial DNA, and nucleotide sequences of expressed and non-expressed segments of cellular DNA. It turns out that such analyses are often as ambiguous as, or more ambiguous than, those based on morphological data. Character state polarities are hard to determine, homoplasies are frequent and very difficult to detect, and 'characters' often disappear; there is often little long-term conservation at the nucleotide level.

However, with these caveats noted, results from the analyses of genetic character states resemble closely those of protein immunology. Humans are close to African apes, while orangutans, gibbons, Old World monkeys, and New World monkeys are successively more different. Until recently, it has proved difficult to find unambiguous agreement on which two out of chimp, gorilla, and human form a monophyletic group. Two recent studies seem now to have tilted the balance in favour of a slightly closer relationship of humans and chimpanzees.

The nucleotide sequence of a portion of mitochondrial DNA (about 1000 base pairs long) has been determined in hominoids, and suggests that humans and chimps are slightly closer to each other than either are to gorillas. Orangutans and gibbons are clearly distinct. The second study involves DNA-DNA hybridization. In this technique (ideally) all the single copy DNA (present only once per genome) plus one copy only of repetitive DNA for a particular species is heated, causing the two helices to separate. There are almost 2×10^9 nucleotides in such a DNA preparation for most mammals, including primates. The DNA is then sheared into short (approximately 500) base pair segments, and then sheared DNAs of two species are mixed together. One species is radioactively labelled. Double helix hybrids are formed, despite the fact that there will be mismatches between non-complementary bases where no hydrogen bonds will form. The hybrids on reheating will then separate into complementary strands and this separation is carefully monitored in such a way as to provide an indirect measure — in the form of the differences between the temperatures at which the homoduplex and heteroduplex separate — of total nucleotide sequence difference between two species.

These measurements summarize the entire single copy portion of the genome. They integrate differences in expressed (no more than 10 per cent) plus unexpressed (at least 90 per cent) segments and do so for an immense (over 10^9) number of nucleotides. Although any particular small portion of DNA (exons of a gene, for example) may show fluctuations in evolutionary rate along or between lineages, when summed over enormous numbers of such segments these differences will cancel each other out. The average rate for the entire genome will be, theoretically, monotonically steady through time, and will be the same for similar genomes. As judged by the rate test, such data for hominoids are remarkably uniform, implying an underlying mechanism that is indeed extremely regular.

Both DNA sequencing and hybridization show that humans and chimps are more similar than either is to gorillas (Fig. 2.5). This can be used to infer relatedness because all evidence points to an underlying mechanism capable of being used as a 'clock'.

Some concerned with phylogenetic analysis, particularly morphologists, have argued that only character analysis can give unambiguous patterns of relationships, and that natural selection will cause evolutionary rates to fluctuate to such an extend that a clock is an impossibility. Again, one can only point to the rate test which appears to show that even a single protein, albumin, but more especially an assay of the entire genome, DNA-DNA hybridization, yields very regular patterns of phenetic distance, strongly implying a very regular pattern of change.

However, nature is not to be deciphered too easily. We still cannot expect a genetic 'clock' to give us temporally linear readings, because convergences and reversals will always make the actual number of mutational events which happened less than the number of differences we can actually see. Although this reduction should often follow a regular pattern, we can only hypothesize what it is. However, sensible hypothesizing and a judicious use of the fossil record should allow us to produce a synthesis of the molecular and palaeontological which throws new light on evolutionary patterns, especially those of hominoids.

One lesson to be learned from the last 20 years' work in evolutionary biology is that it is sometimes easy to misread morphology in reconstructing phylogeny. Few in 1950 would have predicted that orangutans, not humans, were the outliers among large hominoids; fewer still in 1980 that gorillas were the outgroup among 'African' hominoids. Since both patterns have become generally accepted, we have seen some new morphological analyses supporting the newer phylogenies. Yet it is clear that, in some cases at least, genetical and morphological patterns can appear out of step. Chimpanzees and gorillas are morphologically very

DNA-DNA   DISTANCES   (Sibley 1985)

	Ch	Hu	Go	Or	Gi	Ba
Chimpanzee	—	1.6	2.2	3.6	4.8	7.4
Human		—	2.3	3.6	4.8	7.3
Gorilla			—	3.5	4.8	7.1
Orangutan				—	4.8	7.4
Gibbon					—	7.1
Baboon						—
Column average		1.6	2.2	3.6	4.8	7.3
Column range			0.1	0.1	0	0.3
% Ra / Av			4	3	0	4

UPGMA   ch   hu go   or   gi   ba

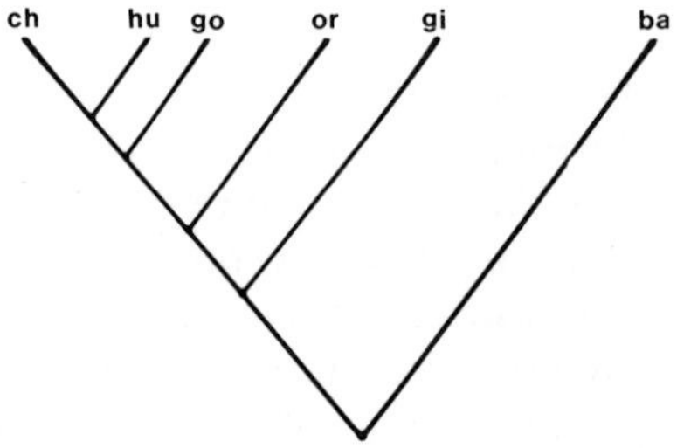

Fig. 2.5.   Matrix of DNA hybridization distances in catarrhine primates (C. B. Sibley, pers. comm.). The distances are equivalent to percentage sequence differences between species pairs. Note how much more uniform the data are than for albumins (Fig. 2.3). The branching sequence is derived from the matrix by the unweighted pair group method of analysis (UPGMA). (J. Sept.)

similar, humans very different — in brain size, limb proportions, hair distribution, to pick just three features. However, in every structural protein looked at, humans are very similar to the two apes. Thus, the differences are in a broad sense regulatory — in the way the building blocks are assembled, in the control of developmental timing and rates. Small genetical changes can produce big morphological differences. As an important corollary, morphological features, although they are all we have in the fossil record, are not always obvious indicators of genetical and therefore evolutionary relationships. They need careful reading.

3   Primates

Humans, primates, and mammals

Before we list important events in human evolution and take a closer look at several of them, we shall discuss the Primates, the mammal group to which modern humans (*Homo sapiens*) belong. By looking at our close relatives we can highlight critical differences and similarities between them and us, and thus guide our study of human evolution (Table 3.1).

Homo sapiens is one of about 180 living species in the order Primates, which is itself but one of 17 orders of mammals. Placental mammals or eutherians can be characterized under three main kinds of features: morphological, physiological, and behavioural. Eutherians have high metabolic rates and generally high activity levels. They are endothermic, with efficient insulation in the form of fur and body fat. Blood circulation is divided completely into systemic and pulmonary, and there is a four-chambered heart. The respiratory system is similarly efficient. Mammals have evolved effective adaptations for harvesting and digesting a broad range of foods, including differentiated and specialized dentition, and complex guts. They also have a very diverse array of movement patterns related to food harvesting and predator avoidance, and these are reflected in a broad range of skeletal adaptations. Placental mammals often have extended associations between mothers and offspring. Mothers suckle their young with milk from specialized mammary glands, and the contact between mother and infant is sometimes intense and prolonged. They exhibit varying degrees of paternal care. Generally, eutherians show complex patterns of social behaviour and organization and they have large brains, especially in the evolutionarily newer integrative cortex. They are capable of learning complicated tasks, often have high intelligence and good memory, and have well-developed senses — sight, hearing, smell, touch. The Eutheria are subdivided into 17 major clusters of species or orders, including groups such as Carnivora, Rodentia, Cetacea (whales, dolphins), Proboscidea (elephants), as well as the arrogantly named Primates to which *Homo sapiens* belongs.

Among mammals the primates are a somewhat difficult order to characterize. In many ways they differ from the others because of what they lack, rather than because of what they possess. They lack, for example, obvious anatomical specializations, such as the hooves characteristic of ungulates, or the persistently growing incisors of rodents.

TABLE 3.1
Classification of Primates

Suborder	Infraorder	Superfamily	Family	Subfamily	Genus	Common name
Strepsirhini		Lemuroidea	Cheirogaleidae		*Allocebus* *Cheirogaleus* *Microcebus* *Mirza* *Phaner*	Hairy-eared dwarf lemur Dwarf lemur Mouse lemur Coquerel's dwarf lemur Fork-marked lemur
			Daubentoniidae		*Daubentonia*	Aye-aye
			Indriidae		*Avahi* *Indri* *Propithecus*	Woolly lemur Babakoto Sifaka
			Lemuridae		*Lemur* *Varecia*	Lemur Ruffed lemur
			Lepilemuridae	Hapalemurinae	*Hapalemur*	Gentle lemur
				Lepilemurinae	*Lepilemur*	Sportive lemur
		Lorisoidea	Lorisidae	Galaginae	*Euoticus* *Galago*	Needle-clawed galago Bushbaby
				Lorisinae	*Arctocebus* *Loris* *Nycticebus* *Perodicticus*	Angwantibo Slender loris Slow loris Potto
Haplorhini	Tarsii	Tarsioidea	Tarsiidae		*Tarsius*	Tarsier
	Platyrrhini	Ceboidea	Callimiconidae		*Callimico*	Goeldi's marmoset
			Callitrichidae		*Callithrix* *Cebuella* *Leontopithecus* *Saguinus*	Marmoset Pygmy marmoset Golden lion tamarin Tamarin

Suborder	Infraorder	Superfamily	Family	Subfamily	Genus	Common name
Haplorhini	Platyrrhini	Ceboidea	Cebidae	Alouattinae	*Alouatta*	Howler monkey
				Aotinae	*Aotus*	Owl or night monkey
				Atelinae	*Ateles* *Brachyteles* *Lagonthrix*	Spider monkey Woolly spider monkey Woolly monkey
				Callicebinae	*Callicebus*	Titi monkey
				Cebinae	*Cebus*	Capuchin monkey
				Pitheciinae	*Cacajao* *Chiropotes* *Pithecia*	Uakari Bearded saki Saki
				Saimiriinae	*Saimiri*	Squirrel monkey
	Catarrhini	Cercopithecoidea	Cercopithecidae	Cercopithecinae	*Allenopithecus* *Cercocebus* *Cercopithecus* *Erythrocebus* *Macaca* *Miopithecus* *Papio* *Theropithecus*	Swamp monkey Mangabey Guenon Patas Macaque Talapoin Savannah baboon Gelada baboon
				Colobinae	*Colobus* *Nasalis* *Presbytis* *Pygathrix* *Rhinopithecus*	Colobus monkey Proboscis monkey Langur Douc langur Golden monkey
		Hominoidea	Hominidae	Gorillinae	*Gorilla* *Pan*	Gorilla Chimpanzee
				Homininae	*Homo*	Human
			Hylobatidae		*Hylobates*	Gibbon
			Pongidae		*Pongo*	Orangutan

From: Richard (1985).

Most primates have a rich repertoire of behaviour and live in social groups organized according to relatively complex patterns of interaction. Primates are also very diverse. The order includes many species and they show a wide range of adaptations.

The living primates are usually divided into two groups. Until recently the subdivisions have been the 'lower' primates or Prosimii (Fig. 3.1) and the 'higher' primates or Anthropoidea (monkeys, apes, and humans; Fig. 3.2). The division is based above all on relative brain size: brains are no bigger in 'lower' primates than in most other mammals, but they are several times larger, relative to body size, in 'higher' primates. An

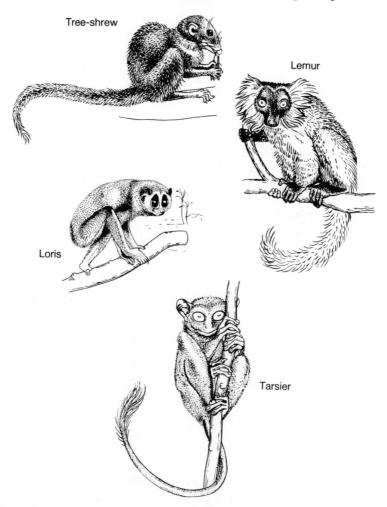

FIG. 3.1. A close relative of primates: tree-shrew: and three prosimians: lemur and loris (Strepsirhini) and tarsier (Haplorhini).

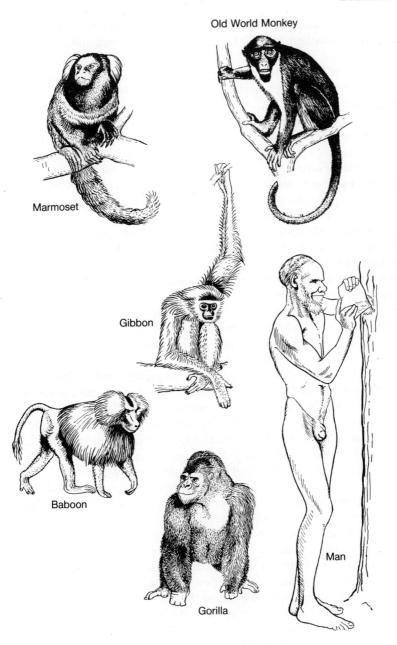

Fig. 3.2. Anthropoid primates.

alternative classification has been more popular recently, in which the tarsiers are classified not with prosimians, but with anthropoids. The subdivisions are then labelled Strepsirhini and Haplorhini. This second classification is a cladistic one, reflecting the hypothesis that small-brained *Tarsius* is more closely related to Anthropoidea than to other small-brained prosimians.

The living prosimians such as lemurs and bushbabies are the survivors of what was, 40–50 million years ago, a larger and more diverse group. The anthropoids appeared as new adaptive types 35 million years ago or more. In addition to larger brains they had redesigned teeth and skeletons, implying important changes in the types of food they preferred, and in the ways that they obtained and processed it, as well as parallel developments in social organization and interaction.

The living anthropoids fall into two natural groups, platyrrhines from South and Central America, and catarrhines from the Old World. Judging from their genetic similarities, they shared a period of common ancestry after diverging from the prosimians. Each anthropoid group is diverse and heterogeneous and difficult to characterize in a few words. Both contain species which have large brains and are intelligent, and exhibit an unusual richness of interactional behaviour within social groups (although it is not living in social groups as such that is particularly unusual — many other mammals do this). Both anthropoid groups show a broadly similar amount of behavioural complexity, and a similar range of diets, movement patterns, and so forth. A zoologist would describe them as sharing a similar 'grade' of organization.

The platyrrhines or New World monkeys are arboreal (tree-living) forest species, all basically vegetarians, with some species emphasizing leaves in their diets and others fruits, with greater or lesser amounts of insects. They are all agile climbers, clamberers, leapers, and acrobats in the trees, and live in groups varying from a few individuals to those containing many adults of both sexes. They are interesting in their own right, but they are especially interesting to palaeontologists because they offer good living models or analogues for our early ape-like ancestors. Some platyrrhine features (wrist and elbow joints, for example) are convergent on the same region in extinct hominoids. We can thus turn to the New World monkeys in addition to closer relatives for clues to the ways that our extinct ape-like ancestors moved and fed.

The Old World catarrhines are also a varied group, but one in which there are two main clusters, monkeys and hominoids, reflecting a division which occurred more than 20 million years ago. The monkeys of Africa and Asia, the cercopithecoids, live both in the trees and on the ground, in both forest and more open country. Despite these different adaptations, the cercopithecoids are a morphologically much more homogeneous group than either the platyrrhines, or their fellow catarrhines, the hominoids. All cercopithecoids are quadrupeds like dogs or cats, moving with the body parallel to the ground or a branch. They are also adept climbers, and have grasping hands and feet that are dextrous and mobile.

The other catarrhine group, the Hominoidea, is harder to characterize. They subdivide genetically into three groups: gibbons, orangutans, and chimps, gorillas, and humans. None are quadrupedal like cercopithecoids. All show traces to a greater or lesser extent of descent from an ancestor which was an accomplished climber, hanger, arm-swinger, and biped with long arms and legs, mobile joints, broad thorax, and short lumbar region. Unlike monkeys, they have few adaptations for processing mature leaves or unripe fruits. They depend for food on high quality foliage (young leaves), ripe fruit, insects, and occasionally mammals.

Classification and naming

As we noted in chapter 1, the basic biological unit is the species, and because it is the largest normally interbreeding population it is defined both inclusively and exclusively. All other taxa in the hierarchy are more or less arbitrary. Related species are grouped into genera. Thus, for example, the common baboon, *Papio cynocephalus*, and the hamadryas baboon, *Papio hamadryas* are put in the same genus *Papio*. Genera are almost always, in practice, monophyletic. A species thus has a name or nomen (plural nomina) made up of two words, both italicized: a generic name, capitalized, and a trivial name which is not, the two together making the species name.

Genera usually comprise species which are very similar, perhaps differing in body size or some relatively minor feature of body proportions, behaviour, coat colour, and so forth. Genera are in turn clustered into families. Families contain species that are of broadly similar adaptive type: all omnivorous–frugivorous cercopithecids versus leaf-eating colobids, for example. Families are subdivided into subfamilies or clustered into superfamilies and there can be honest differences of opinion about the levels of classification when one deals with 'family group' categories (thus one person's colobids and cercopithecids might be another's colobines and cercopithecines).

Family group nomina also follow a definite pattern. All family names end in -idae, hence, Hominidae (formally capitalized, colloquially non-capitalized as hominids). Superfamilies end in -oidea (Hominoidea, hominoids), and subfamilies end in -inae (Homininae, hominines). There are, in addition, categories between order and superfamily (suborder, infraorder), and between subfamily and genus (tribe, subtribe). It is important to remember that, except for species, these other categories are arbitrary.

The broad outlines of primate classification have been well established for many years and reflect to a considerable extent the broad pattern of primate phylogeny. Most phylogenetic judgements were made almost exclusively on the basis of comparative morphology. The fossil record played little part, and it was always possible to fit the mostly fragmentary and ambiguous fossil record to several alternative evolutionary frameworks.

In the last 25 years, 'molecular' studies, discussed in the previous section, have become widely used as additional sources of information about evolutionary relationships. In many cases the patterns of similarities derived from these studies have matched morphological ones, but not always: hominids and their relationships to the other hominoids are a case in point. However, careful reanalysis of both morphological and genetical records generally produces agreement. Some problems remain, for example, at the level of subordinal relationships: are tarsiers closer relatives of anthropoids, all prosimians, or some prosimians? However, the essentials are agreed.

Nomenclature is also widely agreed. Although there may be disagreements about which taxonomic level should be used, there is less disagreement about co-ordinate status of taxa. There is one exception: Hominidae. Hominids are of compelling interest to other hominids. Judged from a cladistic perspective, hominids are no more distantly related to chimps than are baboons to macaques or gibbons to siamangs, and less so than chimps are to gorillas. Family status for *Homo* and its ancestors cannot be justified on grounds of degree of relatedness. However, there is a plausible counter-argument that lineages along which a great deal of important change occurs should be recognized, and that taxonomies should balance cladistic and phyletic 'interests'. We shall continue to include *H. sapiens* and earlier non-ape hominids in Hominidae.

Primates: traditional definitions

Primates have traditionally been defined mostly by anatomical features and, as noted already, there is no defining character present in all species. Primates then are a particularly good example of a polythetic set: a group of individuals defined by the presence of a list of characters, not one of which is present in all individuals of the set. Most sets in the living world are polythetic, as opposed to the monothetic sets more typical of the inorganic: groups in which individuals are defined by characters all of which are present in all individuals.

It has often been said that primates can be defined by 'trends'. Thus, Le Gros Clark summarizes decades of thinking as follows.

Broadly speaking... it is now possible to define the Primates on the basis of the prevailing tendencies which dominated their evolutionary development. Using such criteria we may say that they form a natural group of mammals distinguished from other groups by the following prevailing evolutionary tendencies: the preservation of a generalized structure in the limbs, associated with free mobility of the digits (especially the thumb and big toe) and the replacement of sharp compressed claws by flattened nails, the elaboration of their visual powers and a corresponding reduction of the olfactory apparatus, the shortening of the snout or muzzle, the preservation of a relatively simple pattern of molar teeth, and the progressive development of large and complicated brains.

These features have usually been seen as 'generalized' or 'unspecialized', and the 'success' of the primates was believed due to their unspecialized arboreal way of life as opposed to the 'narrow specializations' supposedly associated with life on the ground. Manual dexterity, stereoscopic vision, and high agility were all believed adaptations to life in the trees. Implicitly and sometimes explicitly, being generalized was seen as somehow superior to being specialized. Terms like specialized and generalized are usually difficult to define and often can only be applied in retrospect; unless absolutely necessary they are best avoided.

Recently, it has been proposed that the basic primate anatomical and behavioural patterns evolved in nocturnal animals which hunted insects by stealth requiring high degrees of hand–eye co-ordination. This would explain forwardly directed eyes and stereoscopic vision, and also the great manual dexterity. A refinement of this idea implicates plant as well as insect food. The early primates began to diversify 50–60 million years ago at roughly the same time that many groups of flowering plants also began to diversify. These primates became highly selective feeders, during twilight and at night, on both animals and plants, accordingly evolving elaborate patterns of vision, olfaction, hand–eye co-ordination, and movement.

These ideas represent a more recent trend in modern evolutionary biology, namely that food and its distribution are very important in determining many attributes of a species' biology. Other important factors would be predator avoidance and the ability to locate mates.

Perhaps a better way to think about primates is in terms of their ecology and behaviour, and only then of the morphological consequences of primarily behavioural adaptations.

Primates as mammals: behaviour, physiology, anatomy

Skull

The form of the skull differs between prosimians and anthropoids (Fig. 3.3). Prosimians have larger and more projecting faces, partly related to

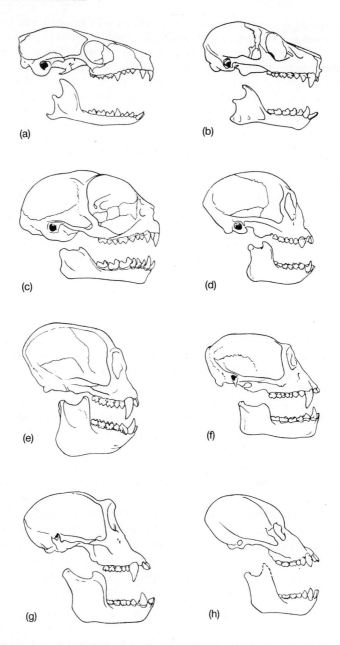

F<small>IG</small>. 3.3.   Living primate skulls: (a) is the tree-shrew *Ptilocercus*; (b) *Galago*; (c) *Tarsius*;
(d) *Cebus*; (e) *Presbytis*; (f) *Hylobates*; (g) *Pan*; and (h) the Miocene hominoid *Proconsul*.
(After Le Gros Clark.)

the importance of olfaction and the relatively large peripheral sense organs, and partly to the biomechanical requirements of tooth use in mainly insectivorous forms which use their projecting incisors and canines as dental combs or gum scoops; and partly reflecting a small brain. The orbits are surrounded by a bony ring, are quite large, and face more forwards than sideways.

In anthropoids the skull is more globular, the braincase being larger and rounder while the face is less projecting and deeper from top to bottom. The two sides of the mandible are fused at the symphysis, the back wall of the orbit is formed by bone, and the orbits face directly forward. These changes in the face are related, in part, to reduction in the olfactory apparatus, but also to biomechanical factors that involve moving the dentition backwards under the orbits. This allows larger, and differently oriented, chewing muscles (masseter, pterygoids, and temporals). Chewing forces are greater and these forces can be dissipated more quickly through the vertically deep face.

In hominoids, the cranium is atop a vertebral column which is often more vertical (orthograde posture) than horizontal (pronograde). At least in the smaller species, with relatively smaller faces, the foramen magnum — through which passes the spinal cord — is placed underneath the skull. In hominids the combination of facial flattening and bipedal posture has resulted in a foramen magnum relatively far forward.

Ecology and diet, teeth, and gut

Primates are principally tropical, mainly forest, mostly arboreal creatures, and for the most part are highly selective feeders on predominantly plant food. Most species live in permanent social groups. Although primates are mainly plant eaters they almost all consume animal food as well, particularly in the form of insects. Diets must include proteins (important as 'growth' food), carbohydrates (energy), and fats (mostly energy storage), vitamins, minerals, and trace elements. From the animal's perspective foods provide nutrients, but they also pose problems in harvesting. They may be widely scattered or randomly distributed, high in trees or buried below ground. They may be covered in spines or be otherwise resistantly packaged. They may contain secondary compounds such as alkaloids and tannins which are generally metabolically costly to ingest.

Primate foods include both ripe and unripe fruits (although these pose different 'problems' and require different 'solutions'), seeds, leaves and stems (mature versus young leaves contain different nutrients and present different problems for the animal), grass parts, underground plant parts, gum and sap, and animals. Species have been sorted into dietary categories according to what they eat: for example, frugivores (fruits), folivores (leaves), insectivores, carnivores, omnivores (animals and plants are eaten), graminivores (seeds), grazers (grass), and browsers (leaves and stems). These are not exclusive and non-overlapping categories. Such classifications do help in communicating information succinctly, but they are also deceptive. Most primates are eclectic feeders and eat a wide range of food items. While these may be predominantly of one general kind (leaves, say) virtually all species eat other items too. Each species has a somewhat different dietary repertoire. Thus, two frugivores, although having predominantly fruit dominated diets, may eat different kinds of fruit, and one may eat additional leaves while the other prefers supplementary insects. The two diets may well pose different problems — harvesting of clumped nuts with tough coats, or widely scattered softer ripe fruits — and also provide contrasting combinations of nutrients. Diets with items having similar physical properties may, in fact, contain rather different foods. Hence, though a useful communicative tool, labels such as these need to be treated with proper caution.

Primates are mainly individualistic herbivorous foragers, carefully selecting many different and particular parts of a wide range of plant foods. They also eat some animal food. Humans are an exception in being truly omnivorous. They rely on substantial amounts of both plant and animal food; in particular they exploit underground plant resources in ways other primates do not, and they frequently transport, elaborately prepare, and share food.

Primates have evolved a broad array of adaptations to their diets, including features of the hands, jaws, teeth, masticatory muscles, the alimentary canal, and of course behaviour patterns. Primate hands are generally five-fingered, and a range of grips involving fingers to palm and fingers to fingers are used in harvesting and processing food items. For example, a particular part of a fruit or leaf might require careful selection, or sand might be brushed from the food. Apes, with long, curved metacarpals and phalanges, reflecting their primary adaptation to arboreal climbing and arm-swinging, have well developed thumbs which permit grips between thumb pulps and the sides or fronts of other digits. One cercopithecoid monkey, the gelada baboon *Theropithecus*, is descended from species that were already part terrestrial with shorter fingers than their arboreal cousins. This reduces stresses on the digits during palmigrade quadrupedal walking. Geladas are now exclusively terrestrial animals, and often feed on grass corms and roots which they harvest by digging with their hands, using extended fingers together so that their hand becomes a kind of trowel. Because digits are short they can also pick grass blades carefully using index finger and thumb. Thus, hand form is often influenced by the demands of feeding, although other factors are also important, e.g. grooming and locomotor behaviour. Hand structure is, at any time in an animal's evolutionary history, a compromise.

The dentition of primates is perhaps more varied than hand structure. Primates have retained the four kinds of teeth found in early mammals: incisors, canines, premolars, and molars. All tooth classes show variability. Incisors are important in harvesting food as well as in its preparation. The needle-shaped procumbent (forwardly projecting) incisors of most prosimians and a few New World monkeys are used in scraping gum, and also in scraping other primates during grooming. In most anthropoids incisors are rather chisel-like and more vertically implanted than in the majority of prosimians. In addition, the two halves of the mandibles arc firmly fused at the midline symphysis in anthropoids, and the faces are shorter from front to back (flatter). This promotes greater force in incisal biting. Frugivores have relatively larger incisors than folivores, imply a different kind of incisal preparation.

In most primates, canines are tusk-like projecting teeth, although in virtually all prosimians and in hominids canines have altered in shape and function, and become essentially incisors. Almost all anthropoids are sexually dimorphic (males and females differ in an anatomical feature) in canine size and shape: males have projecting tusks while female canines are much shorter. The canine is a social tooth, using in fighting, bluffing, and other interactions, particularly between males. As a correlate, male faces are longer and larger, neck muscles and the rear portion of the temporal muscles are better developed, and the bony crests to which they are anchored are prominent. In gibbons and siamangs canines are more similar in males and females, this being associated with the fact that their basic social group is a monogamous pair.

The cheek teeth, premolars and molars, vary in number and morphology among primate groups. Premolars are generally simple with two cusps, the most anterior of the lower premolars usually being an elongated single-cusped tooth occluding with the enlarged upper canine. Molars are multi-cusped and larger. Cusp relief and the sharpness of cusps and interconnecting crests varies between species, generally depending on the physical properties of foods and therefore on diets. Insectivores have generally high sharp cusps and crests, while fruit-eaters have on average lower and rounder cusps.

Most primates have retained the primitive mammalian number and disposition of cusps — four on upper molars and five on lowers — although there is considerable variation on the basic theme. The shape of the tooth surface (occlusal surface) is controlled by the shape of the underlying, internal, dentin core of the tooth (dentin is a hard substance similar to bone, though denser), and the thickness and distribution of the coating enamel, a harder, mineralized tissue that is highly resistant to wear. Enamel thickness can vary markedly in closely related species. Humans have thick enamel, chimps and gorillas thin enamel, laid down on a dentin surface that is basically similar in shape. One species of macaque and another of capuchin have thicker enamel than their congeners. Exactly why thickness varies is still poorly understood, but it may be related to the presence in food of unpredictable small, hard items. There may also be advantages to having thin enamel which wears to expose dentin circles on the occlusal surface; the rim formed by the depression of softer dentin below harder enamel might be an effective shredding mechanism.

Cercopithecoid monkeys differ from hominoids, which have retained a more primitive pattern, in the derived nature of their molar teeth. Cusps are paired, linked by crests or lophs running from side to side. Particularly in colobids, cusps and crests are high and self-sharpening (because of the interlocking nature of upper and lower teeth). This 'bilophodont' pattern in cercopithecoids is probably an adaptation to efficient reduction of leaves.

As already mentioned, humans differ from African apes in having thick enamel, although the underlying dentin surface relief is rather similar. Orangutans have enamel of intermediate thickness, and the dentin surface shows little relief, so the occlusal surface itself has low cusps which rapidly become worn flat.

Proportions of different teeth vary according to broad dietary categories. Frugivores have relatively large incisors and often smaller cheek teeth, folivores smaller incisors, while species eating tougher or more abrasive food often have larger cheek teeth. Face shape also varies. Colobids have faces that project less and are deeper from top to bottom than those of cercopithecids, and this is related to greater effectiveness in incisal biting. With a flatter face the bite point is closer to the jaw joint, and the muscles that close the jaw (masseters, pterygoids, and part of the temporals) have average vectors closer to the bite point and further away from the fulcrum; mechanical advantage of biting is higher than with a more projecting face. The relative size of masticatory muscles also varies. In animals that chew powerfully, masseters and pterygoids, and often anterior temporals, are enlarged.

The alimentary canal is relatively unspecialized in most primates. The most derived species are the colobids, predominantly eaters of leaves. This food source is high in the structural carbohydrate cellulose which is tough and indigestible, and also in alkaloids and other secondary compounds. Colobines have evolved complex sacculated stomachs, convergent on those of ruminant artiodactyls such as cattle, in which gut bacteria break down cellulose and potentially toxic substances. Cercopithecines have not evolved (or have lost) these features, but have developed cheek pouches in which food is stored and softened for preliminary digestion. One possibility is that the pouches are used in feeding situations where there is intense competition from conspecifics; the pouches are storage organs allowing an animal to harvest, move away, and digest in solitary peace.

Hominoids have no such clearcut specialized features. Gorillas have enlarged large intestines, related to their predominantly folivorous diet. Relative to chimps, humans have smaller stomachs and large intestines, and larger small intestines, presumably as adaptations to more omnivorous diets. However, there is still important work to be done in understanding the anatomical and physiological differences between humans and their closest relatives.

A good deal of the individual, and thus collective, behaviour of primates is shaped by food and its distribution. One example out of many will serve as illustration. *Pan troglodytes*, the common chimpanzee, has an unusual pattern of grouping behaviour, in which the larger social group or community often fragments into units made up of mother plus dependent offspring, or of small numbers of males and females. On occasion, these units coalesce into larger groups. The degree of fragmentation or coalescence seems to be related to the availability and pattern of distribution of their favoured food, ripe fruit. This is generally quite widely dispersed, and is so in frequently unpredictable ways.

As a generalization, primates having diets in which food items are unpredictably distributed spatially and temporally tend to be both large bodied and large brained. As noted earlier, food and its distribution are critical factors in moulding adaptations of several kinds.

Positional behaviour

Effective movement within and between feeding, sleeping, and resting sites is one of the most important activities for any animal. Positional behaviour and its anatomical correlates play a central role in survival and reproduction, and so are closely linked to fitness. Positional behaviour is a useful term covering what an animal does when it moves (locomotor behaviour) and when it does not (postural behaviour). In both cases anatomy must be appropriate to allow the behaviour to be effective.

In only one primate, *Homo sapiens*, can positional behaviour be accurately described with a single simple label, bipedalism. Of course, humans are also capable of being quadrupedal, tripedal, and of swinging and hanging by the arms. We are not normally particularly good at such behaviours, whereas we are effective bipeds. Other primate species have much broader positional repertoires. Thus, baboons are normally quadrupedal in trees and on the ground, but can also sit upright or stand on hind limbs while feeding, and occasionally hang by the arms. However, the dominant activity is quadrupedalism, and baboons can conveniently be labelled quadrupeds.

Chimpanzees are also frequently quadrupedal, but move in a very different way from baboons. The hands, instead of contacting the substrate on the palmar side of fingers and palm are used with flexed fingers so that the backs of the middle digits contact the substrate. This is described as 'knuckle-walking'. Chimpanzees also hang and swing by one or both arms, occasionally propelling themselves though space (leaping) using arms alone. This has been called 'brachiation'. Furthermore, chimps are bipedal in a variety of circumstances. They also climb vertical supports using a combination of arm pull and leg push that is very familiar to humans. Here it is less clear that a single label (knuckle-walker, modified brachiator) can do adequate justice to chimpanzee positional behaviour.

If we turn to the anatomy underlying positional behaviour we can see that a species such as the chimpanzee which has a varied positional repertoire shows an essentially compromised anatomy. The anatomy is not designed biomechanically solely for any particular activity within the repertoire. Each activity is, of course, performed effectively within the animal's habitat and, overall, the compromise anatomy will be efficient, but it is a total efficiency within the context of the animal's total repertoire within its real habitat. On the other hand, if a single activity comes to dominate the repertoire, human bipedalism for example, anatomy will itself shift to reflect this. The pattern could be described as committed rather than compromised. We can imagine positional behaviour evolving by shifts in the proportions of various behavioural activities, and corresponding modifications in the underlying anatomy.

Primates are, or were, predominantly arboreal creatures, designed to deal with a habitat which is markedly three-dimensional, and generally irregular and disconnected. This has selected for mobile limbs and dextrous hands. Manipulative skills were also shaped by the demands of highly selective feeding, and by other activities like grooming and tool use. Primates climb by grasping, critical adaptations being their general possession of flat nails rather than claws, and opposable, prehensile first fingers and first toes.

Given caveats about the inevitable simplifying that goes with single word or single phrase labels for positional repertoires, and with focusing on a few descriptive anatomical features, the positional behaviour of living primates can still be quickly and usefully summarized. Prosimians can be subdivided into three basic categories. The first group are quadrupeds such as *Lemur catta*, moving for the most part rather like other small mammalian quadrupeds such as cats, but climbing in the particular way made possible by having prehensile extremities rather than claws. The trunk is long, especially the lumbar region, and acts rather like a springy bow. Hindlimbs are somewhat longer than forelimbs, and there is a tail. A second group are the 'vertical clingers and leapers' like *Indri* or *Galago*, species which habitually cling to and leap from vertical supports. Hindlimbs are longer and forelimbs shorter than in prosimian quadrupeds, and some joints — hips and knees for example — are better adapted for the habitual motions involved in powerful springing. Yet a third prosimian group are 'slow quadrupeds' such as *Perodicticus*, the pottos, which have no leaping phase in their deliberate and careful locomotion.

Each of these positional patterns can be plausibly linked to diet or to predator avoidance. Thus, the slow quadrupedalism of pottos is correlated with the careful stalking of animal prey, unnecessary behaviour for the more agile, unrestrained, vegetarian lemurs. Ripe fruit doesn't have to be ambushed. *Indri*-type leaping permits rapid escape from predators in *Indri*-type forest habitats.

Turning to anthropoids, the Ceboidea of south and central America are arboreal forest species which are predominantly quadrupedal, although with many variations on that theme. Some smaller species such as marmosets have an active springing component to their quadrupedalism and, as might be predicted, they have low fore-to-hindlimb length ratios. Larger species such as the capuchins of the genus *Cebus* are more 'typical' quadrupeds. The largest New World species, the howler (*Alouatta*), spider (*Ateles*), woolly (*Lagothrix*), and woolly spider monkeys (*Brachyteles*) show modifications to varying degrees on the basic (by inference, ancestral) quadrupedal patterns. All have long, strong prehensile tails which function as fifth limbs when the animals are suspended beneath branches or propped against trunks. In addition to quadrupedalism these species have varying amounts of arm-swinging behaviour in their positional repertoires. They hang with one arm, two arms, arms and one or two legs, tail with or without limbs, indeed almost any combination of limbs and tail. They are also bipedal in a variety of contexts.

Morphological correlates, for example in *Ateles*, of a positional repertoire, including important components (arm-swinging, bipedalism) in which the body is more vertical than horizontal, are long and limber arms and long legs. The thorax is broad from side to side and shallow from front to back, and the lumbar region is short with a reduced number of elements. These are correlates of positional situations in which the centre of gravity frequently acts parallel, rather than normal, to the vertebral column. Because the arm is frequently used with the hand above the head and the shoulder extended, the scapula is shaped and placed on the thorax in such a way as to facilitate these movements. These larger ceboids have been labelled 'semi-brachiators' or 'New World semi-brachiators'. Again, it is important to remember that these are merely useful labels, and descriptions of neither evolutionary history nor the full positional repertoire.

The two main groups of Old World anthropoids have very different positional repertoires and anatomies (Fig. 3.4). Cercopithecoid monkeys are predominantly quadrupedal, in the trees and on the ground, and their repertoires are basically rather unvarying. With their narrow, deep chests and long lumbar regions, Old World monkeys are similar to many other quadrupeds. Such variation as exists lies mostly in the extremities. For example, terrestrial species like baboons (*Papio*) or patas monkeys (*Erythrocebus*) have forelimbs longer than hindlimbs, and hands and feet with relatively short digits.

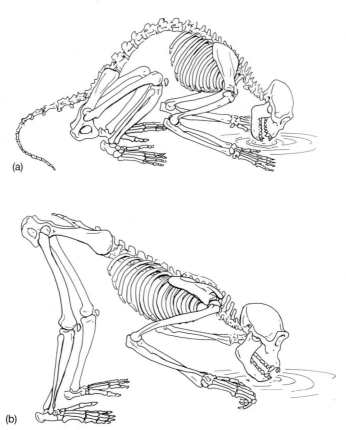

(a)

(b)

FIG. 3.4.   A cercopithecoid monkey (a) and chimpanzee (b) drinking. Note the long flexible lumbar region, deep thorax, and short pelvis in the monkey, and contrast with the chimpanzee with its short stiff back, broad shallow chest, and long pelvis. (After A. Schultz.)

Hominoids are more heterogeneous in positional behaviour than cercopithecoids, and this is mirrored in their anatomy. All hominoids have broad, shallow chests and relatively short lumbar regions. This probably reflects a common ancestor which climbed, hung, swung, and stood with the trunk more vertical than horizontal. Since this (hypothetical) ancestral stage, each main hominoid lineage has followed rather different paths.

We have already discussed chimpanzees. Gibbons, labelled brachiators, are actually capable of a varied range of positional behaviours, including not only arm-swinging and vertical climbing, but a significant component of bipedalism. As corollaries, hind limbs are long — a fact often overlooked because of the gibbon's very long arms which are adaptations to spectacular brachiation, and there is some muscular control of pelvic and truncal tilt associated with balancing the body during bipedal walking along branches where the feet cannot be placed far from a rather narrow path, and hence there is a need to control tilting of the body when only one foot is in contact.

The chimpanzee as noted also has a varied repertoire: knuckle-walking quadrupedalism, arm-swinging, vertical climbing, bipedalism. The other African apes, *Pan paniscus* (the bonobo or pygmy chimpanzee) and *Gorilla gorilla* (the best known populations being the eastern mountain gorillas, which differ in a number of ways from the western lowland subspecies), have a broadly similar range of behaviours. Bonobos occasionally brachiate, and adult gorillas never do, but otherwise both positional behaviours and anatomies for the two chimpanzee species and gorillas are quite similar. All have relatively long arms, hands, and fingers, with joints adapted for suspension. Superimposed on these features are those associated with forelimb use as a strut during quadrupedal knuckle-walking. The thorax is broad and flat, the lumbar region reduced (there are typically only three or four lumbar vertebrae), the pelvis both broad and deep. Hindlimbs are relatively longest in bonobos, shortest in gorillas. Limb proportions and body size are intimately linked to efficiency of climbing, arm-swinging, and walking. The African apes provide good examples of a compromised post-cranial morphology, reflecting a varied range of positional behaviours which often involve 'conflicting' pressures: arborealism and ground-walking, for example.

The fourth large ape, *Pongo pygmaeus*, the orangutan, lives today in the forests of Sumatra and Borneo. It is a predominantly (adult males) to exclusively (females and young) arboreal species and is a slow, deliberate contortionist. Orangs have very long arms, with long and curved hands and fingers, and much shorter legs with long and curved feet and toes. All limb joints allow considerable mobility, so that these heavy animals, by hanging on to branches usually with three or four extremities, can distribute their body weight adequately. As with the African hominoids, the lumbar region of the vertebral column is reduced, to four vertebrae.

The non-hominid hominoids — large and small apes — are diverse anatomically and in their positional behaviour, but all share certain behavioural patterns: suspension by the arms in a variety of contexts; sitting, climbing, standing with the trunk relatively upright; and a particular pattern of vertical climbing. These general features were most probably present in the common ancestor of all hominoids. The living apes have built upon this early pattern in a diversity of ways, but still retain its traces. Hominids, although greatly modified now, also retain some traces.

Body size

Primates range in body weight from a few grams to over 150 kg. Size is a very important influence on life-history patterns of any species, and many important biological features are correlated with overall body size. For example, maturation rates are quite strongly correlated, as are longevity, age at first reproduction, gestation time, interbirth interval, birth weight, adult brain weight, home range size, and a host of other variables.

There is a good general relationship between gross dietary categories and body size in primates. The smallest primates, below 1 kg, are specialists on insects and gum. Larger species, up to 10 kg, feed on fruits of various kinds because they are unable to catch enough insects to subsist entirely on them. They do, however, supplement their diets with insects, a source of protein. Still larger primates, averaging over 10 kg, add leaves to the fruit, leaves also being a source of protein; some of these larger primates feed predominantly on leaves and herbs.

Body size also has an influence on positional behaviour, although other factors are involved too. It is easier to be a small arboreal quadruped, running along branches, than a large one, and larger arboreal primates are more likely to be suspensory. However, two species of the same size can have radically contrasting positioning behaviours, just as they could have quite different diets. Differences in diet are linked with differences in metabolic rate. With body size similar or with the calculation done allowing for the effects of body size, fruit-eating species have higher metabolic rates than leaf-eaters.

Hence, there are clear links between body size, life-history patterns, diet and other ecological parameters, brain size, and positional and ranging behaviours. A frequently observed phenomenon in evolutionary change is size change: an ancestral species evolves into a larger or smaller descendant. Two descendant species derived from the same ancestor might be different in body size. So many other features are linked to body-size differences, and body-size changes seem to pose at least no more of a morphological problem than more specific adaptive shifts, that size changes allow organisms to enter new niches relatively easily.

Brain and special senses

Brain size relative to body size is large in primates, especially in anthropoids. Primates, again particularly anthropoids, are relatively intelligent animals. This is reflected in their puzzle-solving and manipulative skills, although those reflect abilities honed in social contexts at least as much as in environmental contexts. Some hominoids exhibit mental qualities that are best described as 'conscious'. Primates have a high degree of visual acuity with binocular and frequently colour vision, and their auditory sense is also well developed. Olfaction is especially significant among prosimians, and although reduced remains important in anthropoids. Neuromuscular co-ordination is highly developed, and tactile stimuli, espectially from fingers and hand, are of considerable importance.

Most non-human primates, and essentially all monkeys and apes, live in year-round social groups. Interactions between individuals and groups are complex and subtle, and depend on sophisticated communication systems. Primate communication heavily involves vocalization, and in consequence both vocal and auditory systems are highly developed. In addition, body and tail postures and, particularly, facial expressions convey important information, and hence neuromuscular control of facial expression and body position, and visual acuity are subject to strong 'social' selection pressures.

This array of sensory, motor, and communicative skills can be clearly and closely related to the facts that most primates are arboreal creatures, highly selective feeders on high quality (mainly plant) foods, and live complex lives within social groups made up of well known and often related individuals.

Almost all parts of the brain are enlarged in primates, but this applies particularly to the cerebrum relative to the midbrain and brainstem. The cerebellum is also important. There is a good deal of discussion and some disagreement about exactly how brain size varies within the order, but anthropoids do have larger brains than prosimians by a factor of at least two, relative to body size, and the large living hominoids have larger brains than the living monkeys.

The cortex is divided topologically into several distinct regions. Lying around the edges of the cortex is the limbic system, or palaeocortex, which is involved in behaviours such as aggression, sex, feeding, fear, and a range of attentional biases and behavioural predispositions; namely, the physiological substrates of a wide range of emotional behaviours critical for survival and for interactions with other individuals. Primates have well developed limbic systems, and they have also elaborated other so-called neocortical regions.

The neocortex receives input information from sensory systems such as vision, the spectrum of auditory frequencies, the skin surface, and muscles, tendons, and joints, and sends output instructions to the muscular system. Electrical stimulation of primary sensory and primary motor regions shows that the parts of the body are represented like a map on their surfaces: the proportions of the map vary depending upon the importance of the particular body part to the species. Thus, in humans relative to a macaque monkey, tongues, lips, and thumbs are 'over-represented'. The primary visual centres are in the occipital lobe, those for

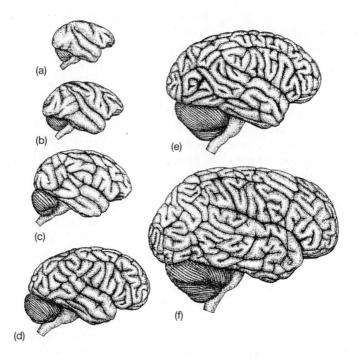

Fig. 3.5. Living anthropoid brains: (a) *Saimiri*; (b) *Macaca*; (c) *Mandrillus*; (d) *Pan*;(f)
Homo. The extinct early hominid *Homo habilis* is reconstructed in (e).
(Courtesy of Dr T. Deacon.)

auditory stimuli in the temporal lobe, and somatic sensory in the parietal
lobes, separated from primary motor cortex by the central sulcus. Each
primary sensory and motor area has adjacent to it an association area,
receiving input only from the primary areas. In non-human hominoids
there is a parietal association area receiving projections from all the
association areas. This is expanded greatly in humans.

The brain is made up of functional cells called neurons, which have
excitable surfaces and which send out signals. They are complex in shape,
with long appendages known as dendrites which generally receive
stimulation from other neurons, and axons which primarily stimulate
other cells. At the gross level the brain is heterogeneous, with groups of
neurons of common morphology and function, and bundles of connecting
axons which carry stimuli between them. Neuron cell bodies and their
dendrites are called grey matter, axon bundles are white matter. There are
two typical kinds of grey matter. The first are nuclei, usually deep
structures which receive inputs from one direction and have outputs exiting
from the other. Nuclei typically collect, relay, co-ordinate, and integrate

input (sensory) and output (motor). They generally mediate between periphery (sensory and motor), and the organizing and analysing functions of the cortex. A second kind of grey matter is located on the surface of the brain, forming the cortex. Because it is a surface structure, input and output connections enter and leave from the same side.

The surface of the brain increases in area rapidly as its total volume grows; hence larger brains have surfaces which are folded to produce deep valleys known as fissures and sulci (separating major brain segments) (Fig. 3.5).

Connections between different parts of the brain can be studied in non-humans by injecting tracer chemicals into neurons and seeing where they are carried by the axons. The mammalian brain is as conservative as the skeleton, in that similar circuits linking homologous nuclei and cortical areas are quite stable over a diverse range of species. However, relative size of functional units and circuits varies greatly (as with the mammalian skeleton: compare human hand and horse hoof) and so, therefore, do functions.

Careful measurement of various portions of the brain makes possible studies of the relationships between parts of the brain, the whole brain, and body size. Although studies are still at an early stage, it is clear that the mammalian brain is organized in regular and predictable ways. Larger animals have larger brains, but the various parts do not vary in mathematically identical ways, so species with large brains do not preserve the exact internal proportions of those with smaller brains. Differences in proportions of nuclear and cortical structures, and of circuitry, are correlated with differences in function. An important issue is the extent to which the greater brain size in bigger animals, for example of large apes versus cercopithecoids, reflects merely greater body size or includes some evolutionary increase. This is debated still, although it is clear that the human brain is roughly three times its expected value for a hominoid of that body size (Fig. 3.6).

The quality and complexity of the special senses, manipulative, and communication systems of especially hominoid primates reflect their eclectic diets, their varied positional repertoires, their object-using skills, and their elaborate social milieu. The sense of smell is important for prosimians in the recovery and selection of food items and in social interactions and marking behaviour. Lemurs and lorises have a naked, glandular rhinarium, a segment of skin connecting external nose with mucous membrane binding lips to gums. In anthropoid primates the olfactory system is somewhat reduced relative to body size, and greatly reduced relative to overall brain size. However, the sense of smell is still important, in particular in social interactions. The role of pheromones in monkey, ape, and human behaviour needs further exploration.

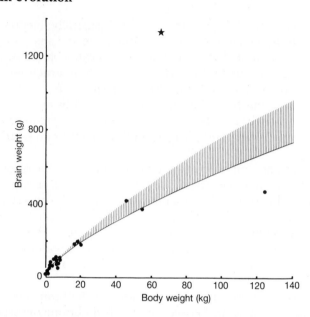

Fɪɢ. 3.6.   Brain weight plotted against body weight for living anthropoids. The star is *Homo sapiens*. The solid line is the regression for all species except *H. sapiens*; the hatching shows the regression line shift with the addition of *H. sapiens*. (Courtesy of Dr T. Deacon.)

The visual system is elaborate in all primates. Eyes are relatively large, directed forward, and, particularly in monkeys and hominoids, adapted for stereoscopic, diurnal, colour vision. Peripheral adaptations of the eyeball and retina are matched in the high fibre content of optic tracts, in the complexity of midbrain and cortical connections, and in the size of visual association areas. It is now believed more likely that early primates evolved their particular combination of visual adaptions as part of a visual–olfactory–manipulative system involving feeding behaviour, rather than simply arboreality as such. The elaboration of vision in anthropoids, along with enlargement of the brain, may also be causally linked to the demands of finding highly specific food items (mostly plant parts) in an environment which is spatially and temporally both heterogeneous and unpredictable (among other factors).

Communication

The temporal lobe is the region in which auditory stimuli are processed, sorted, and analysed and where behavioural sequences are comprehended; order and context are important. Complex sequential behaviours, especially involving vocalizations, are recognized and interpreted. Vocalizations in non-human primates are entirely controlled by limbic and midbrain areas, there being no neocortical involvement.

Communication systems in primates are complex, in part because the mainly social contexts in which they are used are complex. Information is conveyed between individuals within and between social groups — information about an animal's emotional state, degree of arousal, sexual interest, attitude towards status interactions, and so forth. Regardless of debates about whether or not apes are capable of learning human-like languages of various kinds, it is clear that real communication systems of anthropoid primates, especially hominoids in their natural habitats, are more elaborate than had been previously believed. While they may not have qualities typical of human language such as displacement and arbitrariness, for example, at least in the case of apes and some monkeys information which is highly context dependent can be transferred, along with information about things remote in space and perhaps in time too. Information is conveyed by vocal signals, by postures, gestures, and facial expressions (made easier in anthropoids by the free upper lip and elaborate facial muscles).

The precise nature of primate communication in the wild is a problem only now beginning to be tackled, and our knowledge of underlying neuroanatomical and neurophysiological subtrates is likewise at a still immature phase. As noted above, vocalizations are controlled by connections to the limbic system and midbrain, although the temporal lobe is implicated in deciphering vocalizations. In monkeys, presumably apes as well, there is a region of frontal cortex adjacent to the motor region homologous to the area in humans known as Broca's area, implicated in language production in humans. In non-humans it controls face and mouth movements in facial expression and feeding, and is connected by circuits to the temporal and parietal regions as well as the limbic area. Connections as well as areas are homologous in humans, monkeys, and apes, although overall function or output of the systems differs radically.

Tools

Tool use is rare among non-human primates. Objects are used for the purpose of obtaining food, or in display behaviours, by only a few primate species. Baboons have been reported to roll stones at or drop branches on human observers. The New World capuchin monkey *Cebus* has considerable object manipulative skills as revealed in laboratory tests, as do all the hominoids. The best known tool user is the chimpanzee, *Pan troglodytes*, which uses a range of natural objects in aggressive displays and in feeding. Tools are important in expanding the range of what can be called extractive foraging. The careful preparation and use of termite-fishing sticks, fig-gathering hooks, stone or wood nut-cracking hammers require forethought, manipulative skills, and the ability to control with precision and sometimes force objects which are extensions of the body.

The capacity to plan such extractive activities implies a brain that is organized differently from those of other primates.

Only humans, the tip of a lineage now totally dependent on tools for survival, surpass chimpanzees in the range and sophistication of tool-making skills and goals.

Life-history patterns

Primates are slow maturers and they live long. This is a function both of their large body size and their generally large brain size. Brain tissue matures more slowly than other tissues, and it has been suggested that this constrains many other growth and maturation rates. Primates have long gestation times. Almost all anthropoids produce one, large, large-brained, and dependent offspring per birth. They reproduce relatively late and infrequently, and for their body sizes show surprising longevity.

Reproduction

Primates mature sexually at a slow rate. Sexual cycles are best understood in females, although males of some and perhaps many species have physiological and behavioural cycles related to reproduction. However, females are more important in the production of offspring because their biology places constraints on reproductive rates.

The sexual cycle in females consists of up to three linked phenomena. (i) A menstrual cycle, of variable length, but averaging around 30 days in anthropoids, involving the gradual build up and then rapid shedding of the uterine lining, generally with blood loss. (ii) An ovarian cycle, which may or may not be of similar length, involving the release of generally one ovum or egg from the ovaries. (iii) An oestrus cycle consisting of the rise and fall of mating-associated behaviour, often including changes in female receptivity and attractiveness. Frequently, though not invariably, oestrus and ovulation are linked. There is often a period of lowered fertility in young females after they commence menstrual–oestrus cycling, probably because ovulation is irregular (so-called 'adolescent sterility').

Gestation lasts 6–9 months in anthropoid primates, long for animals of their size. Infants are normally born with brains roughly half the volume of those of adult females; they are dependent on their mothers for long periods after birth, and are not weaned for anywhere from 1 (some monkeys) to 4 years (chimps, gorillas, some human hunter–gatherers). The transfer of energy in the form of milk is very efficient, but it imposes nutritional costs on the mother. It has been variously estimated that primate female energy requirements are raised somewhere between 20 and 50 per cent during lactation, and pregnancy itself is clearly similarly demanding metabolically.

During lactation and prior to weaning, sexual cycling is first totally and then partially suppressed. This is the period known as post-partum or lactational amenorrhoea. Following weaning or after death of an infant, a female will commence cycling again quite quickly.

Individual and social behaviour

Primates are intelligent animals and capable of complex behaviours involving learning and the integration of past experiences, prediction of future conditions both physical and social, social manipulation, and communication. It would be unwise to partition behaviour into learned and innate or instinctive components, or in the case of humans into biological and cultural (human or cultural behaviour is a result of our species biology). A more useful approach is one which sees behaviour patterns as ranging along a spectrum from relatively open (modifiable during development) to relative closed programmes. We can think of many anthropoid behaviour patterns as relatively open, with internal behavioural states and external behavioural acts well integrated into complex and usually modifiable behavioural programmes.

Most primates and almost all anthropoids are social and live in permanent (that is, year round) social groups which usually comprise at least one adult male and at least one adult female, plus the female's offspring (which may also be the male's offspring). A few species are monogamous, for example the gibbons (*Hylobates*), but most are polygamous, generally with more than one adult male in the social group. It is of interest to note that monogamous species exhibit minimal dimorphism in body or canine size (though not all non-dimorphic species are monogamous), while in most polygamous anthropoid species males are larger and have much longer canines than females.

At the risk of some over-simplification, primates can be broadly categorized as living in two kinds of social and reproductive groups: those which are 'male-bonded' and those which are 'female-bonded'. These terms describe which sex forms the core of the social group, related males or related females, respectively, and which sex transfers from its natal group and moves to others before finally settling down. Most primate species are female-bonded.

Within groups of cercopithecoid monkeys, for example, the females can generally be ranked in a hierarchy (often called a dominance hierarchy) on the basis of priority of access to food, grooming, sleeping areas, the ability to prevail in aggressive interactions, and so forth. Female offspring tend to rank behind their mothers and ahead of other females in the hierarchy. Female hierarchies are stable. Male hierarchies are less stable, and males rely more on coalitions with other males which can be quite fluid. Age and length of time spent in a group, perhaps reflecting familiarity, have an effect on male 'poise' and 'confidence', and these in turn affect dominance interactions and status. Relationships in hominoids are different, and in groups containing several females the animals are normally not kin, and transfer between groups. In chimpanzees, males are the related sex. In all social anthropoids, relationships between the sexes are generally complex and subtle. It is slowly becoming clear that anthropoid behaviour depends greatly on who the actors are: how they are related, how well they know each other, whether or not their relationships are usually friendly. Primates are intelligent, emotional, long-lived animals, and understanding them requires both long-term field studies and a recongition of the subtleties of their behaviour. There is some behavioural differentiation within primate social groups along age and sex lines, but except in humans, there is little significant non-reproductive division of labour.

Baboons, apes, and humans

We have reviewed some general anatomical, behavioural, and ecological characteristics of primates and now move on to discuss specific primate species. There is insufficient space to describe all, most, or even 'representative' species of all major primate categories, and we shall look more closely at just a handful.

Baboons

In addition to the hominoids, the savannah baboons of Africa are of considerable interest. They were initially studied at least in part because, despite their greater phylogenetic distance from us, their open country habitats are believed to be more similar to those in which early hominids lived than are those of the more forest-living apes. Regardless of these issues, the baboon genus *Papio* is now one of the best studied primates, and it is known in considerable detail in a range of habitats. Four closely related and very similar species of *Papio* are found in the woodlands and savannahs surrounding the west and central African rainforest. Indeed, it is possible that these populations represent a single species. We shall treat *Papio papio, Papio cynocephalus, Papio anubis*, and *Papio ursinus* as separate species, but similar enough for morphological, behavioural, and ecological descriptions to be interchangeable.

Baboons are markedly dimorphic, males weighing 20–30 kg, females about half that. Brain volumes average around 175 cm^3. Baboons are quadrupeds and are at home on the ground, in trees, and climbing on cliffs. Their habitats range from arid savannah to forest, though their modal habitat is bush or wooded grassland. Their forelimbs and hindlimbs are of approximately equal length. Fingers are relatively short, thumbs long, and they have considerable dexterity. Trunks are long with long lumbar regions (six or seven lumbar vertebrae). There is a short tail. In these skeletal features, *Papio* is basically similar to other cercopithecoids (Fig. 3.4). In the dentition, molars are bilophodont.

Baboons are eclectic feeders mainly on plant food which is frequently dispersed in patches; they also eat insects, which they catch using their hands. Some groups hunt, being predators on small game such as young gazelles or small monkeys. The extent to which groups hunt in part reflects local competition from carnivores, and normally meat forms only a very small proportion of the baboon diet. In one well studied area carnivores have been removed and male baboons have become infrequent, but regular hunters. Meat is a prized item; it is not shared between males, but is shared with oestrous females and those females with whom the hunters have 'special relationships' or 'friendships'.

Social groups (troops) normally range in size from 30 to 150 individuals, and contain several to many adults of both sexes, plus offspring. Baboons are, like most cercopithecoids, female bonded: daughters remain in their natal troops while sons migrate out. Males often transfer several times before settling down in a troop. The female dominance hierarchy is clearcut and stable. Where individuals can be identified and carefully monitored, daughters are generally ranked in the hierarchy immediately below their mothers and in order of decreasing birth rank. In contrast, male hierarchies are less stable and much more fluid. Male behaviour is 'flashier' than female and for that reason earlier researchers tended to concentrate more on the males. It took some time to recognize that female behaviour, though more muted, was as important as that of males in modulating group behaviour. Male hierarchies are often not linear, and coalitions are important. Rankings according to different criteria — access to oestrous females, ability to win fights, and so forth, are often not concordant. The dynamics of inter-individual behaviour depend on many factors: fighting ability, social skill, familiarity with other animals.

Males become sexually mature at around 6 years, but generally do not become active breeders until they have reached adult size when they are 9 or 10. Even then they may not have as much access to ovulating and therefore fertile females as older and more familiar males. Females become sexually mature at around 4 years and give birth for the first time at about 5. Both male and female choice are important in determining who mates with whom; this is critical at those points of the oestrous period during which the female is fertile.

Gestation takes 6 months, which is long for a small animal, and this reflects the relatively large brain size. Infants are at first very dependent on their mothers and only gradually begin to play a more independent role in the life of the troop. Weaning is usually complete by just over a year, and within a few months the mother begins oestrous cycling again, normally becoming pregnant within a few cycles. The average interval between births is 2 years, and the length of the interbirth period and the duration of post-partum amenorrhoea are linked among other factors to the intensity and duration of suckling during the third to eight months of infant life. Adult females are normally pregnant or lactating, which explains why females feed as much as the significantly larger males, since heavy metabolic demands fall on them.

In most environments all individuals, even the apparently carefree and playful infants and juveniles, in fact spend the bulk of their waking time feeding or moving. Less than 25 per cent of time is spent resting or in social behaviour. Much of infant social behaviour involves play, which serves a variety of functions including one of simulating situations in later adult life. A good deal of interaction among individuals involves grooming, one animal using the hands to clean another's fur. This has dual functions: hygienic and social. Grooming is actively solicited and animals will attempt to groom others. It is important in a varied range of social contexts and can be seen as an essential part of networks of social exchange.

A previously unreported phenomenon has recently been noted in some baboon groups, so-called special relationships or friendships. These are relationships between a particular male and a particular female marked by periods of proximity, grooming, and infant care by the male, during non-sexual times as well as oestrus. Infant care occurs whether or not the male is the infant's father. Friendships are somewhat more common with resident males than with newcomers, even though the frequently younger and more aggressive newcomers may win fights over low priority items. Females are more cautious of newcomers, perhaps fearing attacks on infants, and perhaps feel more relaxed with familiar males.

These patterns have only emerged from long-term studies in situations where individuals can be identified. Even a 10-year study, extremely costly in resources, energy, and commitment, samples only a fraction of one generation.

Basically similar behaviour patterns, including friendships, are seen in several other related monkeys, particularly species of the Asian cercopithecid *Macaca*.

Apes

Considerable advances have been made in the past decade in our knowledge of the behavioural ecology and evolutionary biology of the hominoids (Table 3.2). Hominoids are normally subdivided into three groups: the lesser apes *Hylobates* (gibbons), the hylobatids; the large apes *Pongo* (orangutan), *Pan* (chimpanzee), and *Gorilla*; and hominids — humans and their close relatives. Because humans and African apes are

TABLE 3.2
Living hominoids

	Gibbon *Hylobates*	Orangutan *Pongo pygmaeus*	Chimpanzee *Pan troglodytes*	Bonobo *Pan paniscus*	Gorilla *Gorilla gorilla*	Human *Homo sapiens*
Distribution	S.E. Asia	Borneo and Sumatra	Tropical Africa	Tropical Africa S. of Zaire Riv.	Tropical Africa	Global
Habitat	Forest	Forest	Forest and woodland	Forest	Forest	Almost all
Body weight (kg)	5–10 (both sexes)	84 (m) 38 (f)	40–60 (m) 30–50 (f)	45 (m) 35 (f)	150 + 75 + (f)	65 (m) 58 (f)
Cranial capacity	100–125	425 (m) 370 (f)	400 (m) 355 (f)	350 (m) 345 (f)	535 (m) 460 (f)	1400 (m) 1300 (f)
Social unit	Male and female plus offspring	Females plus offspring. Solitary males.	Community. Females plus offspring. Males in groups. Mixed.	Community. Mixed male–female groups.	Adult male plus females plus offspring. Some solitary males.	Families, clans, tribes.
Diet	Mostly leaves and fruits; also insects.	Predominantly fruit; some leaves and bark.	Fruit, leaves, shoots, buds, insects, mammals.	Fruits, stems, leaves, shoots, insects, mammals.	Vegetarian; young leaves, stems, roots, fruit.	Omnivorous
Female age first birth (years)	6 +	10–12	12–14	?	9–11	16–20 +
Gestation period (months)	7	8	8	8	8	9
Average longevity	30	40–45	40–45	?	40–45	80–90

more closely related to each other than any of them are to *Pongo*, the large apes are no longer normally grouped formally in Pongidae.

The hominoids and cercopithecoids diverged, probably in Africa and probably 20–25 Ma, the split perhaps being related to a combination of differences in feeding behaviour, habitat preference, and positional behaviour. Morphological contrasts between the groups are marked in all body parts. The ancestors of gibbons and orangutans evolved in the early or middle Miocene (24–12 Ma) and spread to Asia. The ancestor of humans and African apes probably remained in or close to Africa, differentiating into separate lineages during the late Miocene, 12–5 Ma. The large hominoids live longer and mature more slowly than monkeys and gibbons because they are larger and have bigger brains. Hominoids are now less diverse than they were, probably because the areal extent of preferred habitats has declined since the Miocene. They are under particular pressure from humans now.

Lesser apes

Gibbons and their close relatives the larger siamangs (approximately 6 and 10 kg, respectively) are small arboreal Asian forest apes. Smaller species are frugivores while the siamang is more folivorous. They live in monogamous pairs which vigorously defend tiny forest territories. Sexual dimorphism is minimal and both sexes have projecting canines, presumably because females sometimes fight in territorial disputes. Offspring are expelled as they mature. Hylobatid positional behaviour is richly varied, and includes brachiation which is probably an adaptation for the rapid monitoring of territorial boundaries. In addition gibbons have a variety of orthograde postures and locomotion involving climbing, hanging, standing, and walking. Anatomically, they differ from cercopithecoids and the presumed catarrhine common ancestor in having broad flat chests and five lumbar vertebrae. Forelimbs and hands are elongated — derived characters, and hindlimbs are also long, although arms are 65 per cent longer than legs. They are adept bipeds and have a simple muscle system for balancing the trunk during bipedal walking. Both thumbs and big toes are large and well muscled, important in climbing more vertical supports.

Great apes

The large apes are all too big to be successfully territorial, because individuals must range too far in search of food. Their social organizations are quite varied, and they are united by a negative feature: none are female-bonded.

Pongo pygmaeus, the orangutan, is a forest ape found today in Borneo and Sumatra. In the late Pleistocene it was more widespread throughout Indonesia, south China, and probably Indochina. Orangutans are large and very dimorphic, males weighing over 80 kg, females a little under 40 kg. Brain volumes average 425 cm^3 for males, 370 cm^3 for females. Females and juveniles are almost entirely arboreal, while the much larger males can spend up to half their time moving on the ground. They are contortionists, with a broad range of careful positional behaviours in the trees, in which they climb deliberately and suspend themselves using varying combinations of their limbs. They have very long arms and hands, very short legs with long feet which are rather hand-like in having long curved fingers and toes, and a small first digit. Arms are almost 80 per cent longer than legs, and joints permit a great deal of mobility and flexibility. The lumbar region is short and inflexible, and there is little gap between pelvis and ribs. On the ground, orangutans are quadrupedal, and walk with their fists and feet in varying positions, and with fingers and toes tucked under.

They feed on a wide range of plant foods, including fruits which are often dispersed and frequently hard. Cheek teeth have moderately thick enamel and low cusps, and are moderately large relative to body size. Canines are very dimorphic. An adult female and her offspring occupy a small feeding area roughly $\frac{1}{2}$ km^2 in size. One older and high ranked male ('resident') often has a territory overlapping those of a few females, perhaps two to four, and these females normally mate with their resident male and turn to him for protection against other males, for themselves or for their infants. Other mature and subadult males are solitary and tend not to be tied to particular areas.

Females become sexually mature at 9 or 10 years and first give birth around 12. They have an oestrous–menstrual cycle, and oestrus lasts 5 or 6 days. Females exercise some choice in mating and they generally prefer courtship by the resident dominant adult male rather than by one of the migratory males. During oestrus, females generally form consortships with a male lasting 2–6 days, during which they copulate only three times on average. Gestation time is almost 9 months. Infants remain close to mothers for long periods and are not weaned until about 4 years of age; interbirth intervals are long, varying between 3 and 7 years.

The number of oestrous females available in a particular area will be low and, since the large resident males are not particularly mobile, this means that males may copulate barely more than a few times each year. The most frequent as opposed to preferred type of mating is that forced on a female by a large non-resident male. Oestrous females and their offspring are thus protected by remaining close to resident males. Female offspring remain close to their mothers throughout life while sons leave and become wanderers before some establish resident territories.

Orangutans raised in captivity are highly sociable and friendly as long as they are well supplied by humans with food. However, on return to the wild they begin to forage on their own and social bonds break down. Thus, individuals in the wild are basically solitary because of food distribution, foraging patterns, and body size. (Large animals feeding on mostly dispersed foods don't aggregate because they can't.) Females are therefore vulnerable and seek high ranking males. Orangutans rarely use tools in the wild, although captive animals learn to use tools with skill and alacrity under experimental conditions.

Gorillas are African forest apes. There are three sub-species: western lowland, eastern lowland, and eastern mountain gorillas. Almost all that we know about gorillas is based on the not necessarily typical mountain gorillas of Zaire and Rwanda, *Gorilla gorilla beringei*. They are large and mostly ground-living apes, males weighing over 150 kg, sometimes much more, and females only half that. Male brain volumes average 535 cm^3 and female 445 cm^3. Like the two species of chimpanzee, the gorilla spends time in the trees and on the ground, climbing, swinging, and hanging with body relatively upright in the trees, moving quadrupedally or very occasionally bipedally on the ground. Knuckle-walking is the preferred pattern of terrestrial locomotion. Forelimbs are some 16 per cent longer than hindlimbs. Fingers and toes are short. The lumbar region is short and inflexible, and the chest very broad.

Mountain gorillas feed mainly on the ground on up to 100 species of plants, mostly leaves and shoots. They are highly selective and non-competitive feeders. (However, lowland gorillas of west Africa eat considerably more fruit.) Gorillas live in social groups made up of a highly ranked mature male (a 'silverback' from the grey-white saddle pattern on the back), two to six adult females plus their infants, and sometimes with other younger males which are probably related to the silverback. Females migrate out of the group as they mature. Relationships among group females, who are generally unrelated, and between males, who generally are related, for the most part are relaxed and tolerant.

Solitary males and groups of (mostly unrelated?) males live apart from mixed groups, and are often in conflict with the resident silverbacks. Groups are sometimes raided for females, and infants can be killed by outside males. Groups can ideally be thought of as comprising a dominant male and several females plus his infants, all of whom stay close to the male for protection. These groups are cohesive and long lasting. They feed together peacefully and with little competition, because food resources are abundant and evenly distributed. Sexual behaviour is rather like that of orangutans. Oestrus lasts 1–2 days, and copulation rates are very low: one bout every 1–2 years per female. Gestation length is 8 months and the interbirth interval about 4 years.

The common chimpanzee, *Pan troglogdytes*, is the most widely distributed of the African apes, ranging from Senegal in west Africa to Tanzania in east Africa. There are three recognizable subspecies — eastern, central, and western, the first two partially overlapping with current gorilla distributions; the overlap was probably considerably greater in the past. The body weight of chimpanzees is difficult to determine for wild individuals — a general problem for primates. There is some evidence to suggest that the central and eastern subspecies differ. In the former, males average 60 kg and females 47 kg, while in the latter males weigh 43 kg and females 33 kg. Thus, sexual dimorphism is considerably less marked than in gorillas. Brain volumes are 400 cm^3 for males and 355 cm^3 for females. Preferred habitats range from forest to very open savannah with less than 5 per cent tree cover; trees, though, are essential.

Chimps are both arboreal and terrestrial, their diets consisting mainly of ripe fruits which are located in trees. Over 50 per cent of time is spent feeding or resting in trees. These food sources are widely scattered and, even in forest, moving between food sources involves mainly movement on the ground, over 90 per cent of such movements in fact. Chimpanzee positional repertoires are extremely varied: they are proficient climbers in the trees, swinging by their arms and using them in climbing, as well as moving on the ground. They are quadrupedal knuckle-walkers, and quadrumanous climbers and scramblers; they engage in bimanual suspension and arm-swinging, leaping and diving, and bipedalism — both in the trees and on the ground. Their anatomy reflects this diversity; knuckle-walking is clearly a reflection of an arboreal ancestry, on which is superimposed the effects of terrestrial life. Chimp-type quadrupedalism (and that of other apes) is about 50 per cent more expensive energetically than other kinds of primate quadrupedalism, perhaps reflecting the long 'arboreal' hands and feet, and forelimb musculature, stout back, and short legs of a large climber (Fig. 3.4).

In addition to ripe fruits, common chimpanzees supplement their diets with nuts, leaves, insects, and small mammals, some of which are caught by groups of males. Animal food is not a large part of the diet, but in some areas it is a regular part. Meat is frequently shared, sometimes between males, or with oestrous females. Plant foods are also sometimes shared.

The best studies of chimpanzees are of the eastern subspecies, especially populations at Gombe, east of Lake Tanganyika. The well studied communities are not necessarily completely typical of all chimpanzees. Indeed, this is unlikely, given their wide geographical range and broad ecological tolerance. With that caveat in mind, all chimpanzees live in communities containing up to 100 individuals, community members sharing a home range, but living and foraging alone or in small groups, and without permanent large group associations.

Most adult females are immigrants from other communities, and most males do not migrate, so communities tend to be made up of related males and unrelated females. Females are more solitary, generally ranging in small territories with their young, though these territories overlap completely with those of other community females. Although female–female relationships are generally not aggressive they are only passively tolerant; aggregations occur at rich feeding sites, because of the attractions of food, not other females.

Males are more social and spend more time together than do females. They range widely, defending community boundaries, seeking individual females or groups of females. Females newly transferred into the community often associate with males, probably for protection against other females. Relationships among the males, who are often related, are both competitive and co-operative. Dominance interactions are important, and hierarchies are fairly clearcut, though fluid. Coalitions are important, and social manipulative skills are at a premium.

Hence, chimpanzee groups can be characterized as loose aggregations of individual females, monitored by a group or groups of males. Dispersal is common because most foods are dispersed, and because of feeding competition — it being less efficient feeding in large groups than when solitary or in small groups. The co-operative behaviour of males is difficult to explain in terms of feeding, and it is suggested that the closed social network of males is better explained as related to territorial defence and boundary patrolling, because with territory go females: so sociality is related to reproductive competition. Females become sexually mature at around 10 years, and have their first birth around 13–15. Oestrus itself lasts 9 or 10 days. Females come into oestrus unpredictably. Males are not particularly interested in anoestrous females (generally those that are pregnant or lactating), but are strongly attracted to oestrous females. Within a community, at least one female will be in oestrus over 50 per cent of the time and oestrous females are very attractive to males. Mating is promiscuous, but consortships are often formed. Males are more successful in attracting females the more they share food with, groom, and spend time with them early in the oestrous period. Ovulation occurs in the last day or so of the period, and the most dominant male has a very high percentage of copulations during that time. Pregnancy lasts about 8 months, and infants closely associate with their mothers for a long time. Lactating females experience amenorrhoea, and do not copulate. Interbirth intervals average 5 years in the well-studied populations.

Interactions between communities involve a range of behaviours from avoidance to outright aggression. On several occasions the males from one community have been observed to stalk, attack, and severely injure or kill the males of another, eventually taking over territory and females, some of whom were also attacked.

Chimpanzees exhibit a wide range of tool-using and even tool-making behaviours. Leaves are chewed and used as sponges, twigs and stems trimmed for termite fishing, branches modified as hooks to garner fig-laden branches that would otherwise be out of reach. In west Africa, chimps have been observed using wood and stone hammers, often against wood and stone anvils, to crack nuts. Both hammers and anvils are carefully selected and transported to potential feeding sites, and sometimes even cached there. Learning the skills necessary for tool use takes some time: up to 6 years for termite fishing, and even more for nut-cracking. There are some interesting sex differences in tool use, females being more skilled and persistent than males, possibly because of a combination of co-ordination, persistence, concentration, and more appropriate strength.

Pan troglodytes social organization reflects the coexistence of powerful, co-operative males and less powerful, less co-operative, and more tool-competent females. Both sexes are highly intelligent and emotionally labile, capable in the wild of considerable behavioural flexibility, planning, and forethought. Dispersed food sources and complex interindividual relationships are two out of no doubt many factors contributing to the behavioural competence of chimpanzees. It is clear, for example, that chimp communication in the wild is complex, subtle, effective, and poorly understood by us. Controlled laboratory experiments show that while chimpanzees cannot be trained to speak or to produce human-type non-vocal languages, they do have the ability to symbol, are clearly self-aware, and are highly and recognizably intelligent. Our challenge is to understand them on their terms, not to consider to what extent they do or do not achieve human capabilities.

Pan paniscus, the so-called pygmy chimpanzee or bonobo, is confined to lowland rain forest in Zaire, south of the Zaire River. Although still not as well studied as *Pan troglodytes*, our knowledge of this species is expanding rapidly. Despite their vernacular name, bonobos are not dwarfs or pygmies. Their body weights and statures are essentially the same as those of eastern common chimps (males 45 kg, females 33 kg). Relative to *Pan troglodytes*, *Pan paniscus* has shorter arms and a somewhat longer truck; males, especially, appear robust in build. Heads, faces, and teeth are all smaller than in *P. troglodytes*, and overall skeletal and dental dimorphism is very low despite differences in body weight between males and females. Brain volumes are also lower: 350 cm^3 in males and 345 cm^3 in females.

Bonobo positional repertoires are essentially the same as in forest *troglodytes* groups; they have a high component of arm-swinging, and leaping and diving, and are agile and adept in the trees. In addition to quadrumanous climbing in the trees, they also move bipedally and by quadrupedal knuckle-walking both in the trees and on the ground.

Their social organization into communities is apparently similar to that of common chimpanzees, and although data are not yet fully adequate it looks as though females transfer between communities. The subgroups within communities are larger than in *troglodytes* and are usually mixed, with the sex ratio approaching 1. Relationships between female bonobos are more affiliative and tolerant, and less solitary than in common chimps, and animals feed near each other and frequently share food (a rare occurrence in *troglodytes*). Male bonobos on the other hand are somewhat less social with other males, and share less food. The two sexes aggregate more closely together. Unlike common chimpanzee females, female bonobos have sexual swellings for longer in each oestrous cycle, and begin cycling much sooner after giving birth, approximately 1 year later. Females mate essentially throughout the oestrous cycle (as they were known to do in captivity). It might be expected that interbirth intervals would be less than in *troglodytes*, although this is as yet unknown.

As we already noted, bonobos are exclusively lowland forest creatures. Their diets consist of both plants and animals — primarily insects, although small game is also taken. Plant and animal food is shared, more frequently by females. Bonobos prefer ripe fruits, and feed both on widely dispersed and on relatively clumped kinds. In addition, they frequently feed on the ground, particularly on arrowroot and ginger, plants which are generally abundant in tropical forests though not in woodland or savannah. These are also favoured foods of gorillas, and of common chimpanzees when available. However, they are absent from much of *troglodytes* ranges, and where present gorillas are, or were, significant competitors. Gorillas and bonobos, however, do not overlap geographically.

Thus, greater aggregations in bonobos relative to *Pan troglodytes* are made possible by more abundant and evenly dispersed food resources permitting more tolerant feeding, especially by females. Males are as important protectors of females and their infants as in the other apes, and males themselves gain protection by associating with others since single males are vulnerable to attack. In general, levels of aggression are lower, and affiliative and sharing behaviours higher in bonobos than in other apes. This, coupled with certain anatomical features (longer legs, slightly shorter arms) and the increased sexual receptivity of females has made them a species of great interest in speculations about early hominids. In addition, bonobos are at least as smart as common chimpanzees and show if anything greater skills at acquiring artificial languages which mimic in some features human language.

More studies are essential of all the apes before they become extinct in their natural habitats — or before those habitats disappear.

Humans

The fifth surviving species of large hominoid is *Homo sapiens*, genetically one of the African 'apes', and doing very well. Humans are today exceptionally abundant and they have an exceptional impact on ecosystems. They are found in most terrestrial communities, and generally live at relatively high population densities, frequently in population concentrations. Humans live mostly in 'complex' societies with particularly elaborate economic systems and technological patterns. However, prior to the shift which began 10–15 thousand years ago from food gathering to food production, humans lived very different and rather less varied lives as hunter–gatherers. Very few people today live this traditional way of life which typified all humans for at least 30 thousand years, from the emergence of modern humans late in the Pleistocene.

Hunter–gatherers are normally nomadic people and live in small groups where 'face-to-face' relations are the norm. Social relationships tend to be less differentiated and more egalitarian than in most human societies. Food is a mixture of hunted animals and gathered wild plants. The differences between them and most of us in social, economic, and political organization, and ritual behaviour flow from economic-technological changes over the past 10 thousand years, not from biologically determined behavioural changes. It is to human hunter–gatherers that we address our attention, both those few alive today and those known in the ethnographic present in enough detail to be used to infer the patterns of earlier humans prior to the food production revolution.

Humans are constrained in many ways by their primate heritage; similarities to other hominoids are particularly marked. Thus, humans have similar sexual cycles and give birth to single, large, dependent young. Diets are omnivorous. Social groups are composed of individuals who have varied, intimate, and complex long-term relationships. Within and between groups relationships apparently resemble those of other hominoids; normally, females marry out, males are initiators of inter-group conflicts. It is unclear to what extent such similarities are strictly homologous.

However, many features of difference do stand out, and these we now review briefly. Humans exhibit only moderate body size dimorphism, males averaging 60 kg, females 50 kg; brain volumes average 1400 cm^3 for males, 1300 cm^3 for females. (These values are for Australian Aborigines and are representative of the many other human populations which could have been cited.) In distribution of breasts, body hair, muscle, and fat, patterns of human sexual dimorphism differ from those of other primates. Humans are not particularly hairy, although hair follicles are abundantly present. Pigment cells (melanocytes) are in the epidermis rather than the dermis, and there are abundant eccrine (sweat) glands in the dermis. Humans thermoregulate mostly by sweating; they sweat profusely, and are highly dependent on water which must be consumed frequently and regularly. These and other physiological features suggest that human ancestors were tropical; indeed, we carry tropical microenvironments with us everywhere in the form of clothing and housing.

Humans are habitual, upright, striding bipeds, and bipedalism dominates our positional repertoire. During bipedal walking and running, the body has frequently only one support at a time, unlike the situation in quadrupeds. The centre of gravity is moved in as undeviating a line as possible during locomotion, to minimize energy expenditure. Human bipedal walking is energetically no more costly than that of typical mammalian quadrupeds (and less costly than that of *Pan*). Bipedal running is twice as costly, but humans are good endurance runners and the cost of locomotion does not vary with speed. All speeds up to maximum are equivalent, and in this humans differ from other mammals in which certain speeds are linked to certain gait patterns which minimize energy costs.

There are many anatomical correlates of bipedalism, including body proportions, muscle mass and its distribution, joint structure and orientation, and muscle orientation. Important are control of pelvic tilt and pelvic rotation to minimize centre of gravity displacement, a linked pattern of hip, knee, and ankle flexion and extension, and a particular pattern of weight transmission through the foot. The pelvis and hip region has been radically remodelled, producing low and broad iliac blades with outwardly oriented small gluteal muscles permitting controlled rotation and tilt of the pelvis. The hamstrings have a short moment arm. The foot has a parallel and stout big toe, short and straight lateral toes, and a wide heel (Fig. 3.7). Human bipedalism makes many things possible — carrying food and tools, use of weapons, generally improved manipulative skills, but the extent to which one or any of these played a role in the early evolution of hominid bipedalism is unclear.

Females have a 28-day sexual cycle which is basically similar to other hominoids although ovulation is more 'concealed', there being no obvious oestrous period (although there are changes in cervical mucus and some individuals monitor ovulation by feeling the ovum being released, so-called 'mittleschmerz'). There is no sexual skin, unlike *Pan*, but like *Gorilla* and *Pongo*. Relative to other hominoids, human males have large penises and moderately large testes.

Human newborns are large and have large brains, and are born at a more immature stage than other hominoids, particularly with regard to the nervous and immune systems. Considering the general mammalian pattern linking maternal brain and body weight to gestation time, human pregnancies are a couple of months shorter than expected. In most mammals, newborn brain volumes are roughly half those of adults, whereas in humans the ratio is about one-quarter. This pattern is linked to obstetrical constraints imposed by the maximum biomechanically feasible size of the maternal pelvis. Thus, human infants are highly dependent on mothers and other family members, often including fathers.

Breast-feeding in hunter–gatherers is usually frequent, on demand, and continues for several years. Human milk promotes growth and also contains antibodies — immunoglobulin A's — that are specific to infant gut pathogens. A feedback loop exists between infant and mother whereby infant pathogens stimulate secretion of specific antibodies in maternal milk. Prolonged suckling both stimulates milk production and inhibits

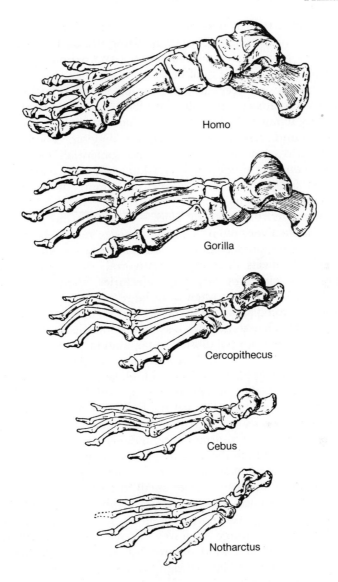

Fig. 3.7.   The foot skeleton of a series of living anthropoids and the Eocene prosimian
Notharctus. (After W. K. Gregory.)

ovulation (via nerve-hormone circuitry linking nipple, brain, ovary, and breast). The suppression of ovulation and menstruation (amenorrhoea) is a contraceptive, and together with cultural factors (abstinence, abortion, infanticide) limits births to 4–5 year intervals, as in apes. It is interesting to note reproductive differences between Kung hunter–gatherer women and women in a typical industrial society. The cumulative duration of lactational amenorrhoea is 15 years in Kung versus 1 year in the U.S.; Kung females menstruate cumulatively for approximately 4 years while U.S. females menstruate a total of 35 years.

Human hunter–gatherers are omnivorous, eating animal food, which is mostly hunted by males, and wild plant food, gathered mainly though not exclusively by females. Included in plant foods are tubers and roots which are underground resources and require digging. Food is generally transported and often shared within and sometimes between families. Foods are frequently prepared, by cooking, soaking, mashing, and it is often stored for days or in the case of hunters in strongly seasonal environments for months.

Humans forage from a central place — a camp or home base — which may not be moved very often. They forage over very wide areas and have even lower population densities than the large apes: 1/km^2 as opposed to 1–5/km^2. This reflects their more carnivorous diet. Males and females generally follow different foraging trajectories, though both return to camp each night. Humans often hunt large game, and this may involve complex and detailed planning involving several individuals and spanning several days. This has sometimes been called logistic hunting. The degree of carnivory varies from population to population, and often from time to time in the same population, depending on both the availability of game and the abundance of plant food. Most high latitude hunter–gatherers are heavily dependent on animal products, while in the tropics some populations at some times may get the bulk of their diet from plants.

Patterns of work vary: some groups work long hours to meet subsistence needs, while others may in times of abundance work much less. Humans are most similar to *Pan* in the little that is known of digestive anatomy and physiology, but show adaptations to a diet higher in meat (smaller stomach and large intestine, larger small intestine). Canines are smaller and incisor-like in both sexes and this, combined with the flatter face makes possible very powerful incisive biting. Cheek teeth have very thick enamel and are also arranged in a way that maximizes the effectiveness of the masticatory muscles, so that powerful chewing is possible. As with bipedalism, the original adaptive significance of these features is unclear.

Relationships between individuals and groups are varied and complex, even more so than in apes, and this reflects differences in communication (how much information is stored and transmitted), and in the extent to which external and internal worlds can be ordered and organized symbolically. Of particular importance is speech, a peculiarly human form of vocal language.

Human groups have as their basic unit the nuclear family, consisting usually of one father, a mother or mothers, and offspring. Families can be monogamous or polygamous, polygynous (usually) or polyandrous (rarely). All families are based on marriage and on exogamy, one sex almost always moving from its natal group. Marriage involves reproduction, sex rights, mate guarding, economic reciprocity, politics, and other factors. Adult males, usually biological fathers, are integrated into nuclear families to variable extents, and are involved in infant care and protection to variable extents. These differences from non-humans are made possible by and are apparently heavily dependent on language. The human family structure performs or satisfies several functions; resource defence, sharing, co-operation, reproduction, mate and infant guarding and protection, and alliance building. Which, if any of these, was initially important is unclear, strongly debated, and probably less tractable a question than is the case for systems such as subsistence or positional behaviour.

Families are clustered into larger units (up to tribes or nations) the smallest of which are communities within which marriages are usually contracted, averaging up to hundreds of individuals. In hunter–gatherers, nuclear families sometimes disperse to forage alone, and sometimes aggregate into groups of fluctuating composition. In hunter-gatherers community ties between families are maintained mostly through males. Within-community relationships emphasize sharing and reciprocity, and are relatively egalitarian at low population densities. Differential wealth and power is minimized, and there is an egalitarian ethic if not complete reality. Inter-community relationships in humans are often competitive and hostile, and male-initiated aggression is common.

Population density may be very high in non-hunters, up to $1000/km^2$ or more, and this is made possible by economic-subsistence differences. Division of labour and social stratification are, in part, consequences of higher population density, as are extreme variations in the control of economic and political resources — wealth and power.

Human brains are three times the size of those of apes (Fig. 3.6) and contain more neurons (25 per cent more in the cortex) which are more complex. Although there seem to be no literally new structures in the human brain, the relative sizes of nuclear, cortical, and 'circuit' structures are different, and this underlies the difference in output. Perhaps the most fundamental difference between human and non-human minds flows from our ability to generate and understand language. Human language is vocal, it uses words in strings or sentences, and word order affects meaning. Language permits communication about things, including things remote in time and space. Furthermore, we can communicate about conceptually remote things — abstractions. Language is referential, and the objects and concepts to which words refer are arbitrary and changeable and hence symbolic. They must be learned, but they are learned within a system which is clearly dependent on highly species-specific neural circuitry. Language involves movements, of body and face, as well as sounds, and these characters can be manipulated, strung together, and recombined with infinite variety.

Although the neurological bases of language are far from being fully understood, it is clear that certain portions of the cortex are essential for language comprehension and production. Broca's area in the prefrontal cortex is relatively greatly enlarged compared to its homologues in monkeys and apes (Fig. 3.8). In macaques, and probably other catarrhines, Broca's area equivalent is involved with face, mouth, and tongue movements during feeding and lip–face actions; vocalizations are

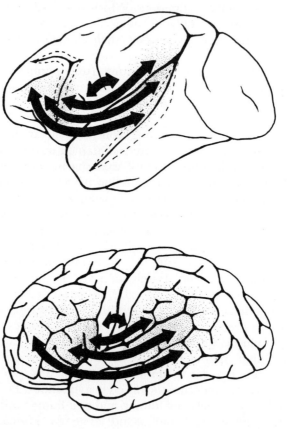

Fɪɢ. 3.8. Monkey brain (top) and human brain (bottom) drawn to same size. Interconnections of cortical areas in the monkey brain were determined by axon tracer techniques, and interconnections of cortical areas in the human brain were inferred from electrical stimulation experiments. Anterior and posterior language areas in the human brain are stippled, and approximately correspond to Broca's and Wernicke's classic speech areas. The stippling on the monkey brain indicates structurally homologous areas to human language zones. Although posterior temporal areas in the monkey brain are involved in vocal communication, the anterior zones have not been found to play critical roles in vocalization. (Courtesy of Dr T. Deacon.)

under limbic control. In humans, among other differences, the limbic connection is continued and expanded, but vocalizations, tongue, lip, and face movements are also integrated into neocortical circuitry. This allows language production to be expanded to non-emotional (non-limbic) motivations. Broca's area is adjacent to frontal motor cortex which is also differentially enlarged in humans, especially portions for the hand. This presumably reflects selection for tool-making and craft skills. Portions of the limbic system are relatively enlarged too, so the human brain reflects both our intellectual and our emotional richness. Which, if any, of these factors was involved in initial brain expansion is unclear. That they are all linked seems likely.

The functions of language now — whatever the original function(s) — are many and pervasive, and are primarily responsible for the enormous gulf in behaviour separating humans and non-humans. Language places constraints on friendly and non-friendly social interactions. It is important in mediating aggressive interactions non-physically, through negotiation or the invocation of authority. Language permits formalized systems of marriage, kinship, and incest avoidance, and here it is the rules as much as the behaviours themselves that differ from non-human situations. Language induces belief, encourages sharing, checks cheating and makes it possible, allows inheritance (sharing over generations), division of labour and co-ordination of complex economic activities, pooling and redistribution, peaceful crowding, complex alliances, warfare, etc.

'Culture' is a favourite anthropological term and it is used in several different ways. For example, stone and other kinds of tools are often described as 'material culture'. Sometimes the term means 'learned behaviour', and some non-human species can then have 'culture' or 'proto-culture'. However, we need something more to capture the uniqueness of human learned behaviour. At least as good as any other is Clifford Geertz's description: culture is 'a set of control mechanisms — plans, recipes, rules, constructions, programmes for the governing of behaviour'. These control mechanisms and the forms they impose on behaviour are learned, arbitrary, symbolic, and only possible because the human brain is a language (specifically speech) producing brain. It is true that non-humans learn and may have traditions. However, they do little to generate institutions — group-specific systems of social relationships which take the same form regardless of individuals. Human culture identifies and restricts group membership because unfamiliar individuals find it difficult or impossible to adopt new roles by learning new rules.

Past and present

Many accounts of human evolution can be analysed on two levels: what the author thinks is going on, and what is actually going on. It seems, or seemed, clear to those of us interested in reconstructing the human past that the primary evidence came from the fossil record — the only direct or 'hard' (as it sometimes is confusingly and misleadingly called) evidence we have. The fossil and archaeological evidence is used to build up a story of the past, starting with older and proceeding to youngest.

It does not work that way. Explicitly or implicitly, our knowledge and understanding of the present is needed to unlock the past. We must study the past as though it were a kind of present, albeit a different one, but one which operates according to the same ecological-behavioural-physiological 'rules' or 'principles' operating today. Our challenge is to comprehend and elucidate those rules. We can then proceed to do our best, given the very limited features of past systems that have been preserved, to reconstruct extinct species, and to do so as much as possible as though they were alive today. That is, we need to pretend that we are watching from a time machine.

Thus, one of several important reasons for making a detailed comparative study of the present is to understand the important rules or factors governing the interactions among ecological, behavioural (individual and group), and physiological-functional features. Which adaptations affect success in the search for food, mates, and protection? What generalizations can we make about them? Which are the important features? We can make a short list of significant ecological parameters; body and brain size, and dimorphism; life history patterns; positional behaviour; ranging behaviour; diet and subsistence patterns; adult grouping behaviour; relationships among females, among males, between adults and young, between groups; equipment use and skills; communicative behaviours. Furthermore, what are the physiological-function correlations of important behaviours? Remember, the fossil record preserves only morphology and we have to use that judiciously in order to infer as much as we can about behaviour. Finally, and obviously, studying the present defines the current differences between humans and non-humans, and, once we have articulated these, the significant adaptations which characterize humans give us our agenda: the features whose evolution must be explained. There is one important caveat; functions and roles change in evolution, and a feature important now in a particular role may not originally have been shaped by natural selection for that role.

Try to imagine a species as a system; a system is a collection of objects which interact or behave together as a unit according to certain rules and which produces or generates external behaviour which is measurable or predictable. Systems are convenient abstractions, useful for analytical purposes, and can be anything one wants them to be. A university can be analysed as a system, and so can a car or a television. Once certain critical rules are known, the behaviour of each element in the system is characterized or understood. Changes in the interactions of elements, following some changes in 'internal' rules, may cause changes in the output or 'external' behaviour of the system as a whole. Sometimes changes in one of the elements will have that effect. Small internal changes can produce small effects, but sometimes the effects are very large — as with so-called threshold effects (blowing a balloon up beyond the critical bursting point, for example, changes the state of the balloon system — irreversibly). Thus, if we can understand how sometimes small or subtle variations of the same interactional rules can produce different configurations or outputs, it is possible to imagine how one system can be transformed into another which looks quite different. We are trying to understand living species as variations on themes: we seek the themes. Once we have them we can begin more realistic reconstructions of past, different species, as though they were alive today.

As a first attempt at analysis of living primate species we will keep things simple, for example, by focusing mostly on food and its distribution and worrying only secondarily about predators. This is a good example of a situation where simplification is useful and reasonable in the initial stages of analysis, because it helps explain enough of the variation in which we are interested to permit useful generalizations.

Ultimately, we want to understand how current species 'work' as systems, in the light of evolutionary principles, particularly of natural selection and inclusive fitness; how past species worked; and how past and different systems became transformed into those of the present. We are particularly interested in human evolution, and within that in the nature of pre-human hominids, those whose behaviour would still have been recognizably non-human, albeit different from any extant non-human primate.

What we are deliberately not doing is using a particular known species as a 'model'. We are not interested in arguing by analogy: just because baboons and early hominoids shared the same habitat is no reason to reconstruct early hominids like baboons. Because chimpanzees are genetically most similar to humans does not mean that early hominids resembled chimpanzees in most, some, or indeed any critical features. We start by laying out some simple and simplified guidelines, assuming that all animals behave in ways which promote fitness. We may not understand why a particular behaviour exists, and not all behaviours at all times will promote fitness, but we assume that 'net' behaviour does. To survive to leave grandchildren requires that animals successfully find food and mates, and avoid predators.

What can we say about food and feeding behaviour? It is clear that there are relationships between what is eaten and how it is dispersed (large, ripe, dispersed fruit versus mature leaves, insects versus young leaves), body size, positional behaviour, and ranging patterns, and these factors — and more — interact with each other, and are linked by problems of getting and using energy. Large mammals, like gorillas, could not form aggregates or groups if they lived in trees like orangs because sufficient food does not exist for them. Orangutans need large fruits; these occur in discrete and dispersed packages, so orangutans are virtually solitary. Male orangutans do not range widely like male chimps because they are too big and they have no adaptations making terrestrial trekking effective.

Animals need mates to reproduce. But the sexes have different requirements. Put perhaps too simply, females have to carry the foetus and generate milk for the infant, both metabolically costly activities, and also provide most of the primary care and protection. For reproduction, they need a male only occasionally. Males, on the other hand can in principle inseminate any number of females at virtually any time. Thus, although both males and females compete for the opposite sex, the limiting resource is females rather than males.

Both sexes need to avoid predators, and this can include con-specifics. In the case of the large and sometimes co-operative hominoids, members of the same species may present greater hazards than carnivores. Males especially may attack females and infants that are clearly not theirs, and also other males. Even females occasionally attack other females or their infants. Males may be very large (gorillas) or usually move in groups of at least two, as in chimps, and females and infants almost invariably will be close to a male or males for protection.

We can start our simple model with resources. Food and its distribution will affect primarily the distribution of females. Male distribution will tend to monitor those of females, although the two patterns may interact, and of course predators will be important as well. For example, baboon females (*Papio papio, cynocephalus, ursinus,* and *anubis* species) can aggregate into groups because their food resources tend to be of high quality clumped in patches large enough to support several females. Although females will be in competition with each other for food, they gain more from associating because large groups can displace smaller ones at favoured places. Groups are made up of related individuals to minimize the genetic costs of competition. Female-bonded groups recruit males as protectors and — up to a point — the more males the better. *Papio hamadryas* in contrast lives in small groups with only a single male to a handful of females: resources are sparse and scattered.

Among hominoids, the dispersed ripe fruits favoured by *Pan troglodytes* keep females relatively solitary and this is also the case for orangutans. However, orangutans are almost completely arboreal, less mobile, and so less dispersed and spread out. Gorillas and bonobos prefer terrestrial herbaceous vegetation, an evenly and abundantly distributed resource; females can therefore aggregate into groups. Male distributions both map and influence those of females. Male orangutans tend to be solitary because of their large body size, and either overlap the ranges of a few females, or are migratory. Male chimps are relatively smaller than male orangutans (only 30 per cent larger than females rather than 100 per cent or more) and range widely together to monitor females. Gorilla groups have a single very large breeding male as protector, while bonobo groups have almost equal numbers of males and females. Bonobo males and females are closer in body size; both sexes cover wide ranges, and energetic costs are reduced by keeping body size within limits.

We can trace a number of anatomical features to these behavioural patterns. For example, chimpanzee cheek teeth are small and have thin enamel, adaptations to their fruit and herb diet. Orangutans have larger and thicker-enamelled teeth because their diet includes harder and tougher items. The highly arboreal orangutans have many post-cranial adaptations to arboreal feeding. Chimps and gorillas use both ground and trees for feeding and trekking, and show correlated features such as the range of knuckle-walking compromises. Degree of dimorphism, particulary male body size, interacts with both positional and ranging behaviour: compare more mobile and smaller male chimps with male orangutans. It also reflects male–male competition, particularly within groups: contrast very dimorphic gorillas and orangutans with less dimorphic and more co-operative common chimpanzees and bonobos.

We can use these interrelationships to make some educated guesses about the last common ancestor of humans and apes. It was probably a forest or woodland species which was omnivorous though mainly vegetarian, showing some sharing behaviour. Grouping behaviour would have depended on diet and habitat. Females in a social group were probably not usually related, and usually transferred out. Male relationships were mildly to markedly tolerant within groups — if containing more than one male — but intolerant to aggressive between groups. Sexual relationships might have varied from relatively exclusive (gorillas) to more promiscuous (bonobos), and it is quite likely that there would have been 'friendships'.

This very sketchy framework needs to be filled out more, and we can return to it later when we have reviewed the relevant fossil record. As we shall see, there is plenty of room for ambiguity and disagreement even in the case of those systems where we have deceptively 'hard' data: post-cranials, teeth. In the case of more 'intangible' aspects of behaviour, *caveat emptor*!

4 Primate evolution

The fossil record is our only direct source of information about past states which were different from extant ones. We analyse the characters of past species for two principal reasons: to infer relationships and hence to build up evolutionary sequences (trees or bushes), and to reconstruct patterns of behaviour. If we can order the fossil record as a dated sequence we can build up a picture both of the species in which we are particularly interested — in this case the evolving hominoids and especially hominids — and of their ecological and community relationships.

Fossils

Fossils are the remains of living organisms, and mostly are the mineralized remains of hard parts or biomineralized tissue such as bone, teeth, and shells. To become fossils, these have to be buried in appropriate depositional environments so that they are not destroyed too soon and can be mineralized. Most vertebrate fossils are parts of jaws and teeth, and fragments of limb bones. Other kinds of fossils do form given the right conditions, e.g. feathers, leaves, pollen, burrows, and footprints.

Rocks

From our point of view the most important kinds of rocks are sedimentary and igneous. Sedimentary rocks are created by the erosion, transport, and deposition of small rock particles, and by chemical precipitation mainly of calcium carbonate. The agents of transport and deposition are water and wind. Today sediments are being deposited in oceans (marine rocks), especially along continental margins, in deltas, and on beaches. On land (continental rocks) sedimentary rocks laid down by water form in lakes, rivers, and deltas; those by wind in deserts and beaches; and those by wind and water in caves. Sedimentary rocks usually bear the 'signature' of the depositional environments in which they were formed (continental shelf, delta, cave, lake) in features such as grain size, mineralogy, sedimentary structures, and fossil content. This information is used in reconstructing palaeogeography, palaeoclimates, and palaeoecology ('palaeo-' here means past).

Igneous rocks such as lava flows and ash falls (which form widespread and fine-grained blankets called 'tuffs' spread by wind or water) are formed in a molten state. They are important for dating purposes.

Time and dating

Since the early nineteenth century the study of long fossiliferous sequences has been used to show, first, that immensely long periods of time are involved in the formation of sedimentary rock sequences and in the generation of observed changes in fossil forms, and, secondly, to show the temporal sequence of life itself. The major divisions of the geological time scale reflect the major developments in the fossil record of plants, invertebrates, and vertebrates.

There are three principal ways of dating the record of sedimentary rocks and the fossils they contain: by faunal correlation, radiometric dating, and palaeomagnetic stratigraphy. Faunal correlation is based on changes observed in the fossil record of long continuous sequences. The method relies, in part, on the principle of superposition which states that older rocks were laid down before younger ones and therefore sedimentary sequences, when undistorted, are temporal sequences with the oldest rocks at the base of each stratigraphic section. Changes in faunal composition and morphological changes within individual lineages together produce temporal patterns which permit correlation between fossiliferous sequences. In particular, temporally short or fragmentary sequences can be tied in to longer and more continuous ones. Faunal correlation is a form of relative, as opposed to so-called 'absolute' dating; it specifies the order of events, but not their absolute ages.

Radiometric dating relies on the inherent instability of certain isotopes which decay into other isotopes of the same element, or into another element. This decay proceeds at a constant and known rate, usually measured as the time it takes an amount of an element to decay to half that amount; this time is known as the element's 'half-life'. The ratio of 'mother' to 'daughter' elements, together with knowledge of the half-life, allows calculation of the age at which the process began. If this can be associated with a clear event — the eruption of a lava or ash containing minerals of appropriate composition — the age of the event can be determined. As long as such events have a known association with fossils, we have an estimate of 'absolute' age.

Examples of dating techniques widely used in palaeoanthropology are carbon^{14}; potassium^{40}-argon^{40}; argon^{40}-argon^{39}; and fission track dating using uranium^{238}. Carbon^{14} dating uses wood, charcoal, or bone, and is usually good back to 40 thousand years. Potassium-argon dating uses potassium-rich minerals in volcanic rocks and is capable of dating rocks from hundreds of thousands of years to thousands of millions of years. Fission-track dating utilizes volcanic glasses and other uranium rich minerals, and can be used to date events from yesterday to thousands of millions of years. All radiometric dating techniques have sources of error and even the very best samples that are very carefully analysed in the laboratory will always have a range of ages within which the actual age will lie.

Palaeomagnetic stratigraphy relies on the fact that the earth's magnetic field has switched direction or polarity many times during geological history, because of internal changes in the earth's core. At present the earth's field runs from south to north, described as 'normal', but in the past the field has frequently been in the opposite direction, and this is described as 'reversed'. The polarity of rocks can be determined using a magnetometer, and both igneous and sedimentary rocks, marine and continental, can be measured for polarity. In many cases where the polarity of rocks is known the age is too, based mostly on potassium-argon dating. From these dated samples, a composite calibrated reversal pattern has been built up which is a worldwide pattern (because reversals of the earth's field will have a worldwide synchronous effect). This is known as the magnetic polarity time scale (MPTS). Each reversal is dated, the younger ones with great precision, the older ones with progressively less precision. Local sequences can often be matched to the MPTS, if the sequences themselves are known to be long and complete and if the age is approximately known. Providing the reversal(s) in the local sequence can be matched to those of the MPTS the local sequence can then often be more accurately dated.

Changes in global climates

Climates can be divided into global, regional, and local; although the three levels are linked, each is affected by a somewhat different set of factors. Our knowledge of past climates is expanding rapidly, and this is important for the insights we gain into understanding more about the biological structure of past communities and the adaptations of species. For a variety of reasons we know more about global climates than we do about those of more restricted areas.

Global climates are reconstructed mostly from analysis of the marine record. This is obtained from deep-sea cores, long sequences of the mud and organic ooze at the ocean bottom reclaimed by punching a hollow tube into the sediments (coring). Because these sequences are often relatively complete over long periods of time, and because climatically informative data sources are abundant in cores, we now have a remarkably complete record. In contrast, knowledge of local continental climates is more fragmentary because land sequences are generally less complete and usually represent only snatches of time. Also, data that yield climatic information are normally less well preserved on land than in the oceans.

The primary source of marine data from which climatic inferences are made are Foraminifera (forams), micro-organisms living in the ocean or on the ocean bottom, with hard shells or tests made of $CaCO_3$. Forams take up carbon, oxygen, and hydrogen from the oceans, and incorporate it into their tests. The ratios of stable isotopes (^{18}O, ^{16}O, ^{13}C, ^{12}C, D/H) can be measured in fossil forams and used to determine the isotopic ratios of ancient sea water. Since these ratios reflect a combination of water temperature and the total volume of ice on earth, being able to monitor past isotope ratios through forams in cores gives powerful insights into past climates. In addition, particular species and combinations of species can be sensitive indicators of past temperature and ocean salinity.

For continental deposits, stable isotopes can sometimes be determined in carbonate-rich rocks, and a variety of sedimentological and geochemical indicators also give clues about climate. In addition, fossil plants — mostly pollen, but sometimes leaves and fruits — are valuable indicators of local climate.

These and other sources of data make possible our current reasonably comprehensive understanding of the climates of the Cainozoic Era, covering the last 65 million years. This is the period of major mammalian radiations, including that of the Primates (Fig. 4.1). Cainozoic time can be subdivided into four main periods, each characterized by a broadly typical climate. Between 65 and 37 Ma climates were on average warm, with little longitudinal zonation. Within this period, the early Eocene was

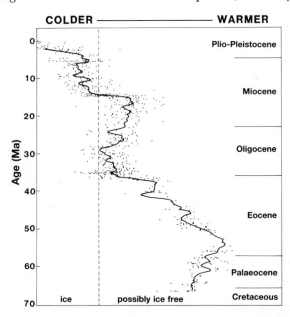

Fig. 4.1.   Climatic record for the last 70 million years based on oxygen-isotope analyses of deep ocean Foraminifera. Individual points and average trends are shown.
(After K. G. Miller and R. G. Fairbanks.)

warmest, and the latest Eocene saw a fairly abrupt cooling. Between 37
and 15 Ma temperatures were a little above those needed to maintain large
permanent Antarctic ice sheets. The early Miocene, around 20 Ma, was
the warmest time of this second period. About 15 Ma a grounded ice sheet
formed in Antarctica and temperatures oscillated around an average some
5°C warmer than today. This condition remained until a little over 3 Ma,
although between 15 and 3 Ma there were a number of cooling episodes in
which ice may have formed on northern continents, but the earth was still,
essentially, in a preglacial mode.

Around 3.2 Ma low-amplitude fluctuations with a frequency of around
100 thousand years began in response to moderate continental glaciations
in the northern hemisphere (Fig. 4.2). The scale of these glaciations
increased around 2.4 Ma and there was a further intensification at 0.8 Ma
leading to a climatic mode which has continued to the present. Since 3.2
Ma, climates have been dominated by a 100-thousand-year cycle of cold-
warm-cold, the cold times (glacials) being characterized by continental ice
sheets, lowered sea levels, colder temperatures, and more marked
longitudinal zonation and seasonality. The warm times (interglacials) are
relatively free of northern continental ice, less seasonal, and much briefer.
Of the last 800 thousand years, less than 5 per cent has been as warm or
warmer than it is in the present interglacial.

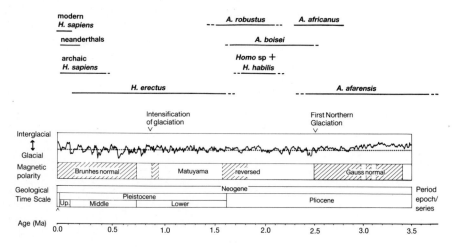

FIG. 4.2.   General diagramatic summary of climatic change over last 3.5 Ma, showing
geological age ranges of hominid species. (In part after N. Roberts.)

These climatic changes and cycles are driven by a variety of
phenomena, both extrinsic and intrinsic to the earth. The major shifts
punctuating the Cainozoic are probably driven more by intrinsic factors

such as continental position and mountain systems. Changes in positions of the continents have changed considerably the geography of the earth. Continental configurations have altered and major new mountain systems have appeared. The separation of Australia from Antarctica, and the isolation of Antarctica and development of circum-Antarctic currents played a leading role in events that led to Eocene–Oligocene cooling. The collision of Afro-Arabia and India with Europe and Asia, beginning in the late Oligocene and completed by middle Miocene, radically changed continental outlines, built major new mountain systems from Alps to Himalayas, altered atmospheric and oceanic circulation patterns, and ultimately triggered a cooling trend which resulted in an earth with significantly more ice and a new climatic mode driven by fluctuations in ice volume.

Careful analysis of deep-sea sequences prior to 3 Ma show that in addition to these very long-term and non-periodic trends there are other shorter and much more regular fluctuations. In the last 3 Ma (Pliocene and Pleistocene) these have a wavelength of 100 thousand years. Although earlier time periods have been looked at in much less detail, regular and periodic fluctuations are also present before 3 Ma, but with wavelengths of 50 thousand years or less. In the Pleistocene the climatic fluctuations are apparently intensified by feedback effects resulting from the presence of very large ice sheets. Waxing and waning of these ice sheets is mostly under the control of variations in the structure of the seasonal radiation cycle: the distribution of total solar energy. These variations are caused by alterations in the earth's orbital geometry, namely in eccentricity (400 and 100 thousand years periodicity), obliquity (41 thousand years), and precession (23 and 19 thousand years). Each changes in a narrow, and characteristically specific, way. When integrated together, the results are gross fluctuations from warm to cold over a roughly 100 thousand years cycle.

In the Pleistocene (last 1.6 million years), these fluctuations are now well known and dated on the basis of very detailed analyses of continuous deep-sea cores. The more fragmentary land sequences rarely reveal the true and full complexity of the climatic record. However, slowly the small pieces of time usually represented in continental sequences are being matched up, and this is greatly helped by the rare, but valuable long land sequences.

The notion that there were a small number of high-latitude glacials (four or five), separated by long interglacials, correlated with low-latitude pluvials and interpluvials has died hard. However, it is now clear that the climatic story is actually much more complicated, and inter-regional correlations of land sequences are more difficult than had been previously believed. As far as the Pleistocene goes, probably the best strategy is to break it down into three broad blocks (Fig. 4.2), each of which can then be further subdivided: early Pleistocene (1.6–0.7 Ma), middle Pleistocene (0.7–0.125 Ma), and late Pleistocene (0.125 Ma to present).

Habitat change

Knowledge of Cainozoic habitats and patterns of change is also growing. Palaeoanthropology has been dominated historically by rather simple notions of the links between shifts in climate and habitat, and evolutionary change. The story usually goes that early hominids exploited the newly expanded savannahs as the climate cooled, while *Homo* responded to the challenge of the ice ages by evolving a bigger brain. This is too simple, both as far as the trajectory of climatic and habitat change is concerned, and the nature of the biotic response.

The boundary between the Cainozoic and the last segment of the Mesozoic, the Cretaceous, is marked at 65 Ma by major biological events. For example, abrupt shifts in carbon isotope ratios in deep-sea sediments suggest that the productivity of surface waters plunged precipitously in a time too short to be geologically measurable. The cause is still debated, though impact by a large extraterrestrial object looks increasingly likely, the effect of which was to radically change climates and habitats. In the Palaeocene and Eocene, climates were generally warm and there is considerable evidence from plant fossils that habitats were predominantly forests and woodlands of various kinds. This inference is supported by animal remains, the bulk of which also suggest forest. Beginning at the end of the Eocene, there is evidence for a shift in the relative proportions of habitats, an increase in woodland and grassland, and a proportionate decrease in forest, although this proceeds at different rates in different continents. The shift from a system with the bulk of plant biomass in the form of trees to one with more biomass as shrubs, grasses, and below ground plant parts reflects changes in the annual distribution of temperature and precipitation (increased seasonality). Plants change as seasonality increases, and change further when the cold season drops below freezing.

The animal and plant fossil records of the Oligocene are rather poor, but for the Miocene they are now quite good. Forests and dense woodlands were still abundant in Africa, Europe, and Asia, but there is evidence too for more open habitats, including grasslands. These became more abundant especially after about 10 or 11 Ma. From 2.4 Ma on, the intensification of the glacial–interglacial cycle drives fluctuations of habitat. In the tropics and subtropics the rainforest blocks of Africa and South America fragmented during glacial times, perhaps even disappearing as communities, and grasslands and deserts expanded. This implies that for the majority of the last 2.4 Ma habitats in tropical regions have been significantly more open than they are today.

Faunal change

Animal communities do change as their individual components evolve. Although there is a low level of appearances and extinctions at virtually all times, they are often quite stable in general composition over hundreds of thousands or even million of years. However, major periods of change involving evolution, immigration, and extinction occasionally punctuate the record, and these periods can usually be correlated with some major climatic event. How the two sets of events are causally linked is a matter of lively debate. There are at present a number of hypotheses concerning the nature of lineage and community (clusters of lineages) change. Would evolution occur in the absence of environmental change? Do new species usually arise because lineages are fragmented by environmental change, or because new niches open up permitting adaptive radiations to occur? These and other interesting questions are under active consideration, but we need not go into details here. It will suffice to give the broad patterns of community evolution.

Palaeocene and Eocene communities are dominated by 'archaic' species, those that are very different in many ways from forms living today — particularly the medium and large herbivores, and carnivores. The overall faunal aspect indicates adaptation to predominantly forest or woodland habitats. Oligocene faunas in part continue these patterns, but include new herbivorous species with higher crowned teeth, resistant to abrasion by grasses, and with limbs adapted to rapid running — both indications of open habitats. Unfortunately, little is known of South America in the first 30 million years of the Cainozoic, and effectively nothing of Africa where Oligocene faunas are known only in Egypt and contain forest-adapted species including primates.

Miocene through Recent faunas in South America are best known from relatively open country areas, the forest being under-represented. South American mammals were for the most part highly indigenous, except for rodents and primates, until the Pliocene when the connection to North America through the Panamanian isthmus was established. The 'great faunal interchange' which followed resulted eventually in the replacement of many South American forms by northern species.

In Africa, early Miocene faunas are known only from east Africa where both forest and woodland communities are sampled. Between 15 and 18 Ma there was a substantial faunal turnover, particularly in the more open components, which involved an exchange of species with Asia. This is consequent to the final link up of Afro-Arabia with Europe and Asia, and final closing of the Tethys Seaway between what are now the Mediterranean and Indian Oceans. Newer species mark these middle Miocene faunas,many showing resemblances to living forms and including more open-country groups. There is a further pulse of faunal turnover at around 10 Ma, again marked both by immigrants and new indigenous species, and this is followed by a period in which the faunas of at least eastern Africa begin to take on a structure which is essentially modern. This process is well under way by the middle of the Pliocene (3–4 Ma). Because the African fossil record is still poor during middle Miocene through Pliocene times (14–4 Ma) the exact pattern of faunal change is unclear and so is its linkage to the climatic changes which occurred 10 and 5–6 Ma. Around 2.4 Ma there are changes in the large herbivore faunas in particular, but also in other faunal components, which may well be linked to the intensification in glacial cycling known to have occurred then.

In Europe and Asia the pattern is clearer because it is better sampled, and more complex than that of Africa, possibly because it is better sampled. Faunal communities show both a north–south and east–west patterning. More forest-adapted communities are found in a southern fringe from Spain to South China in the early Miocene. To the north, particularly in Asia, are more open communities. Through the Miocene climates cool and become more seasonal, and — a consequence of continental drift — shallow inland seas steadily disappear and major mountain systems develop trending roughly east–west.

The mixture of appearance of new species, immigration, extinctions, and the shifting of inter-community boundaries can be exemplified by a well known South Asian sequence, the Siwaliks of Pakistan and India. Early Miocene faunas suggest warm forested conditions. Middle Miocene faunas reflect faunal exchange with Africa and Europe and some climatic cooling. A major faunal pulse at 10 Ma and another more diffuse one from 7 to 5 Ma parallel late Miocene phenomena observed or inferred in Africa. After 5 Ma there is a sequence of community shifts involving south Asia, south China, South-east Asia, and the Indonesian archipelago during which forest community boundaries progressively shift as those habitats shrink toward the south-east.

Many of the pulses of faunal change are broadly synchronous across the Old World. Those at about 17, 10, 7, and 2.5 Ma may well be widespread responses to the same phenomena. It also seems likely that a major faunal change at a little under 1 Ma correlates with a climatic change towards intensified glaciations: open faunas spread, and there is evidence of migration into and out of Africa. The climatic cycles of the last 800 or 900 thousand years had a marked effect on northern faunas triggering regular extinctions (Fig. 4.2). Their effect is less clear in tropical regions because the record is so much less complete, but is likely to have been less marked.

There are some particularly interesting changes in the carnivorous species of Africa over the past 2 million years, as relative numbers changed of stalking carnivores (lions or cheetahs for example), running carnivores (hunting dogs, hyaenas), and scavengers (jackals, vultures). The hominids were also shifting their trophic position during this period, starting as herbivores, but emerging as full fledged and extremely effective predators by at least the time of modern *Homo sapiens* 40 thousand years ago. These shifts were perhaps correlated, linked to patterns in the herbivore faunas, in turn probably driven in part by climatic and habitat change.

A final series of major faunal changes, marked by the local extinction of many large mammals, occurred in most parts of the world around the time of the last deglaciation some 12 thousand years ago. Whether or not the hunting of humans was implicated and to what degree is unclear and hotly debated.

The Primate radiations to the hominids

Primates are not known for certain in the fossil record until the Cainozoic. Early in the Palaeocene they became relatively abundant, and remain prominent members of tropical forest communities until the present. The broad outlines of primate evolution are known. There were at least two early radiations in Palaeocene and Eocene times, the second of which contained many lineages broadly similar to the living lemuroids of Madagascar. Sometime in the Eocene, one lineage evolved new traits including a larger brain, and split in late Eocene or early Oligocene time into platyrrhine and catarrhine lineages. Both groups radiated successfully from the Miocene on.

The Cainozoic record of primates and other interesting mammals is still patchy. The records for North America and Europe are good enough so that absence of evidence of primates can be interpreted as evidence of absence, but this applies less to Asia, particularly southern and eastern Asia, where the earlier half of the Cainozoic is poorly represented. In Africa and South America records prior to the Miocene are poor. This is particularly frustrating because Africa is a likely area of origin for Primates.

Prosimians

There are either two or three main superfamily groups of prosimians, depending upon whether or not *Tarsius* is included. The lemuroids (lemurs, sifakas, indris, aye-aye) live today in Madagascar along with a mammal fauna which is quite different in most elements from that of Africa. Living lemuroids are all small, arboreal, forest species, many of them nocturnal. Within the last few thousands years a number of species have become extinct, and these include larger and more diverse forms, some of which were terrestrial. Madagascar has been separate from Africa for at least 120 million years and it is unlikely — though not completely impossible — that some of the mammal fauna reached there by that time. Thus, dispersal from Africa was presumably mostly across water. Unfortunately, there is as yet no earlier Cainozoic fossil record for Madagascar.

The lorisoids are a second major prosimian group, found today in Africa and Asia. They can be subdivided into the more active leaping galagids or bushbabies, and the slow-moving and deliberate lorisids (including loris and potto). These are small and mostly forest or woodland species, generally nocturnal or crepuscular. Galagids are confined today to Africa and are known there in the early Miocene. Lorisids live in both Africa and Asia, and are also known from the Miocene records there. Lorisids probably evolved in Africa and are known there in the Oligocene, and spread to Asia from there, perhaps as late as the middle Miocene. The tarsioids, represented by a single genus *Tarsius*, are small, nocturnal, arboreal, vertical clinging and leaping primates living in Borneo and parts of Indonesia.

Lorisoids and lemuroids share a number of similarities, including a peculiar modification of the front teeth in which canines and incisors become needle-like and project horizontally to produce a 'grooming comb'. The only African fossil deposits older than Miocene containing definite primates are the earlier Oligocene (31–37 Ma) Fayum beds of Egypt. The only possible prosimians known there are a tarsioid — represented by one jaw fragment, very similar to *Tarsius*, and a lorisoid tooth. Prior to the Oligocene the African primate record is essentially blank, like the mammal record in general. This is unfortunate, because it is distinctly possible that primates originated in Africa, possibly in the Cretaceous, and radiated subsequently on several occasions into Europe, Asia, North America, South America, and Madagascar, with the 'centre of gravity' of the order remaining in Africa.

Very primitive species related to primates, but made up of more rodent-like forms radiated in North America and Europe during the Paleocene. Other than having molars which resemble those of later primates they show few characters diagnostic of living primates. The radiation is best represented by forms like *Plesiadapis*, *Phenacoleur*, and *Microsyops* — small clawed, rather squirrel-like genera (Fig. 4.3).

A series of primate genera of more modern aspect appear in the earliest Eocene of Europe and north America. These show up abruptly as immigrants at a time when there is evidence for major faunal turnover in those areas for which we have reasonable sequences. Sampling from Asia is poor, but generally similar forms are present. One possibility is that they were all immigrants from Africa. This Eocene radiation has been subdivided into two groups: Adapidae, consisting of *Cantius*, *Notharctus*, *Adapis*, and others; and Omomyidae, including *Tetonius*, *Necrolemur*, *Pseudoloris*, and *Omomys* (Fig. 4.3). Their earliest members were similar, but became more divergent through time. The Adapidae were generally lemuroid-like in skull, dentition, and skeleton, although there are some interesting and consistent differences between the two groups. Possibly the resemblances are primitive, reflecting an earlier and pre-Eocene split between an adapid–omomyid lineage and the lemuroid–lorisoid stem (in Africa?). Omomyids, especially later ones, develop a number of derived features resembling those of tarsioids and they have frequently been viewed as the ancestral group from which *Tarsius* evolved. The recent discovery of the tarsier-like jaw (unfortunately only a jaw) in the African Oligocene at least raises the possibility that living tarsioids have a more directly African derivation and that omomyids perhaps evolved tarsier-like features in parallel.

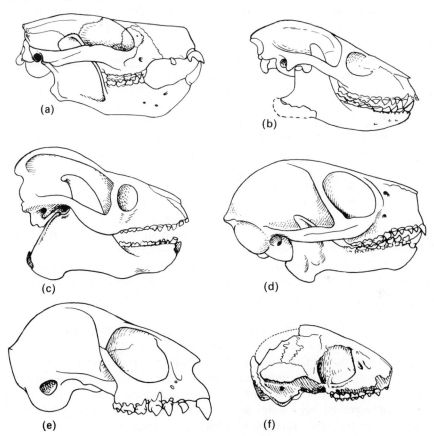

Fig. 4.3. Skulls of selected early primates. (a) *Plesiadapis*, × 1; (b) *Notharctus*, × 1; (c) *Adapis*, × 1; (d) *Necrolemur*, × 1; (e) *Tetonius*, × 3; (f) *Rooneyia*, × 1.

Unfortunately, the Palaeocene and Eocene fossil record is still too poor in critical areas like Africa and Asia to enable us to settle these issues now. Also, genetic patterns among these major living primate groups still need clarifying so that at least relative timings can be better estimated.

The New World Anthropoidea

The two principal groups of living anthropoids are the South American platyrrhines or Ceboidea, and the Old World catarrhines, the hominoids and cercopithecoids of Africa and Asia. They share a number of features not present in prosimians and this together with genetic patterns implies a fairly long period of common ancestry for an anthropoid lineage prior to its divergence into platyrrhine and catarrhine stems.

The ceboids are an exclusively arboreal tropical-forest group with a wide range of body sizes, dietary, and positional behaviours. Fewer than 20 genera are known, and there are five or six anatomical-genetical clusters of genera whose roots can be traced back to the beginning of the radiation. The oldest South American primate fossil is *Branisella*, a latest Oligocene (about 25 million years old) platyrrhine which may not be directly related to any living ceboid. Two late Oligocene or early Miocene genera are said to share derived features with living ceboids: *Dolichocebus* with the squirrel monkey *Saimiri*, and *Tremacebus* with *Aotus*, the owl monkey. Middle Miocene fossils have also been linked with extant genera: *Neosaimiri* to *Saimiri*, *Cebupithecia* to the pithecines (sakis), and *Stirtonia* to the howler monkey, *Alouatta*. These ancestor–descendant links are not universally agreed upon, although some are highly likely.

Despite the fact that the Amazonian rainforests almost certainly fragmented repeatedly in phase with the glacial cycles of the Pleistocene and later Pliocene the ceboids are not known to have evolved terrestrial baboon equivalents — although the fossil record is still poor. Why is unclear because the African and Asian forest primate radiations produced terrestrial species. Possibly terrestrial niches were pre-empted by successful non-primate competitors; possibly habitat structure was inadequate.

The origins of Ceboidea are obscure. The vertebrate fossil record in South America has a yawning gap between late Oligocene about 28 Ma and Eocene rocks around 45 Ma. During this time rodents and anthropoid primates entered the continent; exactly when is unclear and strongly debated. The likeliest sources are north America and Africa, both requiring dispersal over water assuming colonization to have taken place sometime during the Eocene or Oligocene (roughly 55–26 Ma). The earlier consensus favoured a northern origin, but recently Africa has enjoyed a vogue. Ceboid-like anthropoids are known in the Oligocene of Africa, from the Fayum region of Egypt.

The Fayum sequence, mentioned earlier, runs from at least 30 Ma, probably 32 or 33 to 37–40 Ma. Primates are an abundant part of what was a forest and swampland community associated with a large river system and its associated gallery forests. Half a dozen or more genera of arboreal primates inhabited the region. In addition to the tarsioid and lorisoid mentioned above, the primates fall mostly into two groups:

Parapithecidae (*Parapithecus, Apidium*) and Propliopithecidae (*Propliopithecus, Aegyptopithecus*).

The parapithecids are well known from jaws, teeth, and many postcranial elements, and some skull parts have also been sampled. They show surprising and consistent similarities to ceboids, including several features which can plausibly be interpreted as derived; like cebids they have three premolars, and the articulation of skull bones in the temporal fossa resembles cebids. Post-cranial features also match New World monkeys. It is a distinct possibility that they are indeed phyletically or cladistically ceboids. Other Fayum genera discussed below are as plausibly catarrhines. The split between the two groups goes back to at least 35 Ma, perhaps earlier, based on these Egyptian primates.

South American rodents show similarities to African forms too. It is therefore a possibility that a very improbable event before 26 Ma (but how much before we do not know — remember the missing South American fossiliferous sequence) involved the rafting of at least one gravid rodent and one gravid parapithecid from west Africa to eastern South America.

Anthropoidea: Old World catarrhines

The Old World anthropoids, today comprising cercopithecoid monkeys and hominoids, evolved as arboreal, forest primates in Africa, splitting there into the two groups in the early Miocene. Members of both superfamilies spread subsequently into Asia and Europe, and survive today in Asia. Species of both have colonized terrestrial habitats. Many terrestrial species became extinct during the Pleistocene and the majority of living catarrhine species are arboreal.

The earliest known catarrhines come from the Oligocene Fayum of Egypt and date to at least 35 Ma. They are represented there by several species of two genera, *Propliopithecus* and *Aegyptopithecus*. These are very primitive catarrhines, and only a few cranial and dental characters link them to the living forms. Earlier analyses, based on the fragmentary jaws and teeth which were available before Professor Elwyn Simons began a new series of explorations over 20 years ago, suggested that possible ancestors of cercopithecoids, hylobatids, large hominoids, and even hominids could be recognized among the Fayum primates. However, this seems now to be very unlikely.

Propliopithecus species were small, arboreal primates weighing 2–3 kg with teeth suggesting a predominantly fruit-eating diet. Canines were strongly dimorphic and there was noticeable size difference between males and females, at least in jaw size. Post-cranially *Propliopithecus* shows a number of similarities to quadrupedal or actively leaping cebids such as the squirrel monkey, *Saimiri*. *Aegyptopithecus* is now also quite well known and there are several fine fossils preserving parts of the face and head region, including a complete cranium, as well as several post-cranial specimens, and many jaws and teeth. *Aegyptopithecus* was a dimorphic form, the larger males weighing around 6 kg, and both canines and jaws were dimorphic. In this feature *Aegyptopithecus* resembles *Propliopithecus* and most later catarrhines, and differs from prosimians; such a pattern implies permanent breeding groups composed of perhaps several females and one to several males. *Aegyptopithecus* was diurnal judging from orbit size, and small-brained, in fact within the prosimian range of relative brain sizes and below that of anthropoids. However, some features of the endocast (a mould of the inside of the brain case rather than the outside of the brain, though reflecting some details of external brain anatomy) differ from average prosimian patterns in an anthropoid direction: relatively smaller olfactory bulbs, larger visual cortex, and a more derived sulcal pattern. Judging from molar morphology and incisor to molar proportions, *Aegyptopithecus* was basically a fruit-eater. Post-cranially, it preserves a unique mixture of features, some like Eocene and living prosimians, adapids and lemuroids, others like the New World genus *Alouatta,* the howler monkey. A functional interpretation of the limbs suggests an animal that was neither an agile cursorial quadruped running along branches, nor a suspensory clamberer and swinger below branches, but a slow and deliberate climber and branch walking quadruped. Reasonable living analogues would be *Alouatta* or the Malagasy variegated lemur, *Varecia variegatus*.

By the beginning of the Oligocene then, around 35 Ma, primitive primates broadly ancestral to the two main groups of anthropoids were living in North Africa (at least). Clearly, they and related species would have been living elsewhere in Africa, possibly in India too, though almost certainly not in Europe, Asia, or North America. These catarrhines were very primitive and document a phase before the branching into cercopithecoid and hominoid lineages. Fossil samples from Africa, where catarrhines were almost certainly confined, are unfortunately non-existent between 32 and 22 Ma, but by 22 Ma catarrhine fossils from early Miocene sediments in Kenya and Uganda show that considerable evolutionary change had occurred. Although once again only a tiny fraction of Africa is sampled these primate faunas from the early Miocene, 22–17 Ma, are diverse and varied. Afro-Arabia was still separated from northern continents by the Tethys Sea. Judging from gross similarities between north, east, and south African faunas of the earlier Miocene, as well as from geological evidence in deep-sea cores, the Sahara desert did not exist. An African equatorial belt of predominantly tropical forest and woodland was probably flanked by wooded or brushed grassland. It is important not to have too monolithic a view of these palaeoenvironments: all types were probably represented from forest through to savannah, though in different proportions from today.

The early Miocene catarrhines that are well enough known were all arboreal creatures, judging from post-cranial anatomy, although by analogy with many living species they would also have come to the ground occasionally. Until quite recently, many palaeoanthropologists argued that ancestors of the lesser apes and the large apes — sometimes even including hominids, could be found in the middle or early Miocene. One or two features of resemblance to particular living catarrhines would be picked out in a fossil and used to infer a specific ancestor–descendant relationship. This ignored both the overall similar, primitive nature of Miocene species and the overall, derived similarity of many living catarrhine characters, and the excessive extent to which parallism would have to be invoked if ancestries were indeed to be traced back 15 or 20 million years. It now seems more probable, on the basis of comparisons of DNA and proteins, comparative anatomy, and the fossil record, that living catarrhine branching times are approximately as follows: cercopithecoid–hominoid 20–24 Ma; small hominoid versus large hominoid 12–18 Ma; *Pongo* versus other large hominoids at 10–15 Ma; *Gorilla* versus *Pan* and *Homo* around 7–10 Ma; and *Pan* versus *Homo* between 4 and 8 Ma. These are still rather broad ranges, but the true splitting events probably lie within them.

The oldest recognizable Old World monkeys came from Napak in Uganda and Buluk in Kenya, and are 17–18 million years old. They are also present in north Africa about then. Cercopithecoids are much less well known at this time than other catarrhines, perhaps because they were less diverse or perhaps because their preferred habitats are not well sampled in the fossil record. Early monkeys are recognized by their distinctive molar patterns, clearly derived from the more primitive catarrhine tooth design which in its basics has been retained by living hominoids.

The extant cercopithecoids divide into cercopithecines and colobines. The former comprise both arboreal and terrestrial species, many of which are frugivorous or omnivorous feeders. The latter are mostly arboreal and specialized leaf-eaters. The groups differ in a few dental and post-cranial features, but are basically very similar. All are active cursorial quadrupeds with very characteristic post-cranial features: for example, a narrow and deep thorax, and long lumber region with seven or more long lumbar vertebrae. Hands and feet are adapted to palmigrade and plantigrade locomotion, and there are few adaptations to suspensory behaviour. It is possible that the two groups had split in the middle Miocene, and they certainly had by late Miocene time. Some colobids migrated from Africa to Europe and Asia in the late Miocene — possibly 7–9 Ma. In Africa, a number of Pliocene and Pleistocene species became as adapted to terrestrial habitats as some baboons. Some arboreal species became very large. Since the Pliocene most terrestrially-adapted colobids have become extinct: climatic–habit change and/or competition are possible explanations.

Cercopithecoids also radiated out of Africa, but did so later than colobids, arriving in Europe and Asia only in the Pliocene. Cercopithecids are the most diverse group of living catarrhines. There are many arboreal forest species, whose fossil record is only poorly known. The large number of species of *Cercopithecus*, the forest guenons, may reflect the repeated Pleistocene fragmentation of the equatorial African rainforest block. The more terrestrial baboons, both those related to *Papio* and to the even more terrestrially adapted geladas, *Theropithecus*, radiated during the Pliocene and Pleistocene. Their diversity is now greatly reduced and all but one surviving open-country species belongs to *Papio*.

The cercopithecoids are specialized generalists. They have a generalized quadrupedal positional repertoire based on a fundamentally active cursorial pattern. Their post-cranial anatomy resembles in many features that of other vertebrate quadrupeds, but at least some of those features might be secondarily derived. Dentally, they are derived and the colobines evolved specialized stomachs for breaking down cellulose and detoxifying secondary compounds. Both arboreal and terrestrial species are abundant throughout the record, and it is unclear exactly what were the earliest cercopithecoid adaptations. It is possible though that the earliest cercopithecoid was both arboreal and terrestrial, an active quadruped, capable of feeding on leaves and unripe fruits. These are foods requiring chemical as well as mechanical processing for digestion. Cercopithecoids would therefore have been able to exploit more seasonal and more open habitats.

The early hominoids, and the generalized ancestors of hominoids and cercopithecoids seem to have been rather different kinds of animals. The hominoids of the Miocene still retained features which would probably have been present in the common ancestor of both surviving catarrhine groups. This explains why many people have used 'hominoid' to describe species which almost certainly lived prior to the split (*Aegyptopithecus*, for example), even though they are not cladistically Hominoidea. One group of species, *Pliopithecus* and related forms, are probably also derived from catarrhines prior to the diversification of modern lineages. They are middle and late Miocene primates known from Europe and Asia, which originated in Africa and emigrated early in the middle Miocene around 16 Ma. Often considered to be early gibbons because of dental similarities, they are better seen as primitive and basically frugivorous catarrhines, arboreal quadrupeds, but more agile and active than *Alouatta* and perhaps more like *Lagothrix*, the woolly monkey in showing a mixture of above- and below-branch positional behaviours.

More difficult to place are the several genera so abundantly represented in the early Miocene rocks of east Africa: *Proconsul*, *Rangwapithecus*, *Dendropithecus*, *Limnopithecus*, *Micropithecus*. These are more probably phyletically hominoid than not (that is, falling after the divergence of hominoids and cercopithecoids). They are associated with predominantly forest faunas. Another Kenyan early Miocene site, Buluk, dated around 17 or 18 Ma, has a somewhat different fauna suggesting a different and perhaps more open habitat. *Proconsul* and the other genera are absent, while monkeys are abundant (they are exceptionally rare in other early Miocene sites). *Afropithecus* and *Turkanopithecus* are new hominoids which partly resemble both some later hominoids known in Asia (*Sivapithecus*) and some in Africa (*Kenyapithecus*); but they have their own unique features.

Dendropithecus, Limnopithecus, and Micropithecus are all small forms weighing less than 5 or 10 kg. They were arboreal, probably predominantly frugivorous, and sexually dimorphic. *Dendropithecus* post-cranial bones are known and suggest a diverse positional repertoire, quadrupedal, but with clear adaptations for arm-swinging and other suspensory behaviour. One or more of this group might be related to *Pliopithecus* (and would not then be strictly hominoids); alternatively, a link to the living hylobatids is a possibility. It is possible — no more — that hylobatids originated in Africa from one of the known Miocene genera, spread to Asia, became extinct in Africa and south Asia, and survived only in the tropical forests of South-east Asia (including until very recently, China). Middle Miocene small apes, very similar to *Micropithecus* and *Dendropithecus*, are known from fragmentary remains in Pakistan and south China. However, it is very difficult to recognize hylobatid ancestors because in most dental and skeletal characters gibbons are relatively primitive. They are most derived in their very long arms, and their lack of body and canine size dimorphism.

Proconsul-group species are not very well known except for one species. There is some debate on exact placement of the group, but they are probably hominoids and must stand very close to the origin of the superfamily. If not the ancestor, *Proconsul* is probably very similar to the ancestor. The early Miocene remains come from half a dozen major localities in Kenya and Uganda. One place in particular has produced many fossils and is still producing them, Rusinga Island in Lake Victoria. Both *P. africanus*, weighing around 10–15 kg, and the larger *P. nyanzae*, weighing about 30 to 40 kg, are abundant at Rusinga. Recent new material and reanalysis of older specimens has suggested a possible alternative grouping: that *P. nyanzae* and *P. africanus* at Rusinga actually represent males and females of the same species, *P. nyanzae*. (*P. africanus* still exists as a species, but only at other early Miocene localities.) Just two kinds of canines are found at Rusinga, long blade-like 'male' types and short stubby 'female' types, not the four kinds expected from two species; the 'male' canines when associated with other teeth are in *nyanzae* specimens and the 'females' in *africanus* specimens. However, in the average size of cheek teeth and possibly in average body weight, presumed males and females of *nyanzae* are more different than in the most dimorphic living catarrhines like the orangutan. We have several possibilities here: two species are in fact involved, or one species of normal 'modern' variability is present, but sampled in an unusual way, or one unusually variable species is present. Proving the third alternative is intrinsically difficult, since we have no modern equivalents. However, other Miocene sites show similar patterns, and we may be seeing here something characteristic of many pre-modern hominoids.

Judging from tooth morphology and dental proportions *Proconsul nyanzae* was basically frugivorous. Brain volume is known for 'females' and is approximately what would be predicted in an ape of this body size. The external morphology of the brain, estimated from an endocast, resembles that of a gibbon (which is probably primitive for Hominoidea). Recently discovered 'female' post-cranial bones from Rusinga add substantially to a forelimb skeleton found in 1951, and many new 'male' bones have also recently been recovered. *P. nyanzae* was a quadrupedal form with a varied positional repertoire, including some suspensory and some bipedal behaviours; predominant patterns though would have been above-branch, and more deliberate than agile and rapid. It is likely that animals would have come to the ground from time to time, and that sometimes they would have stood and moved bipedally.

Other species similar to *P. nyanzae* are known from Kenya and Uganda, though they are not as well represented. *Afropithecus* and *Turkanopithecus* species from Buluk show that we may still have only a glimpse of the actual species diversity in the African early Miocene.

The exact relationship of these early Miocene species is still poorly understood. A few features, for example the anatomy of the lumbar vertebrae of *P. nyanzae*, hint at specific links to hominoids, but most of them are sufficiently generalized to be ancestors of both monkeys and apes. These forms are clearly close to the divergence. The Buluk material, frustratingly meagre though it is, suggests a somewhat closer relationship to later hominoids than do *Proconsul* group species. For the moment we will assume that the division of hyblobatids from other hominoids happened around 16 Ma, plus or minus 2 million years, while the cercopithecoid–hominoid split occurred at least 20 Ma and probably earlier.

The hominoid fossil record in Africa is very poor from 17 Ma on. That for Europe and Asia is better and improving, and it is now possible to see the outlines of a framework in which fossil data and genetic information can be coherently integrated (Fig. 4.4, Table 4.1).

During the 1960s and 1970s the middle and late Miocene hominoid *Ramapithecus* was widely perceived by palaeontologists as a pre-*Australopithecus* hominid, rather different from its contemporaries such as *Sivapithecus* and *Dryopithecus* which were assumed to be apes (Fig. 4.5). During the last 10 years new fossils have greatly improved our knowledge of these and other Miocene hominoids, and it is now clear that

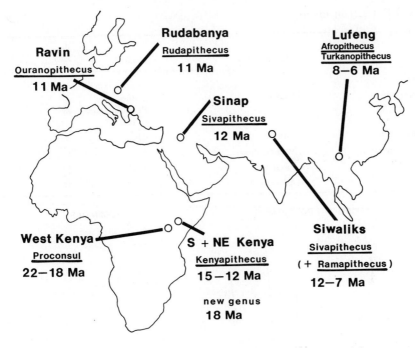

FIG. 4.4.   Distribution of better Miocene hominoid material.

TABLE 4.1
Selected Miocene hominoids

Taxon	Main localities	Country	Age (million years)
Africa			
Proconsul africanus	Koru, Songhor	Kenya	19
P. nyanzae	Rusinga, Meswa Bridge	Kenya	22–18
P. major	Songhor	Kenya	19
	Napak	Uganda	18
New genus	Moroto	Uganda	13–18
Afropithecus Turkanopithecus	Buluk, W. Turkana	Kenya	18
Kenyapithecus	Maboko, Fort Ternan, Nachola	Kenya	15–11
New genus	Samburu Hills	Kenya	9
Europe			
Dryopithecus	Various	Western Europe	14–10
Rudapithecus	Rudabanya	Hungary	10
Ouranopithecus	Salonika (near)	Greece	10

TABLE 4.1 (continued)

Taxon	Main localities	Country	Age (million years)
Asia			
Sivapithecus	Various	Turkey, Pakistan India, Nepal	12–7
New genus	Lufeng	China	8
Indeterminate	Vienna Basin	Austria	15
	Pasalar	Turkey	14
	Ad Dabtiyah	Saudi Arabia	16

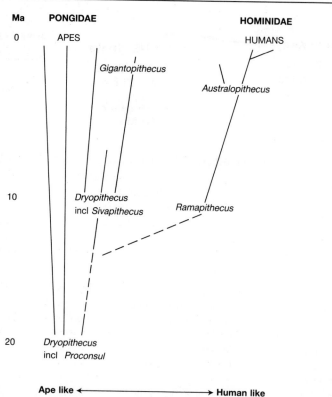

Fig. 4.5.   Diagram showing hominoid relationships believed by some palaeoanthropologists in the 1960s and 1970s. (Compare Fig. 2.4)

fragmentary jaws and isolated teeth are not particularly useful body parts
for sorting out species or for ascertaining phylogenies.

The middle Miocene record documents the expansion of large

hominoids out of Africa into Europe and Asia across what had previously been the Tethys ocean barrier. Between 15 and 12 Ma apes dispersed between eastern Europe and south China; at least four genera and probably nine species were involved. They all seem to have been members of basically forest and woodland communities, and they disappeared when local conditions became more open.

New material of *Sivapithecus* and *Ramapithecus* from India, Pakistan, and Turkey shows that much of the *Sivapithecus* and *Ramapithecus* specimens may represent a single form, *Sivapithecus*, which was, like *P. nyanzae* from Rusinga, highly dimorphic. Post-cranial remains suggest body weights ranging from less than 30 to over 70 kg. *Sivapithecus* was arboreal and more ape-like than *Proconsul* in that its positional repertoire would have included suspension and quadrumanous clambering, as well as quadrupedalism and bipedalism. The foot was adapted for powerful grasping and suggests vertical climbing abilities like those of hylobatids and chimpanzees. The cheek teeth were relatively large and had thick enamel and rather low cusps looking superficially somewhat like hominids, and the mandibles were robust. Canines were ape-like and dimorphic, and the front teeth heavily worn from food and object manipulation. Microscopic tooth wear closely resembles that of *Pan*, a surprising result given that chimp teeth are relatively small and have thin enamel. We can reconstruct *Sivapithecus* as a mostly, but not exclusively arboreal, dimorphic, fruit-eating hominoid which would have moved in recognizably ape-like ways though it was different from any living ape.

What about its evolutionary relationships? In many features *Sivapithecus* is unique. In some, for example the foot, it resembles *Pan*, though there are good reasons to believe that these features are primitive for large hominoids. In others, the dentition for instance, *Sivapithecus* is similar to *Australopithecus*, but when examined in detail the similarities are only superficial; they are probably homoplasies. Neither set of resemblances helps in determining phylogenetic relationships. In a very few features *Sivapithecus* resembles the Asian orangutan, *Pongo*, in particular in the form of the face and details of the nose, premaxilla, and orbit. When these features are assayed across all extinct and living hominoids, each characteristic can be subdivided into a few character states, and their distribution is of great interest (Fig. 4.6). One pattern, A, is shared by *Hylobates*, cercopithecoids, *Proconsul*, and at least one later Miocene hominoid *Rudapithecus*. It is almost certainly primitive. A derived pattern, B is found in *Gorilla, Pan, Ouranopithecus*, perhaps *Kenyapithecus* (both mentioned later), and the early hominid *Australopithecus afarensis*. Yet another derived pattern, C, characterizes *Pongo* and *Sivapithecus* (including *Ramapithecus*). This pattern can plausibly be evolved from A via B; neither A nor B can readily be evolved from C (Fig. 4.7).

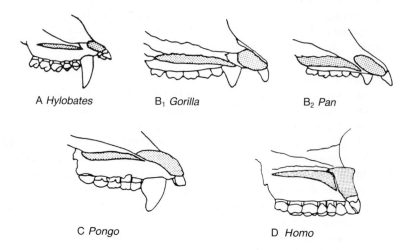

FIG. 4.6. Midline cross-section of lower face in living hominoids showing premaxillary-palatal morphology and relationships. (After S. Ward.)

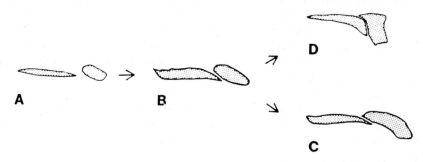

FIG. 4.7. Hypothesized evolutionary sequence in hominoid premaxillary-palatal morphology (see Fig. 4.6).

Hence, an evolutionary relationship between *Pongo* and *Sivapithecus* seems likely (Fig. 4.8). The oldest *Sivapithecus* specimens preserving the relevant body parts are 11–12 million years old. *Sivapithecus* in this carefully defined sense is known to be confined to the regions between Turkey and India, and is bracketed in age between 12 and about 7 Ma. The animal community containing *Sivapithecus* disappears around 7 Ma as forest and woodland forms are replaced by more open country species. Related and possibly descendant species are found in south-east Asian Pleistocene and recent communities, which may strengthen the proposed *Sivapithecus–Pongo* connection. Even if the link is correctly inferred, it should not be forgotten that *Sivapithecus* and *Pongo* are still very different creatures; orangutans have evolved markedly since the late Miocene. For

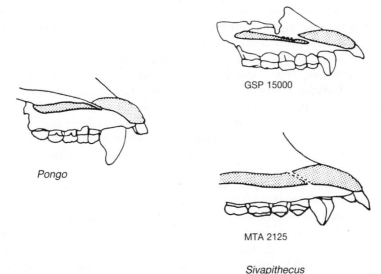

GSP 15000

Pongo

MTA 2125

Sivapithecus

Fig. 4.8. Mid-line sections of lower face in *Pongo* and *Sivapithecus* (GSP 15000 from Pakistan, MTA 2125 from Turkey) showing similarity in premaxillary-palatal morphology. (After S. Ward.)

one thing they have become more committed arborealists. *Sivapithecus* might have had a social organization somewhat more like the bonobo than like orangutan.

What of other later Miocene hominoids? Let us first consider Europe and Asia. The middle Miocene record in Europe is poorly known, documenting not much more than the spread of hominoids into south-eastern Europe by around 14 Ma, south of a large island seaway, the Paratethys. A possible second colonization of west and central Europe north of this seaway when it began to break up into large lakes 11 or 12 Ma is documented by the poorly known *Dryopithecus* and *Rudapithecus* from Spain, France, Germany, and Hungary. Where preserved, their cranio-facial anatomy is relatively primitive; isolated limb remains suggest apes with well developed suspensory abilities, and also relatively long hindlimbs.

Another interesting large hominoid from Greece, *Ouranopithecus*, is known almost entirely from jaws and teeth. The dentition is intriguing: canines are dimorphic but low crowned and they are set next to large, thick-enamelled cheek teeth with crowns which resemble in a number of ways those of African *Australopithecus*. The degree of sexual dimorphism in tooth size is considerable. The structure of the lower face differs from that of both *Rudapithecus* and *Proconsul*, and *Sivapithecus* and *Pongo*, and resembles more the African ape/early hominid pattern (pattern B).

Ouranopithecus and *Sivapithecus* lived in Greece and Turkey at approximately the same time (somewhere between 12 and 8 Ma) yet are as different in facial and dental morphology as *Pan* and *Pongo*. To complete the Euro-Asian sweep, material from south China is being recovered in great abundance from a late Miocene locality called Lufeng, in Yunnan Province. Initially referred to species of *Sivapithecus* and *Ramapithecus*, the hominoids probably represent a single highly dimorphic species of a new genus. Several crania are known, unfortunately crushed, and differ from the *Sivapithecus–Pongo* pattern. These hominoids were all members of forest and woodland communities, and disappeared as climatically and tectonically induced habitat changes brought more open communities in the second half of the late Miocene (along with first colobid and then cercopithecid monkeys). Large apes were once widespread in Europe and Asia, but only the orangutan lingers today in its diminishing forest refuge.

From the African middle Miocene come one or two species of *Kenyapithecus*, a hominoid with thick enamelled teeth rather like *Sivapithecus*, probably markedly dimorphic. The Buluk material is quite similar to this genus. It is fairly abundant at Fort Ternan (14 Ma) and Maboko (15 Ma), also in Kenya. Isolated limb fragments suggest a quadruped with a varied, but basically generalized positional repertoire. New material from Nachola in northern Kenya dated around 11 million years old includes an upper jaw fragment with a facial region like African apes and *Ouranopithecus* (pattern B) and unlike *Proconsul* or *Sivapithecus*. Most enigmatic of the new African finds is an upper jaw piece from the Samburu Hills of Kenya dated tentatively at around 9 Ma. The teeth are gorilla-sized, but with much thicker enamel. Premolars resemble somewhat those of gorillas, and this fragment may represent a proto-gorilla.

These newer specimens of Miocene hominoids allow us to paint the following phylogenetic picture. Somewhere between 17 and 12 Ma the species ancestral to large hominoids evolved, probably a generalized large bodied, very dimorphic, arboreal frugivorous ape with a varied positional repertoire including arm-swinging, active climbing especially of vertical supports, and occasional bipedalism. It fed mainly in trees, but was a less restricted arborealist than *Pongo*. Sometime between 18 and 12 Ma one or more large hominoid species migrated out of Africa. In the European and Asian forests and woodlands these hominoids underwent a modest adaptive radiation, including *Sivapithecus* from west and south Asia which can be linked to *Pongo* the sole living Asian great ape. The oldest well preserved *Sivapithecus* are 12 million years old, and this would then mark the youngest possible age for an 'Asian' (*Pongo*) versus 'African' (*Pan, Gorilla, Homo*) split. The Asian and European hominoids largely disappeared following later Miocene climatic-habitat change, and *Pongo* is a highly derived descendant of *Sivapithecus*.

Our knowledge of what was going on during this time in Africa and relevant adjacent areas is poor. Is *Ouranopithecus* or *Kenyapithecus* related to the 'African' branch, or to the common ancestry of all larger hominoids? Was the 'African' clade always African? We do not know. We can, however, make some general predictions about branching times. Assuming that DNA-DNA hybridization distances are reasonable reflectors of time and assuming that *Sivapithecus-Pongo* similarities tell us about evolutionary relationships, given a *Pongo* lineage diverging at least 12 Ma, gorillas would diverge 7–9 Ma, and chimps and humans 5–7 Ma (Fig. 4.9).

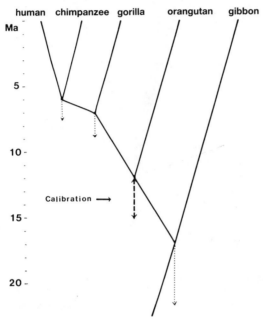

FIG. 4.9.   Hominoid branching sequence based on DNA hybridization data (Fig. 2.5), calibrated assuming *Sivapithecus* is ancestral to *Pongo* and oldest *Sivapithecus* is 12 million years old. (J. Sept.)

The oldest definite hominids, at least well sampled ones, are barely 4 million years old, so as much as 40 per cent of the hominid lineage may be left unsampled. The situation for African apes is even worse, there being no fossil record at all. We would very much like to know what the earliest hominids and prehominids were like, and since it is entirely possible that *Pan* and *Gorilla* are as unlike their ancestors as *Homo* and *Pongo* are unlike theirs, recovering an African ape record is important.

Trying to reconstruct ancestors just from living species is a very chancy business (Fig. 4.10). It is widely assumed at present that chimp and gorilla

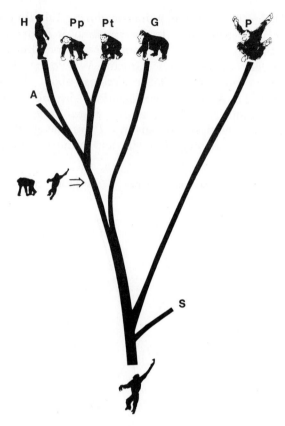

FIG. 4.10. Hominoid tree showing diversity of living large hominoids. Their common ancestor is depicted as a generalized arboreal ape similar to *Sivapithecus*. The ancestry of hominids and African apes is unknown. H, *Homo*; Pp, *Pan paniscus*; Pt, *Pan troglodytes*; G. *Gorilla;* P, *Pongo;* A. *Australopithecus;* S, *Sivapithecus*-like ancestor. (J. Sept.)

ancestors would have been (among other things) knuckle-walking quadrupeds, and that the ancestral hominid would therefore also have been a knuckle-walker. However, none of the known Miocene hominoids or Pliocene hominids show any features suggesting knuckle-walking, and homoplasy has been frequent in primate evolution. Until we have an appropriate fossil record we should perhaps reserve judgement.

Simply considering how different the three extant African apes are suggests how difficult ancestor reconstruction and recognition might be; all might have undergone evolutionary change. Gorillas are probably larger than their ancestors, became more terrestrial, more herbivorous, and more anatomically adapted to terrestrial as well as arboreal behaviour. Their dimorphism is probably retained. It has been suggested that the common chimp pattern is relatively primitive for chimpanzees, an adaptation to scattered food sources typical of woodland and marginal forest, and effectively characteristic of forests given the sympatric presence of gorillas monopolizing more readily available food sources. Compared with the much larger gorilla, *Pan troglodytes* has a radically different social organization. Bonobo social organization presents a third constrasting pattern, possibly derived from the *troglodytes* pattern. Despite the morphological similarity of common and pygmy chimpanzees their behaviour is markedly different in many important features. Given this variability, the difficulties in linking behavioural patterns with morphological patterns, and the apparent lability of the former, we should be very cautious in our reconstructions of the earliest hominids and in our attempts to infer their behaviour.

5   Hominid evolution

The early hominids: Australopithecus

The first well known hominids come from Africa between 5 and 1 Ma.
Australopithecus was a small-brained biped, and there were several species of
the genus. Often known colloquially as 'australopithecines', they are not
now thought to constitute a separate subfamily of Hominidae. They are
much better sampled than the Miocene hominoids. There is no exact
agreement about the number of species of *Australopithecus*, but they fall into
two groups: smaller species averaging about the size of chimpanzees; and
somewhat larger species, with larger cheek teeth and somewhat
bigger brains (Table 5.1). All species of *Australopithecus* are associated with
woodland or savannah communities, and they are clearly less forest
adapted than any Miocene hominoids.

They are found in two geographical areas. In South Africa they have
been recovered from cave sites which cannot be dated radiometrically, but
only by the less precise means of faunal comparisons. However, each site
represents accumulations of a few tens of thousands of years, and several
of them contain remains of dozens of individuals — albeit predominantly
jaws and teeth. We can therefore get some notion of variability in earlier
hominids within brief time periods. The second area is east and north-east
Africa, from Tanzania through Kenya to Ethiopia, where rich fossil sites
are associated with the rift valley system that acted both as a 'sink' in
which bones could be trapped and as a source of volcanic rocks with which
fossils could be dated. The eastern African sequence is extremely well
calibrated, mainly by potassium-argon dating and magnetostratigraphy.
However, in contrast to South African sites most eastern African sites
contain only one or a few specimens, and particular segments of time
hundreds of thousands of years long may have no fossil hominids at all.

These hominids and the earliest *Homo* species such as *Homo habilis* are
often known as 'Plio-Pleistocene hominoids' because they came from
sequences within which the Pliocene–Pleistocene boundary could not be
securely drawn. Plio-Pleistocene hominids have attracted considerable
attention from both public and professionals: they are 'missing links' *par
excellence*. Their study is in itself worthy of study, and it is becoming clear
that several biases have attended their description and interpretation.

Early hominid evolution, like that of later hominids, has until recently
been interpreted implicitly within a gradualist framework. Whether
scenarios were gradualist or punctuated, palaeoanthropologists have been

a little too free-wheeling in their behavioural interpretations of past species and with their explanations for the evolutionary transitions from 'ancestors' to 'descendants'. Behavioural reconstruction has generally been by analogy to particular living species (humans, particular primates, carnivores, beavers) rather than by applying ecological-behavioural 'rules'. Transitions, to be understood, would need to be very much better sampled than they are now before plausible explanations are possible. Also the fact that the evolutionary role of a character can and does change through time is often forgotten. Original and later adaptive utilities frequently differ, and all characters arc not always adaptations. Plio-Pleistocene hominids have usually been reconstructed as being rather like us, either overall or in some supposedly critical features. 'Us' shows a good deal of variation depending on the story-teller, from blood-thirsty plunderer to gentle and co-operative vegetarian. There are many reasons for this, including those already discussed. However, one stands out in particular, and this involves the question of perspective.

Historians about 50 years ago spent a good deal of time discussing approaches to studying history (historiography) and some of these discussions are relevant for palaeoanthropology. Within a spectrum of perspectives on the past, historians selected two as contrasting opposites. They have been called 'presentism' and 'historicism'. Presentism is an approach to the past which seeks likeness between past and present. It interprets the past in terms of the present and only studies the past in order to explain how the present got to be the way it is. It sees historical processes as progressive. Such an approach tends to make reconstructions of the past too 'modern'. The alternative approach, historicism, tries to study the past on its own terms and for its own sake. It tries to see the past as a kind of present, using knowledge of interactions derived from the present but assuming that the past was *un*like the present. This is how we must approach human evolution.

History of discovery of Plio-Pleistocene hominids

The order of discovery of Plio-Pleistocene hominids has (Fig. 5.1) also influenced ideas about them: ancestral species were not found first. The first australopithecine discovered was an *Australopithecus africanus* infant skull from Taung in South Africa described by Raymond Dart in 1925. The infant had a brain roughly the size of a young gorilla's. Dart's inferences about phylogenetic position were basically sound; his ideas about behaviour were imaginative. For a variety of reasons *A. africanus* was not accepted as a hominid. Although many viewed it as possibly more closely related to humans than to apes, it was seen basically as an ape. Implicit in the disagreements were ideas about the derived features which defined Hominidae. In the 1920s having a large brain was a *sine qua non*.

TABLE 5.1
Australopithecus and early Homo

	Australopithecus afarensis	*Australopithecus africanus*	*Australopithecus robustus*
Named:	1978	1925	1938
Main localities	Hadar, Omo (Ethiopia), Laetoli (Tanzania), ?Lothagam, ?Tabarin (Kenya)	Taung, Sterkfontein, Makapansgat (South Africa); prob. not elsewhere	Kromdaraai, Swartkrans (South Africa)
Geological age (Ma)	3.7–2.9, perhaps 5–2.6	Approximately 3–2	Approximately 2–1.5
Principal specimens	A1 288 ('Lucy'), Laetoli footprints	Sts 5 ('Mrs Ples'), Sts 14 (partial skeleton)	SK48 (cranium)
Brain size (cm^3)	400 ±	450 ±	500 ±
Body wt (kg)	?35	?35	?45
Tooth area (mm^2)	460	515	590
Brief description	Cranium long, low large face. Large front teeth, canines reduced in height, but dimorphic, thick enamel. Dimorphic, rel. long arms and ? short legs, bipedal and climbing adaptations.	As for *A. afarensis*, but braincase more rounded and face slightly more buttressed.	As for *A. africanus*, but face flatter and more buttressed. Crests for masticatory muscles.

	Australopithecus boisei	*Homo habilis*	*Homo* sp
Named:	1959	1964	
Main localities	Olduvai, Lake Natron (Tanzania), East and West Turkana, and Chesowanja (Kenya), Omo (Ethiopia).	Olduvai (Tanzania), East Turkana (Kenya), Omo (Ethiopia). Perhaps S. Africa.	As for *H. habilis*
Geological age (Ma)	Approx. 2.5–1.3	+ 2–1.5	?As for *H. habilis*
Principal specimens	OH5 ('Zinj'), ER 406, ER 732	OH7, ER 1470, ER 1590, ER 1481	ER 1813, ER 1805, OH 13, OH 24
Brain size (cm^3)	500 ±	700 ±	550 ±
Body wt (kg)	?50	?50	?
Tooth area (mm^2)	750	500	425
Brief description	As for *A. robustus*, but face deeper and more buttressed. Cranium more robust. Very thick enamel, low cusps.	Larger and rounded braincase, deep and flat face. Large front teeth, thick enamel. Known body parts more like *Homo*.	Smaller and rounded braincase browridges, small face. Small front teeth and cheek. Post-cranium unknown.

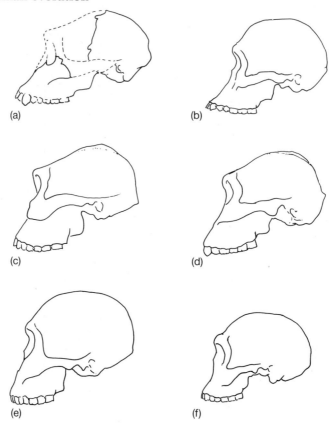

FIG. 5.1.   Pliocene and earliest Pleistocene hominids. (a) *Australopithecus afarensis*; (b) *A. africanus*; (c) *A. robustus*; (d) *A. boisei*; (e) *Homo habilis*; (f) *Homo* sp. (J. Sept.)

In the 1930s and 1940s Robert Broom, and later Broom's student John Robinson, collected more australopithecines from Sterkfontein, Swartkrans, and Kromdraai, including adult skulls and portions of the post-cranial skeleton, and in the late 1940s Dart found more at Makapansgat. By 1947 it was clear that although australopithecines had small brains as adults, within or (so it was thought) somewhat above the ranges of living apes, they had more *Homo*-like teeth with small, blunt canines in both sexes, and a pelvis which was more human than ape-like (Fig. 5.2). Indeed, the defining characters of Hominidae shifted away from brain size to teeth and post-cranial features.

By the 1950s Robinson had begun to synthesize this growing body of data. He reduced the number of species below those originally described and grouped them into two genera, *Australopithecus* and *Paranthropus* — smaller and larger, less and more robust. Initially, he clustered hominids into three lineages, *Australopithecus*, *Paranthropus*, and *Homo*, but later modified this into two, with *Australopithecus* ancestral to *Homo*, *Paranthropus* becoming extinct, and the unknown earliest hominid resembling *Paranthropus*. *Australopithecus* was a tool-user and omnivore, perhaps even doing some hunting, while *Paranthropus* was an 'acultural' herbivore. C. K. Brain had analysed the rocks making up the South African cave fillings and made some tentative inferences about past environmental conditions. The relative ages of the caves was estimated from fauna and unsuccessful attempts made to date them absolutely (which continue).

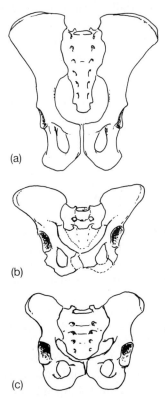

FIG. 5.2.   Pelvises of (a) *Pan*; (b) *Australopithecus africanus* from Sterkfontein, South Africa; (c) *Homo sapiens*. (After J. T. Robinson.)

It is now generally agreed that Makapansgat, Sterkfontein, and probably Taung, which have *A. africanus*, are older (roughly 3–2 Ma), while Kromdraai and Swartkrans with *Paranthropus* or *A. robustus* are younger (2–1.5 Ma). Environments were wetter at Sterkfontein than at Swartkrans, the opposite of what Brain and Robinson originally thought.

During the 1940s and 1950s Dart developed the idea that the smaller australopithecines were aggressive and bloodthirsty predators. This was based on the recovery of broken animal bones associated with the hominids. Dart thought that breakage patterns and other features implied their use as knives, clubs, spears, and scrapers, and that early hominids were human-like predators. The idea is not widely believed now because the patterns of bone breakage and accumulation imply instead the activities of non-primate predators; australopithecines were hunted rather than hunters.

Work in east Africa became very active following the discovery by Mary Leakey in 1959 of an essentially complete australopithecine cranium at Olduvai in Tanzania. Louis Leakey called this *Zinjanthropus boisei*. It was in excellent stratigraphic context and was the first hominid dated by the potassium-argon method, to the surprisingly old age of almost 2 million years. The hominid was associated with simple stone tools which were present at Olduvai in great abundance, and indeed was originally considered to be 'the' toolmaker. Two years later another kind of larger brained hominid of about the same age was found at Olduvai, and called in 1964 *Homo habilis*.

These two hominids became the objects of some controversy. Robinson argued that *Z. boisei* and *H. habilis* were simply east African variants of *Paranthropus robustus* and *Australopithecus africanus*. Few now believe this. Philip Tobias proposed in 1967 that all the australopithecines be placed in one genus, with three species: *A. africanus, A. robustus (Paranthropus)*, and *A. boisei (Zinjanthropus)*. Tobias argued for a three-branched evolutionary tree, with branches to robust australopithecines (*A. robustus, A. boisei*) and to 'graciles' (*A. africanus*) becoming extinct, while *H. habilis* lead to later *Homo*. The unknown common ancestor of all these would, he thought, be a more generalized 'gracile' hominid.

Large-scale multidisciplinary field and laboratory studies of the eastern African sites began in earnest in the 1960s, initially with work at Omo in Ethiopia under the direction of F. Clark Howell, Camille Arambourg, and Yves Coppens, later at East Lake Turkana (earlier Lake Rudolf) under Richard Leakey and Glynn Isaac, and more recently in the 1970s at Hadar in Ethiopia directed by Don Johanson, and Laetoli in Tanzania led by Mary Leakey.

From these studies and others on smaller or less extensive sites we now have an extremely well calibrated sequence for eastern Africa running from about 5 to less than 1 Ma. In addition we have a good understanding of past faunas and a reasonable knowledge of habitats. Hominids are not equally well represented throughout the period, but they are becoming increasingly better known.

We can briefly summarize some of the important newer discoveries from eastern Africa. As more *A. boisei* material was recovered it became clear that the original, Olduvai specimen was male and that sexual dimorphism in the skull was as marked as in *Gorilla* or *Pongo*, implying considerable male–female body size difference. Better *H. habilis* specimens were found (for example ER 1470 from east of Lake Turkana) showing that it had a distinctly larger brain than any australopithecine and was clearly a different species, points that had been in contention. Surprisingly early *Homo erectus* were found at Lake Turkana in dated contexts going back over 1.5 Ma. Certain skulls resembling *H. habilis* (for example, ER 1813) had smaller brains and were thought by some scholars to represent a separate species similar to *A. africanus*. The name *Homo ergaster* was given to a lower jaw which might belong to this species.

The hominid discoveries at Hadar in Ethiopia and Laetoli in Tanzania document the time between around 2.8 and 3.7 Ma which is either poorly or not represented at Omo, East Turkana, and Olduvai. A new and primitive species of *Australopithecus, A. afarensis*, has been described from these sites, and it may also be represented by scraps from Lothagam and Tabarin in Kenya, both close to 5 million years old. *Australopithecus afarensis* may be the ancestor of all later hominids. However, there are some who favour the view that more than one hominid species is represented, an early *Homo* and an australopithecine. Regardless of taxonomy, the footprints preserved in volcanic dust at Laetoli document the presence some 3.7 Ma of bipedal hominids; whether or not the gait is identical to that of modern humans is disputed.

These various species can be sorted fairly easily one from another on the basis of cranial features. Post-cranial remains, however, are a different matter. As these became better known over the last 20 years they were seen to fall into two basic groups. These have been called *Australopithecus* and *Homo* ('early' or 'archaic' *Homo*), the *Australopithecus* group containing specimens of varying size. The early *Homo* group material is similar to comparable parts of *H. sapiens*, although generally more robust, with thicker long bone cortex, sometimes larger joint surfaces, and other features suggesting possible adaptations to higher activity levels than in modern humans. However, the australopithecine post-cranials show a number of differences from their homologues in early *Homo*. For example, in the pelvis the iliac blades are orientated more laterally and the hip joints also face more laterally; the sacroiliac articular area is small. The femur head is small, the neck long and flattened front to back, and the shaft also similarly flattened. Many features of hands and feet differ, as do limb proportions.

Beginning in the early 1970s, Owen Lovejoy argued that most if not all the post-cranial contrasts between *Australopithecus* and *Homo* flowed from the presence in *Homo* of a wider pelvis, a derived feature related to obstetrical problems associated with giving birth to larger brained infants. The more primitive hominid bipedal pattern seen in *Australopithecus* was no less and perhaps more efficient than the *Homo* pattern. These views are being vigorously debated.

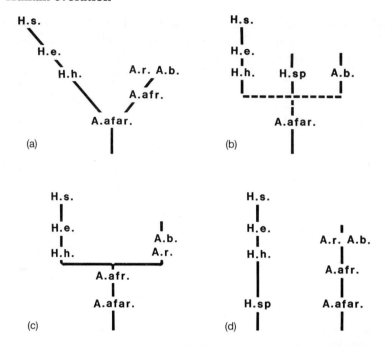

F̶ɪɢ. 5.3.   Alternative phylogenetic trees for hominids. A. afar, *Australopithecus afarensis*; A. afr., *A. africanus*; A. r., *A. robustus*; A. b., *A. boisei*; H. h., *Homo habilis*; H. sp., *Homo sp.*; H. e., *H. erectus*; H. s., *H. sapiens*.

Running parallel with discoveries and analyses of hominids, other fauna, stratigraphy, palaeoecology, and geochronology has been continuing work in the archaeological record. Simple stone tools of the Oldowan industry (named after Olduvai Gorge) first appear a little over 2 Ma. Assemblages consist of sharp flakes and small cores which may be both tools and the sources of flakes. Microscopic edge-wear analyses of a few flakes suggest their use in a variety of ways: cutting plant material, meat, and wood, perhaps to make other tools such as digging sticks. A new tool, the biface — sometimes called a handaxe — appears about 1.6 Ma, a large and more consistently shaped flake tool. This marks so-called Acheulean assemblages, which otherwise are quite similar to the earlier Oldowan.

These early stone tools are found both scattered and in aggregations, with and without animal bones. Some of the bones have cutmarks and toothmarks, some of the cutmarks being produced by stone tools. These concentrations of tools and bones resemble in certain features the remains of camp sites produced by hunting and gathering peoples living today, so it is not at all surprising that the early aggregations should have been interpreted as 'home bases' or 'living floors'. The implication was that space use, ranging and subsistence behaviour, food sharing, division of labour, and co-operation were also hallmarks of these early hominid behaviour systems, and that sharing and reciprocity may have been important triggers of early *Homo* evolution.

After several years of lively debate it is more widely believed now that these early members of *Homo* were certainly behaving in ways different from those of their more primitive australopithecine ancestors and cousins, but that behaviour patterns would not have been like modern humans either. The issues in contention concern the amount of meat in the diet, how it was obtained — by scavenging or hunting or both, and the extent to which significant tasks were habitually focused on one or several places.

Australopithecus afarensis and africanus

Australopithecus afarensis is best known from Laetoli in Tanzania where it is dated to around 3.7 Ma and from Hadar in Ethiopia where it samples a few hundred thousand years around 3 Ma. It may also be represented from the same period at Omo, possibly from East Turkana, and perhaps from Lothagam and Baringo (Tabarin, Chemeron Foundation) where it is older than 4.5 Ma. Material from these localities can only tentatively be assigned because it consists of jaw and teeth fragments which are not the best body parts for taxonomic sorting. However, it is likely that *A. afarensis* existed between 5 and a little under 3 Ma. There is still debate about the number of species represented in the sample. However, we shall take the view that one species is likeliest, while recognizing that more than one species is also a realistic possibility. *Australopithecus afarensis* body weights range from less than 25 to over 50 kg and the species was probably highly dimorphic, certainly more than species of *Pan*, but the exact degree is hard to determine without better and more plausibly sexed material. Brain volume, estimated from endocasts, averaged around 400 cm^3 and was perhaps no larger relatively than that of *Pan*. The interpretation of endocranial features is debated, but it seems likely that they were generally like those of apes, though perhaps resembling hominids in a few features. Overall, the cranium resembled that of a robust chimpanzee or small gorilla, particularly in the cranial base and face, and can be regarded as basically primitive.

The incisors are chimp-like, but not the remainder of the dentition. Canines are dimorphic in size, but not morphology, and in both males and females they are low-crowned and rather incisor-like. Cheek teeth are larger than those of chimps, and this is due, in part, to the presence of much thicker enamel in *A. afarensis*. The dentition is clearly size dimorphic both in canines and cheek teeth, but the extent of dimorphism is unclear because it is difficult to sex jaws; one needs to know sex before the difference between male and female averages — metrical or morphological — can be determined.

Post-cranial anatomy is relatively well-known: the famous 42 per cent skeleton of 'Lucy' is a small and presumably female *A. afarensis*. Arms may have been somewhat longer than those of modern humans while legs were very likely shorter. Some features of hand, wrist, elbow, and shoulder joints suggest better climbing and suspensory abilities than for *Homo*. Proportions of the hand and fingers and wrist joints imply a hand less adapted to arboreal climbing or knuckle-walking than a chimpanzee, and capable of more precise grasping and manipulation. In turn, dexterity and manipulative skill would have been less developed than in early *Homo*.

The hind limb and pelvis show many bipedal adaptations. Broad and short iliac blades, short ischium, approximation of sacroiliac and hip joints, the anatomy of knee and ankle joints, and the parallel big toe, are all more human-like than ape-like and in some cases very human-like. They clearly indicate a positional repertoire in which upright walking and running were important. Other features — for example, an apparently shorter hindlimb, longer foot, and longer and more curved toes — suggest to some that bipedalism would have been different from and perhaps energetically more costly than human bipedalism.

The associated faunal remains and fossil plant material show that habitats in eastern Africa were open: woodland, bushland, and dry savannah, with patches of forest along rivers and streams and near lakes. We know that climates fluctuated throughout the Pliocene and that between 5 and 3.2 Ma climates were warmer and less seasonal than today's, so habitats were probably on average somewhat less open. About 3.2 Ma climates became cooler and the first moderate glaciations began in higher latitudes. Near this time there is evidence from northern Kenya for a brief wetter period during which the central African rainforest extended as far east as Lake Turkana. this may be coincident with the cooling and possibly reflects increased precipitation.

What might *Australopithecus africanus* have been like as a living species? If we could sit and watch a community over a period of months or years how would they have behaved? Which questions would we ask? What would we be interested in? Remember the list of features listed in Table 1.2 and discussed in section 4, adaptations promoting fitness in living species. It included significant ecological features, body and brain size and degree of dimorphism, life history patterns, diet and subsistence behaviour, positional and ranging behaviour and habitat use, adult grouping patterns, relationships among females, among males, between males and females, sexual and reproductive behaviour, intergroup relationships, equipment use if any, and the nature of communication systems.

Assuming that we have sorted Plio-Pleistocene hominids into correct species — still a somewhat questionable assumption — their broad distribution patterns can tell us something about adaptation. *Australopithecus afarensis, A. robustus,* and *A. boisei* all appear to be regional species found in either south or eastern Africa, but not both. *Australopithecus africanus* may also follow this pattern, but that for early *Homo* species is less clear. The first hominid species that is clearly pan-African is *Homo erectus*. More localized species distributions imply a narrower range of adaptations. Unfortunately, moving from what we can say about habitat structure and broad-ranging patterns to the dimensions of realized niche, as in the case of truly living species, is very difficult. Chimpanzees, for example, inhabit a range of habitats from forest to almost open grassland yet focus within those habitats on trees where they find almost all their food. Regardless of habitat, chimps spend well over half their waking time in trees, but a substantial amount of time is also spent trekking between food sources which are mostly in trees. Preferred foods — ripe fruits — have particular physical properties which explain features of the chimp dentition. Their mixed ,arboreal–terrestrial repertoire explains their mixed or compromised anatomy. Thus, ecology, behaviour, and anatomy are linked.

We know now a good deal about *A. afarensis* anatomy and must use it to infer more precisely features of at least potential niche. If *A. afarensis* is one species then it is more dimorphic than chimpanzees and perhaps almost as dimorphic as gorilla. *A. afarensis* had a brain barely if any larger than that of *Pan*. Relationships between adult brain size, body size, gestation length, newborn body size, newborn brain size, and age at first reproduction, among other parameters, show clear patterns in living species, and known adult values can be used to estimate 'unknowns'. Estimates of unknowns always have an error, often substantial, so they must be treated cautiously. The best estimates would be that overall *afarensis* was ape-like rather than human-like in features such as maturation rates, and newborn body and brain size. There are important disagreements over the extent to which obstetrical problems would have affected *afarensis* births, there being advocates both of the view that infants would have been born as in apes without difficulty, and alternatively that maternal pelvic size was already making birth difficult and thereby perhaps selecting for smaller, less mature newborns born after shorter gestation periods.

What about food and feeding behaviour? As with other behaviours it is easier to say what these early hominids did not do than what they did. Remember that *afarensis* incisors were large and chimp-like, while canines were low crowned, blunt, and in part structured like incisors (though still quite size dimorphic). Cheek teeth were large and had thick enamel caps, set in robust lower jaws that were moved by large masticatory muscles. Tooth wear differs little from that in other australopithecines and in turn is similar to wear in living apes, reflecting a diet of softer vegetation including fruits and leaves. However, we are faced with a paradox in that *afarensis*, like other australopithecines and many Miocene apes, had enlarged cheek teeth, thick enamel, and robust jaws, unlike living apes.

Why? Several explanations have been offered: hard food such as nuts and seeds was regularly or occasionally consumed, or food was of low nutritional quality and more had to be chewed. However, we are still unclear about the adaptive meaning of these features. Clearly, *afarensis* was not chewing grass, seeds, meat, or bone, nor probably ripe forest fruits like chimpanzees do. What, if any, is the significance of canine reduction? Does it have implications for social behaviour (males fighting and bluffing less with their canines, more with tools) as many since Darwin have argued? Or implications for feeding? Or both? Possibly food was being prepared with the front teeth even more than is the case with chimps. This might again give us a clue about food type.

Hand anatomy is well known in *afarensis* and differs both from living apes and from *Homo*. The hand is of approximately human proportions, but there are some differences. Fingers are more curved than in *Homo* and suggest powerful grasping. Hand bones imply movement capabilities roughly intermediate between those of apes and humans. Grips would have been powerful with greater precision than in chimpanzees; implications are that objects could be grasped and used more effectively than in apes, though less so than in *Homo*. *Australopithecus afarensis* was almost certainly climbing less in trees and was a good deal more bipedal than any ape, and this may partially explain non-ape features. However, there were positive factors too shaping hand morphology; implications are for more precise and dextrous manipulative skills in feeding and other activities, and probably tool use.

Chimps today use a wide range of natural objects as tools and presumably australopithecines did so too. *Australopithecus robustus* at Swartkrans in South Africa has been found associated with a few bone pieces perhaps used for digging. No clear direct evidence exists for tools associated with *A. afarensis*, but it is reasonable to infer a degree of object use and skill as great as that of apes, without its being intense or patterned enough to have produced recognizable archaeological sites. These australopithecines probably had access to a broader range of plant foods in open habitats than chimps do, and presumably may also have supplemented their diet with occasional animals.

What about positional and ranging behaviour? Here we face immediately a problem of interpretation. There is no disagreement that *A. afarensis* was bipedal a good deal of the time. However, the amount of bipedality in the total repertoire, and its relative efficiency compared to other kinds of bipedalism, is hotly debated. There are at least three positions. First, one species is represented and both sexes are as fully and effectively bipedal as modern humans; indeed, certain features of the hip imply more efficient control of pelvic tilt and rotation than in *Homo*. A second position sees one dimorphic species with smaller females more arboreal than the larger males, neither sex being as fully or efficiently bipedal as *Homo sapiens*. A third view sees two species, one smaller and more primitive, and the other larger and more human.

It will be some time before these different interpretations are settled. We assume here that *afarensis* is indeed one dimorphic species with males much larger than females. While bipedal these hominids were somewhat less commitedly so than *Homo*, climbing in trees more, and post-cranial anatomy was a little more compromised reflecting a positional repertoire broader and more varied than in *Homo*, though less so than in apes. Why were *afarensis* bipedal? They would have stood to feed from low bushes and trees, and they probably trekked bipedally between food sources. They probably also carried things, including their babies, and if they invented slings for babies they would have used them for other objects. However, trying to say what *afarensis* was doing when it was bipedal has to reflect in part one's position on why bipedalism originally evolved, and that is even more obscure, not least because we lack a fossil record for that evolutionary phase.

Differences in male and female body size might have implications for ranging behaviour. Common chimpanzees, in particular, are unusual in that both males and females range widely, males even more than females. Among hominoids living and extinct they are unusually low in degree of dimorphism, and this may be linked to ranging behaviour. The 'smaller' males might be able to range further and more efficiently than would be the case at larger body size. So under this argument a ranging pattern and social organization unlike that of chimpanzees is perhaps implied.

What more can we say about grouping behaviour or relationships between and among the sexes? Again, it is easier to say what *afarensis* was unlike than what it was like. Given its dimorphism it is unlikely to have been monogamous, like gibbons, nor to have had human-type marriage. It might not have had a pattern of fission–fusion characteristic of chimpanzee communities. It probably did not feed on resources concentrated enough to have permitted cohesive groups like those of gorillas or bonobos. High dimorphism, though, implies reproductive units more like gorillas than like bonobos, with bonds between one or a few males and several to many females, rather than more diffuse or paired relationships. If food sources were generally scattered, social groups would have been small, possibly focused around one or few adult males. Intergroup relationships would perhaps not have been very peaceful, especially between males. Our closest living relatives are neither wholly peaceful nor wholly aggressive, but both, depending on context and situation. Presumably, so too were our ancestors.

Communication would have been at least as effective as that of apes. Exactly how effective that is we only dimly perceive: natural ape languages are undoubtedly rich and effective, but poorly understood by us. Judging from brain size and external anatomy which are broadly ape-like, and the likelihood that pharyngeal structure was also ape-like implying a more limited repertoire of sounds than in humans, we can infer that communication systems and the amount of information communicated were broadly equivalent to what is seen in chimpanzees.

In summary, were we able to sit and watch *afarensis* we would probably see them as odd upright apes, basically individualistic plant eaters living in small scattered social groups. What about variations on this pattern? *Australopithecus africanus* is similar to *A. afarensis*. It is best known from the South African sites of Sterkfontein and Makapansgat where skull and body parts are well sampled. Comparative faunal dating suggests an age of 3.0 to 2.5 Ma. *Australopithecus africanus* has also been reported from eastern Africa at Omo between 2.5 and 2.0 Ma although this is based only on teeth. South African *A. africanus* is very similar to *afarensis* except (possibly) for a slightly larger brain and slightly bigger cheek teeth associated with a somewhat more buttressed face. Overall, the two species are hardly more different than the two species of baboon *Papio cynocephalus* and *Papio hamadryas*, or chimpanzee *Pan troglodytes* and *Pan paniscus*. Such a degree of morphological resemblance might be hiding a degree of behavioural difference as in the living species pairs. Or it might not.

Robust Australopithecus

Robust australopithecines are known from South Africa (*A. robustus*) and East Africa (*A. boisei*). For the past 15 years or more they have been classified in *Australopithecus* although there is a growing movement to shift them back to the original genus *Paranthropus*. The two species are similar though clearly distinct, with *boisei* having larger cheek teeth and a more robustly constructed face and braincase. *A. robustus* comes from Swartkrans and Kromdraai in South Africa, between about 1.8 and 1.5 million years old. *Australopithecus boisei* comes from Peninj and Olduvai in Tanzania, Chesowanja, and East and West Turkana in Kenya, and Omo in Ethiopia, ranging from about 2.5 to somewhat over 1 Ma.

Australopithecus boisei is overall better known cranially and dentally. It is the largest australopithecine with the heaviest individuals probably weighing over 70 kg. Unfortunately, post-cranial elements are fragmentary and difficult to identify with certainty. No individuals are known with associated cranial and post-cranial elements, nor are enough post-cranials known to give us information about limb proportions. With those caveats, we can summarize anatomy. *Australopithecus boisei* was a species larger than *afarensis* or *africanus*, with males probably double the body size of females. Compared to *Homo*, forelimbs were relatively longer and/or hindlimbs relatively shorter, like the smaller australopithecine species. Brain volume is over 500 cm^3 in two males and a little under in a couple of females, a degree of difference similar to or less than that between males and females in a dimorphic ape such as *Gorilla*. External brain morphology is different from that in other hominids, although it is unclear what the differences mean.

Cheek teeth were very large with very thick enamel caps and low rounded cusps. Front teeth were smaller than in *afarensis* or *africanus*, and jaws massive and robust. Masticatory muscles were very large, and the front portion of the temporal muscle particularly well developed. The cranium was robust with a flat and deep face, and thickened facial plates acting as buttresses against chewing stresses. Brow ridges and zygomatic bones were well developed and there was a small midline crest toward the front of the braincase in males. These cranial features reflect the large chewing muscles. In addition the cranial base was strongly flexed unlike the condition in *afarensis* or *africanus*.

Many of the cranial differences between larger and smaller australopithecines are probably related to body, tooth, and brain size differences, and this is an appropriate point to discuss briefly the relationships between overall size and relative size or proportions. 'Allometry' is the study of changes in proportions or shape with size. It is clear that as an individual grows, shape changes: compare a child's head with an adult's, or a baby's body with that of an adolescent. Similarly, proportions may change through evolutionary time as species in a lineage change in size. Careful comparisons of related living species allow us to derive some general rules about these shape differences or 'scaling'. Among African apes, for example, both brain size and tooth size increase with body size, but not at the same rate; brain size increases less. As a consequence, large species have relatively larger teeth, faces, chewing muscles, and buttressing than smaller ones, and less globular skulls. In considering differences between fossil species we would like to eliminate as far as possible the effects of size before placing too much emphasis on a functional or adaptive explanation for differences in shape. The australopithecines are an interesting case in point. Cheek tooth size differences and other cranial features can be mostly explained this way. However, cusp shape differs, implying a different function, and front teeth are radically different — smaller in larger species, in ways that cannot be explained by allometry. However, tooth wear in robust *Australopithecines* differs little from Miocene apes, other australopithecines, and living fruit-eating apes, a surprising and 'counter-intuitive' result. Occasional tough or hard foods were probably eaten, but plant foods of various kinds are again indicated.

Both species of robust *Australopithecus* are associated with habitats that were more open than those of the earlier Pliocene. A worldwide climatic shift to cooler conditions a little under 2.5 Ma probably made African plant communities more open so hominids would have been in more seasonal climates with fewer trees. There is some evidence for species change in Africa about 2.5 Ma and perhaps new species of hominid appeared then. However, sampling is poor and while there is a reasonable case to be made out for the view that *A. afarensis* is broadly ancestral to *boisei* and *robustus*, perhaps with *africanus* as an intermediate, when and where the evolutionary transition occurred and how long it took are all unknown.

Equally unclear is the nature of the behavioural difference between larger and smaller *Australopithecus* species. Are the later ones bigger because food was of lower quality? Larger animals, having lower metabolic requirements per unit of body weight, can subsist on lower quality food though they may need to spend more time eating. Is there enough of sufficient quality plant food in seasonal habitats as far from the equator as Southern Africa to keep a large brained, large bodied hominid alive without the regular use of tools? We have no evidence that robust australopithecines used tools, other than perhaps occasional hammer-stones and digging sticks, and, judging from the South African sites, they were hunted by large predators, in contrast to their successor hominids who could out-compete carnivores for cave sites. These creatures, if anything more different than *afarensis* from anything alive today and therefore more difficult to reconstruct, are even more enigmatic. Let us assume though that robust australopithecines either evolved or diversified in response to a climatic change after 2.5 Ma, and that the hominid changes are part of a shift in vertebrate faunas. This was not the only evolutionary response by hominids. Sometime before 2 Ma another new kind of hominid appeared, *Homo habilis*.

Intermediate hominids

Homo habilis

The first *H. habilis*, described in 1964, came from Olduvai Gorge in Tanzania. The original material was fragmentary, sufficiently so to provoke some controversy over whether or not a species distinct from South African *A. africanus* was truly present; *H. habilis* was said to have had a larger brain and smaller cheek teeth. Since then, more material has been found at Olduvai, Omo, and East Turkana where several relatively complete crania were found, including ER 1470 from East Turkana. The age of ER 1470 was originally thought to be in excess of 2.6 Ma, but it is now dated at around 2 Ma.

Homo habilis consists of several reasonable crania, some jaws and teeth, and a few unassociated post-cranials. There are disagreements over the number of species being sampled. If it is only one, there is a considerable range in brain volume. A small number of larger crania cluster around 700 cm^3 or more, while a smaller group average a little over 500 cm^3. These groups differ in tooth size, the former having larger teeth, including canines, and facial morphology differs markedly between the best preserved representatives of each group, ER 1470 and ER 1813, both from East Turkana. If these are males and females the degree of brain, tooth, and face size dimorphism is considerably greater than in any living species. We assume here that two species are probably being sampled, *Homo habilis* and a smaller brained species superficially resembling *A. africanus*. The name *Homo ergaster* was given to a mandible ER 992 from East Turkana which may belong in this second species, in which case the name would be *H. ergaster*. However, mandibles with teeth are difficult to assign between *H. habilis, H. erectus*, and this third species; ER 992 could be a *habilis* or an *erectus*. We shall refer to the third species as *Homo* sp.

Homo habilis specimens are known from a little before to a little after 2 Ma, from East Turkana, Olduvai, probably Omo, and possibly Sterkfontein. Body size is poorly known and degree of dimorphism is unclear, but some individuals weighed 40–50 kg. Brain size ranged between 650 and 800 cm³, and an endocast shows some frontal lobe features not seen in australopithecines and supposedly characteristic of *Homo sapiens*. It has been suggested that these imply speech capabilities. The braincase is rounded, and resembles an enlarged, allometrically scaled version of *A. africanus*. The face was large and deep, with big incisors and canines, and large cheek teeth. Wear patterns on molars and premolars broadly resemble those of *Australopithecus*. Post-cranials are difficult to assign, but an innominate, femur, and tibia are probably *habilis*; hand and foot bones may also be known, but these could be smaller australopithecines or *Homo* sp. The pelvic and leg bones differ from homologues of *Australopithecus* and resemble those of *Homo erectus*, being similar to modern human bones, but more robust. They imply a positional repertoire, gait, and ranging pattern more similar to humans than to australopithecines.

Homo sp. is known from Olduvai, East Turkana, possibly Omo, and perhaps Swartkrans, between 2 and 1.5 Ma. Since it is not clear which, if any, post-cranial elements assigned to other species actually belong to this one, it is impossible to discuss body size or dimorphism. Brain size in probable females is a little over 500 cm³, and one possible male (ER 1805 from East Turkana) approaches 600 cm³, a plausible degree of dimorphism.

Around the time these new hominids evolve — somewhat over 2 Ma — the first archaeological traces also appear. Best known at Olduvai and East Turkana, they also turn up at other places in eastern Africa, perhaps in South Africa as well. Assemblages consist of combinations of broken, concentrated bones and bone parts, and of altered stone. It is often difficult to separate out bones damaged or transported by hominids from those damaged or transported by carnivores (for example, hyaenas) but under some circumstances this can be done. Stone tools are generally recognizable. Patches or scatters of stone alone, of bone alone, and of bones and stone together were scattered around the landscape, although most frequently near lakes or streams.

Under good conditions bone damage clearly produced by hominids can be recognized: cut marks made by sharp stone flakes; crushing as a result of hammering. Stones were sometimes clustered and cached after being carried for several kilometres. Stone was flaked — sharp pieces being knocked off a core using another stone — and the assemblages can be divided roughly into sharp flakes, shaped cores, and irregular cores. Preliminary analyses of edge wear on the flakes show that flakes were used to cut a range of objects: meat, wood, and plants. Some of the shaped cores were perhaps used for cracking bone, or hard nuts and seeds as they are in chimpanzees, and some of the cores were probably carried as a ready source of sharp flakes. Broken bones often would have included at least some marrow, and cut marks are found on both 'meaty' and 'non-meaty' parts. They also occur upon and beneath carnivore tooth marks.

Recently, there has been a lively debate about the meaning of these concentrations and scatters, and it is generally agreed that they do not represent campsites in any human sense; they were not places at which hominids lived and shared, and from which they regularly foraged and hunted. However, they clearly indicate behaviour patterns different from those of any non-human primate. At the very least they show that by 2 Ma hominids were leaving scatters near places where water, protection, and plant food would have been located. Also, they were carrying stones and breaking them to make sharp flakes. They used tools to cut up animals of various sizes and varying degrees of freshness. They were eating meat and marrow, and some of these activities were performed repeatedly at particular places.

It is unclear exactly how meat was obtained — by scavenging or by hunting, and how often — daily, weekly, or monthly. Based on knowledge of how carcasses get consumed today and how bones break, scatter, and weather, these early accumulations represent carcasses to which hominids had access at a variety of times — early to late. Some of them therefore probably came from hunting, while others represented the ultimate scavenged remnants. Parts of the accumulations were due to carnivores, others to hominids. Some accumulations represent brief occupation periods: the scattered pieces of broken bone or stone can be conjoined. Others are palimpsests, repeated accumulations over periods probably much longer than a single life. Both stones and bones with meat or marrow were being transported, perhaps cached, perhaps exchanged or shared.

Clearly, we are seeing a behaviour system based on a wider use of objects than any other mammal and on the consumption of a broader range of resources than most mammals except humans. If we knew what the functions were of the enlarged *H. habilis* brain we would be better able to speculate further. Unfortunately, brains do not fossilize. However, as we learn more about the variation in brain anatomy and physiology in living mammal species, and their links to behaviour, we can make better educated guesses about past hominid brain–behaviour systems. We discussed earlier some recent comparative studies on the brain and these can be used as a guide to thinking about early brain expansion. Prefrontal cortex, including Broca's area, has undergone considerable enlargement in the human brain and, as a consequence, mouth movements and vocalizations are brought under neocortical control. If these areas and others also involved in language enlarged radically under strong selection for a new kind of communication system then overall brain size might increase sharply. If in addition motor and sensory areas involved in complex movements and manipulative patterns were also under selection and became enlarged we have a partial and tentative explanation why the hominid brain increased in relative size by about 50 per cent, perhaps in a rather short time period.

In some ways behavioural reconstruction of these first tool-making hominids is more difficult even than for the australopithecines because, although they are phylogenetically closer to us and in some features more similar, they were clearly unlike us in many ways and we can be easily misled into seeing them as simpler versions of ourselves. But what might they have been like?

Homo habilis was part of the savannah–woodland open country fauna of Africa. *Homo habilis* and *A. boisei* may have used different parts of the habitat, although it is quite conceivable that they could have been completely sympatric. However, judging from the only two sympatric large hominoids, *Pan troglodytes* and *Gorilla gorilla*, the hominids probably avoided each other as much as possible. Many more *boisei* fossils are known than specimens from the non-robust species, and this is certainly because *boisei* teeth and jaws are so robust and easily preserved; but it could also be a consequence of differences in subsistence behaviour. If *boisei* was a herbivore while *habilis* had become more omnivorous the biomass of *habilis* would be lower, with fewer individuals in the community.

We are very unclear about degree of dimorphism in body and brain size in *H. habilis*. At least some individuals were within the range of normal human statures. Since brain size was clearly larger in *habilis* than in any earlier hominid, there would almost certainly have been shifts in life history patterns: slower maturation, greater longevity, bigger newborns with larger brains. Also, perhaps there were greater metabolic demands, especially on nursing mothers and growing infants, because brain tissue is energetically costly.

What about diet and subsistence behaviour? We have anatomical evidence from face and teeth for the physical quality of food being ingested. The teeth are little different from those of *afarensis* and the face is also similarly large. Wear on the molars and premolars resembles that of the australopithecines. Hence, the predominant food was plant food and animal food that differed little in consistency from plant food. As noted earlier, increased brain size perhaps implies a change in either food quality or in predictability or both, because of the high energetic demands of brain tissue. The archaeological record also supports the notion of a dietary shift.

The evolutionary transition to *H. habilis* from something like an *A. afarensis* or *africanus* occurred before 2 Ma, but how long before is unclear. 'Why' questions are impossible to address in our current state of knowledge, but we might speculate on ways in which minor shifts in ancestral patterns might lead to major changes in behaviour. For example, perhaps *habilis* began to soak, bash, and cut parts of animals as they had soaked, bashed, and altered plant parts; naturally sharp flakes preceded artificially flaked edges. New resources became available perhaps following environmental change. Small shifts are always more likely than large ones, but occasionally a small shift will have a major effect.

Pleistocene hominid evolution

The boundary between Pliocene and Pleistocene epochs is dated at 1.6 Ma (Fig. 4.2). The Pleistocene can be conveniently subdivided into three segments: early, middle, and late. The early to middle Pleistocene boundary falls at the last major reversal of the earth's magnetic field from reversed to normal about 700 thousand years ago. The middle to late Pleistocene break is drawn at 125 thousand years, the time of the rapid warming at the end of the penultimate major glaciation and beginning of the last, very warm, interglacial. This transition has left a number of geological 'fingerprints' which make it recognizable in a variety of situations. From about 2.5 Ma on the earth's climate fluctuated between colder glacials and warmer interglacials on a frequency of about 100 thousand years. Around 800 thousand years ago, the amplitude of these cycles increased reflecting a greater severity of glacial climates. Interglacials remained brief, only 10–20 thousand years long, while the intervening cold periods became longer and more intense. *Homo* species were part of open country communities, and open habitats were considerably more extensive for most of the past. As we have noted in Chapter 4, for much of the twentieth century textbooks have been dominated by the incorrect idea that there were four major glacials and three interglacials during the Pleistocene. No attempt will be made here to use these. Although dating techniques are far from precise, it is possible to sort most of the important hominid material into at least broad time blocks.

Homo erectus

A new kind of hominid appeared in Africa in the earliest Pleistocene close
to 1.6 Ma: *Homo erectus* (Fig. 5.4). This species subsequently spread out of
Africa into Asia, and survived relatively unchanged for over 1 million
years, until replaced by *H. sapiens*. The earliest remains are best known
from sites in East Africa, but are also known from southern Africa. A
spectacularly complete discovery in 1984 of a *H. erectus* from the west of
Lake Turkana in Kenya has considerably expanded our knowledge. The
brain volume of early *erectus* ranges from a little over 800 to over 1000 cm^3,
while later specimens have somewhat larger brains. Brain volume is thus a
little larger than in *H. habilis*. The skull is also longer, with flatter frontal
and more angulated occipital regions. Faces are smaller than in *habilis*, the
cheek teeth are reduced, tooth wear is more marked. The degree of
(presumed) dimorphism in face size is interesting, especially in brow ridge

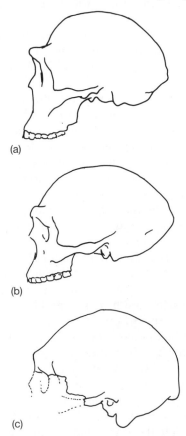

(a)

(b)

(c)

FIG. 5.4. *Homo erectus* crania, ages approximate. (a) KNM ER 3733, Kenya, 1.6 Ma;
(b) Zhoukoudien, China composite, 0.4 Ma; (c) Ngandong, Indonesia, 0.12 Ma.
(J. Sept.)

thickness: as in australopithecines and apes it is more marked than in modern humans (Table 5.2).

The new West Turkana specimen samples most of the skeleton, and although it is a juvenile male, gives us a very clear picture of what at least one individual in one population looked like. The discovery of a relatively complete specimen also allows us to make more sense of the isolated and fragmentary *erectus* limb bones previously known. In Africa *H. erectus* individuals were tall, within the range of larger modern populations (male and female averages in the 5'8" to 6' range), while size dimorphism was probably less marked than in australopithecines. Skeletons were in many ways similar to those of modern humans, although differing in that cortical bone was thicker, and overall skeletal robusticity and degree of muscularity were more marked. Certain features of hand and vertebral anatomy differ from *Homo sapiens*, their significance being unclear at present. Of considerable interest is the narrowness of the pelvis in the West Turkana specimen, even allowing for the fact that the individual was a juvenile male. This implies obstetrical constraints in females, and the evolution of relatively immature newborns. The neck of the femur was long, resembling the australopithecines, while the femoral head was large as in *Homo*.

As noted previously, the few post-cranial bones likely to be from *habilis* resemble those of *erectus* very closely; indeed, all archaic *Homo* are basically similar to modern humans rather than the australopithecines. A positional repertoire which was essentially terrestrial is clearly indicated, involving tool making, carrying, and long-distance walking and running. The origin of *H. erectus* is obscure. The general assumption has been that *H. habilis* was the ancestor, although a link with *Homo* sp. is at least as likely. In that case, a large brain would have evolved more than once in hominid evolution. The pace of the evolutionary transition, its nature, and likely cause are all unknown.

The oldest *H. erectus* from Kenya are probably close to the time of first appearance of the species. Absence of plausible evidence elsewhere in the Old World points to an African origin for the species, where it was confined until about 1 Ma. South, north, and east Africa were inhabited. Around 1 Ma hominids spread out of Africa. An excellent sample of *H. erectus* comes from Zhoukoudian near Beijing in China, mostly crania, mandibles, and teeth, but including a few post-cranial bones. This material can be dated between about 250 and 500 thousand years. Younger and older Chinese specimens are known from Hexian and Lantian. A second Asian collection has been recovered from the Indonesian island of Java, mostly from Sangiran. The bulk of the material is probably between 700 thousand and 1 million years old, and a yet younger set of specimens from Ngandong in Java is probably less than 200 thousand years old. The Chinese and Indonesian material is

TABLE 5.2
Homo erectus

Selected anatomical characters

Brain over 700 and under 1300 cm^3
Long and low skull, sharply angulated occipital
Thick skull bones, large browridges, low frontal
Robust skeleton

Selected specimens

Site and specimen	Remains	Geological age (Ma)	Sex	Endocranial volume (cm^3)
Asia				
Zhoukoudien Layer 3 Loc. 1 (China)	Braincase	0.23	M	1300
Zhoukoudien Layers 8–10 Loc. 1 (China)	Braincase, skeletal parts	0.40–0.46	M	915, 1030, 1225
Hexian (China)	Braincase	0.26	M	1025
Gongwangling (China)	Cranium	0.80	F	780
Yingkou (China)	Skeleton	0.20		
Narmada (India)	Braincase	0.22–0.12		1155–1421
Trinil 2 (Indonesia)	Skullcap	0.90	F	940
Sangiran 2 (Indonesia)	Skullcap	0.80	F	813
Sangiran 4 (Indonesia)	Braincase	1.0	M	908
Sangiran 10 (Indonesia)	Braincase	0.80	?F	855
Sangiran 12 (Indonesia)	Braincase	0.80	?M	1059
Sangiran 17 (Indonesia)	Cranium	0.80	M	1004
Solo I (Indonesia)	Braincase	0.12	F	1172
Solo V (Indonesia)	Braincase	0.12	M	1251
Solo VI (Indonesia)	Braincase	0.12	F	1031
Solo X (Indonesia)	Braincase	0.12	?F	1231
Solo XI (Indonesia)	Braincase	0.12	?M	1090
Africa				
KNM ER3733 (Kenya)	Cranium	1.6	?F	848
KNM ER3883 (Kenya)	Cranium	1.6		804
KNM WT15000 (Kenya)	Skeleton	1.6	M	?920
OH9 (Tanzania)	Braincase	1.2	M	1067
OH12 (Tanzania)	Braincase	0.8		727
Ternifine (Algeria)	Jaws, parietal	0.4–0.7		

morphologically very similar, and both samples resemble African *H. erectus*. There is no good evidence for the presence of *H. erectus* in Europe at any time.

The species is thus documented from South Africa to the eastern and south-eastern margins of Asia, ranging in age from over 1.5 to perhaps less than 0.15 million years old. The extent to which evolutionary change occurs during this time is unclear: we have only a poor understanding of intersexual, intrapopulation, and interpopulation variability. However, there is little evidence for any major change except for a modest increase in average brain volume. The extent of this is difficult to judge because of our poor understanding of the various sources of variability, but it is at most a 20 per cent increase in more than 1 million years (Table 5.2). Other features show a pattern of surprising stability given the amount of time and the geographical range involved.

The time of appearance of *H. erectus* coincides with minor changes in the archaeological record. Some new tools appear, in particular the symmetrical and generally larger forms known as bifaces. They vary in quality and consistency of manufacture, and their degree of apparent sophistication seems to depend to a considerable extent on the nature of raw material and on the intentions of the maker. Industries containing bifaces are called 'Acheulean'. Other than bifaces, the range of smaller tools differ little between the African early Acheulean and the preceding Oldowan. Indeed, many African Acheulean archaeological sites consist of assemblages with few bifaces, and the earliest assemblages in Europe and eastern Asia similarly have few or no bifaces. It is unclear why, but presumably it reflects as yet unknown functional and activity differences.

Acheulean and similar industries lasted from about 1.5 Ma to less than 500 thousand years ago, and showed some change over this immense time period, but as with morphology surprisingly little. During the second half of the period a new technique for making flake tools was invented, 'Levallois' flaking. This involved preparing the final shape of the flake before rather than after a flake is removed from the core. With the development of Levallois technique the number of smaller flake tools increased. Controlled use of fire was also discovered sometime during this period making possible the occasional occupation of caves; it was perhaps the essential factor allowing migration out of Africa. However, there is no evidence for elaborate cooking using fire. It is likely that non-stone tools were also frequently used: wooden spears, nets, bags, carrying slings. Animal remains associated with archaeological sites include sometimes large animals, like hippos or elephants, and occasionally groups of large or dangerous animals (giant baboons or extinct giant bovids).

There has been as much debate over the interpretation of Acheulean assemblages as there has been for earlier and later periods. Do these traces reflect the activities of human-like hunters and gatherers; or of creatures that were in many details of behaviour still unlike us? Again we use both morphological and archaeological evidence in making inferences about past behaviour. Both plants and animals were used for food, and judging from smaller teeth and faces, *erectus* prepared food more than *habilis* did. This food was obtained in generally open habitats, which would often have been quite seasonal. The post-cranial skeleton, basically human-like but robust, implies long-distance walking and running, and perhaps some hunting involved persistent following until prey were exhausted. Males and females seem not to have been very different in size, perhaps because selection favoured larger females than in earlier hominids. Larger females would be at an advantage in carrying and giving birth to infants with larger brains and bodies than preceding species.

There is no unequivocal evidence for planned or co-ordinated hunting of large or herd animals, for killing at a distance, for trapping and ambushing, for transporting of large amounts of meat for sharing, for elaborate food preparation, or for food storage. Technology was simple and raw materials were transported or traded over distances of only a few to tens of kilometres. Caves were rarely used for living although there is some evidence for 'structures' such as huts. Fire could be controlled, but probably not with much sophistication. A mixed subsistence strategy is indicated, perhaps with little interdependence of individuals, probably with little difference between the sexes in how food was obtained, that is, little sharing of large amounts of meat along kin networks. However, females would have high energetic costs during pregnancy and were perhaps provisioned by males. We have as yet too little information about pelvic anatomy, but what we do have suggests that newborn body size would have been constrained by pelvic diameter and babies would have been born, like humans, relatively immature and helpless.

Although females approach males in body size, there are, as we noted earlier, interesting and marked differences between the sexes in face form. Males had bigger faces with more massive brow ridges and this pattern of sexual dimorphism is less like modern humans and more like apes and early hominids. This implies at the least that intra-male competition was more 'biologically' mediated than in humans and perhaps that relationships between the sexes were more like those in non-human species than in humans. The extremely slow pace of technological change over this long time period also strongly implies a distinctly non-human kind of behaviour, one buffered by biological constraints. Learned behaviour patterns were clearly less flexible than in humans, which implies that capacities for symbolic behaviours, including language, were less developed.

We can infer though that *H. erectus* behaviour was mediated more by symbols and through language than in earlier hominids. *Homo erectus* was a longlasting and successfully adapted species. The adaptation involved technological skills significantly advanced over those of apes: the controlled use of fire, omnivory with meat-eating being a significant component obtained by hunting and scavenging, a brain capable of generating some kind of speech and of using symbols to organize internal and external worlds. However, this was not human behaviour. Prolonged anatomical and archaeological stability points to a successful adaptation, but one which in its durability is unrecognizable to us humans. It was one where behavioural flexibility was constrained, where the mix of 'biology' and 'culture' in behaviour was different, the 'biological' component being greater. The challenge is to reconstruct what such a system was like. The problem, at least with *H. erectus*, is the paucity of fossil and archaeological material. The situation is considerably better in the hominids following *H. erectus*, but still preceding modern humans, so-called archaic *H. sapiens*.

Archaic Homo sapiens

The trajectory of anatomical and behavioural change of the genus *Homo* after *H. habilis* has generally been seen as steadily accelerating, with morphology growing gradually more human and behaviour progressively more modern (Fig. 5.5). These later hominids have usually been subdivided between *H. erectus* and *H. sapiens*, the two contrasting mainly in brain size and in archaeological differences implying behavioural advance.

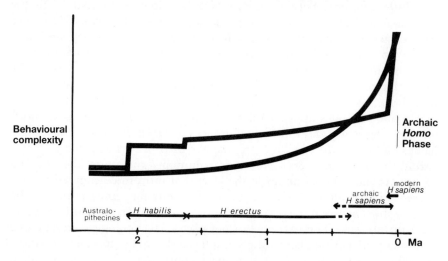

FIG. 5.5. Gradualistic versus punctuated alternatives for hominid behavioural evolution. (J. Sept.)

Supposedly more minor differences separated archaic *H. sapiens* from modern *H. sapiens*. However, we have already noted that anatomical and archaeological patterns changed but little during the time of *H. erectus*. We shall follow a scheme here with three distinct patterns: *H. erectus,* archaic *sapiens*, modern *sapiens*; and argue that *erectus* and archaic *sapiens* differ less than archaic and modern humans. Archaic *Homo sapiens* differs from *H. erectus* in having slightly greater brain size, and in some relatively minor skull features: in particular, the occipital region is more rounded — probably a function of increased brain volume — the cranial base is a little more modern, and skull thickness on average less (Fig. 5.6), but overall skull shape is very similar in the two groups, and post-cranially they are also alike.

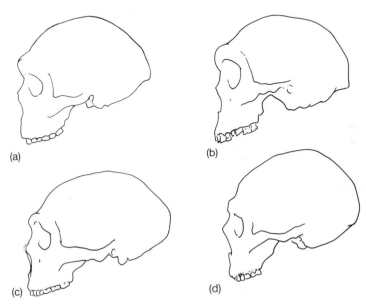

(a)

(b)

(c)

(d)

Fig. 5.6.   Archaic *Homo sapiens* crania, ages approximate. (a) Petralona, Greece, 0.2 Ma; (b) Kabwe, Zambia, 0.2–0.1 Ma; (c) Monte Circeo, Italy, 0.05 Ma; (d) Shanidar, Iraq, 0.05 Ma.

Archaic *H. sapiens* sampling is uneven. Later European and west Asian populations, the neanderthals, are well known while earlier European groups, and especially African and Asian archaic *sapiens* are much more poorly understood. It is unclear how much *erectus* and archaic *sapiens* overlapped in time, but it is likely that the oldest archaic *sapiens* appeared tens or hundreds of thousands of years before the last *erectus*. However, the two differ so minimally that this raises less of a problem than might appear; archaic *sapiens* might originally have been a geographical variant of *erectus*. Use of different names would then be a matter of taste.

The Asian record of archaic *sapiens* is sparse (Table 5.3) and marred by the fact that dating is so poor. A new cranium from Narmada in central India, and similar specimens from Dali and Maba in China can be approximately dated in the 50–250-thousand-year range. They are *erectus-*

TABLE 5.3
Archaic Homo sapiens

Selected anatomical characters

Brain usually over 1000 cm^3
Long and low skull,
Rounded or moderately angulated occipital
Thin skull bones, large browridges
Large face (esp. neanderthals)
Robust skeleton
Long pubic bone (neanderthals only?), short distal limbs (neanderthals only?)

Selected specimens

Site	Remains	Geological age (thousand years)	Endocranial volume (cm^3)
Africa			
Sale (Morocco)	Partial cranium	? 400	860
Bodo (Ethiopia)	Partial cranium	400	
Ndutu (Tanzania)	Cranium	200–400	1050
Ngaloba (Tanzania)	Cranium	120	1367
Kabwe (Zambia)	Cranium, jaw, and skeletal fragments	100–200	1280
Hopefield (S. Africa)	Braincase	300–600	1225
Asia			
Dali (China)	Cranium	130–400	1120
Maba (China)	Partial cranium	75–130	
Shanidar (Iraq)	Several crania and skeletons	45–60	1600
Tabun (Israel)	Cranium, skeleton, jaw	50–80	1271
Amud (Israel)	Cranium, skeletal fragments	35	1740
Europe			
Mauer (W. Germany)	Jaw	400–700	
Vertesszollos (Hungary)	Occipital	400–700 or 200	
Swanscombe (England)	Partial cranium	300	1325
Steinheim (W. Germany)	Cranium	300	1150
Arago (France)	Partial cranium, skeletal fragments	200	1150

TABLE 5.3 (continued)

Selected specimens Site	Remains	Geological age (thousand years)	Endocranial volume (cm^3)
Europe			
Petralona (Greece)	Cranium	200	1200
Ehringsdorf (E. Germany)	Several ind. incl. cranium	225	1450
Neandertal (W. Germany)	Skullcap	45–75	1370
Spy (Belgium)	Crania, skeleton	45–75	1305, 1553
La Quina (France)	Crania	45–75	1350
La Chapelle (France)	Cranium, skeleton	45–75	1626
La Ferrassie (France)	Crania, skeleton	45–75	1689
Gibraltar	Cranium	45–75	1300
Saint-Cesaire (France)	Cranium	30–35	

like in some features — bone thickness, for example — but vault shape is rounder. The Dali face is low and does not project much. In Africa the situation is a little better, although there are also problems with dating. Hominid specimens range from Ethiopia (Bodo) through eastern Africa to South Africa (Saldanha), dating from the later half of the Middle Pleistocene, and document the presence of archaic *H. sapiens* basically similar to those of Asia. These resemble preceding *erectus*, but have larger brains and are as or more robust. Skulls are long and low, faces large, but not markedly projecting, and facial dimorphism is probably substantial. The basicranium is close to modern, being flexed rather than flat. Post-cranial bones are robust, with thick cortex, but they are only poorly known.

The hominid fossil record is substantially more abundant (Table 5.3) in Europe and west Asia than in Africa and the rest of Asia. It is also better dated. The oldest definite hominids appear between 400 and 700 thousand years ago consisting of a mandible (Mauer in Germany) and a skull fragment (Vertesszollos in Hungary). These are clearly not *H. erectus*. More material is known from the second half of the Middle Pleistocene, although all of it is scrappy: jaws, teeth, parts of skulls, unassociated limb bones, and too few more complete crania. What is known suggests that, as in the earlier period, archaic *sapiens* in the circum-Mediterranean region resemble those from Africa. For example, a cranium from Petralona in Greece which is 200 thousand years old or more, closely resembles archaic *sapiens* material from Africa like the cranium from Kabwe (Broken Hill) in

Zambia (Fig. 5.6), but by 150–200 thousand years ago, European populations begin to diverge from archaic *sapiens* populations elsewhere. In the late Pleistocene, European archaic hominids are well sampled as the famous and archetypal cavemen, the neanderthals. Neanderthals are an anatomically homogeneous group of later Pleistocene hominids (200–35 thousand years) from Europe and west Asia that are the best characterized of all non-modern hominids. Entire skeletons are known, and we have excellent knowledge of stature, proportions, brain size, and reasonable estimates of body weight (Table 5.4). Neanderthals differ in certain features from their contemporaries in Africa and Asia. These differences might be due to the poor sampling of non-neanderthals, but are more likely to reflect real and consistent contrasts: faces and basicrania differ, skeletons are hyper-robust, and pelves appear to be different.

TABLE 5.4

Weight, stature, and endocranial volume

	Weight (kg)		Stature (cm)		Endocranial volume (cm^3)*	
	Male	*Female*	*Male*	*Female*	*Male*	*Female*
Early modern Homo sapiens	65.6	53.9	183.6	166.9	1577 (11)	
Neanderthals (Europe, SW Asia)	65.4	54.9	169.0	159.8		
Neanderthals						
(Europe)					1601 (4)	1327 (2)
(SW Asia)					1670 (2)	1271 (1)
Early modern Homo sapiens (SW Asia)					1545 (5)	

* Figures in parentheses indicate sample size.

Neanderthals were relatively short, very powerfully built, and relatively heavy people. They had large brains, but no larger relative to body size than in comparably heavy, though taller, modern humans. Crania were long and low, but with a 'well-filled' look; faces projected substantially. Front teeth were large and often showed heavy wear from use in a range of non-masticatory activities, and this may have influenced degree of facial projection. Basicrania looked more primitive and flatter than those of other archaic *sapiens*; it has been suggested that this implies a more primitive laryngeal position and pharyngeal anatomy, and hence a more restricted sound-producing repertoire. This is distinctly possible; it is also possible that there is a range of bony anatomy for a particular pharyngeal form, the flatter neanderthal cranial base being but the configuration associated with a long cranium and very projecting face. Post-cranial bones had a thick cortex, and joint surfaces were relatively massive. Muscle markings were substantial, even in the very young, and the overall impression is of massive, muscular, heavy, and extremely powerful individuals. This implies high energetic requirements and presumably less effective cultural adaptations than modern humans. Hand bones showed minor, but interesting differences from modern humans, muscle markings and finger proportions suggesting much more powerful grasping abilities along with fine manipulative skills.

Neanderthals had short distal limb segments: tibia short relative to femur, radius and ulna short relative to humerus. This could be an adaptation to cold, reducing surface area and heat loss. The succeeding modern populations have more 'tropical' proportions with longer distal segments, implying that modern *sapiens* has a greater degree of cultural control of environmental factors than did neanderthals.

Neanderthal males and females have pelves with long pubic bones, significantly longer than in modern human populations. Unfortunately, we have no relevant material for comparison with other archaic *sapiens* populations, so it is not clear how typical the feature is of archaic *H. sapiens* as a whole. It implies that the birth canal would have been larger in female neanderthals than in modern humans. Two alternative explanations have been offered for this. Either neanderthals, as short but heavy individuals, had babies which were very large relative to maternal body size. Or neanderthals gave birth to absolutely larger infants than moderns after a longer pregnancy time — 11–12 months; such a gestation time is actually predicted for modern humans on the basis of overall mammalian patterns associating maternal, and newborn body and brain size. If this second alternative is correct, the birth of more mature and presumably less dependent offspring might be linked to other aspects of a non-modern social system.

The archaeological record for the period between 200–300 and 35–40 thousand years ago, which corresponds approximately to the span of archaic *sapiens* populations, is best in Europe and west Asia, but is improving elsewhere in the Old World. Although there are regional variants, the Middle Palaeolithic industries of Europe and west Asia, and the Middle Stone Age industries of Africa, and south and east Asia are made up of a very similar range of tool types. There is little change over this substantial period of time. Many tools are made on flakes, and the size and number of bifaces decreases markedly relative to Acheulean biface industries. Many of these tools would have been hafted and used as knives, cutting tools, or stabbing spears, and many would have been used without hafting.

Fire continued to be important and there is evidence from temperate regions that it was definitely used for cooking, in addition to opening up caves as living areas by keeping out carnivores. The patterning of stone tools within sites and across the landscape suggests that home bases still may not have developed in fully human ways as foci for a wide range of subsistence and social activities. The animal remains in archaeological sites represent mainly parts that were either scavenged after carnivore hunting, or young, old, sick, or otherwise vulnerable animals taken by hunting. This, together with the massiveness of archaic *sapiens*, suggests that they were less efficient than modern humans in planning, organizing, and carrying out hunting or other activities which involved long-distance trekking and endurance running. Modern humans can achieve more with less expenditure in building and maintaining massive bodies, implying better cultural than biological adaptations.

By at least 80 thousand years ago in both Europe and in Africa there is evidence for a shift to increased hunting, although it was still not patterned as in modern humans with elaborate planning and co-operation. Neanderthals buried their dead with some ceremony and took care of injured individuals. They were by no means brutes and they were probably behaviourally advanced relative to *H. erectus*. However, they were not fully human, and the challenge to us lies in trying to imagine a non-human, pre-human behavioural system, based on technology and communicative skills substantially more developed than in earlier hominids or in apes, but which was clearly not like that of modern humans. Relationships between the sexes and within sexes (especially males) probably also differed from modern *sapiens*. Overall, the implication is of a social system in which less information was stored and exchanged, based on a less flexible and variable behavioural programme. Language was clearly involved, but symbolling capabilities were less developed — perhaps considerably less — than in modern *H. sapiens*.

Modern humans

The transition to modern humans had occurred throughout the Old World by a little under 35 thousand years ago, and perhaps even before then humans spread into previously uninhabited areas: the coldest parts of Asia, Australia, and, soon thereafter, the Americas (Table 5.5). This required new cultural abilities to survive in very cold climates as well as to travel substantial distances across water. The first modern humans definitely appear by 40 thousand years ago, probably much before. Early modern *H. sapiens* are difficult to date because radiometric dating

TABLE 5.5
Early modern Homo sapiens

Selected anatomical characters

Brain average 1400 cm^3
Short and high skull, well curved occipital, long curved parietals
Thin skull bones, weak browridge, strong chin
Gracile skeleton

Selected specimens

Site	Remains	Geological age (thousand years)	Endocranial volume (cm^3)
Africa			
Omo-Kibish (Ethiopia)	Partial skeleton	?130	1430
Singa (Sudan)	Braincase	?80	1500
Klasies River (S. Africa)	Skull and jaw frags.	80–120	
Border Cave (S. Africa)	Cranium, infant skel. jaw frag.	90–110	1507
Jebel Irhoud (Morocco)	Crania	50	1305, 1450
Asia			
Skhul (Israel)	Crania, skeletons	?40–50	1554, 1450
Qafzeh (Israel)	Crania, skeletons	?50–70	1568, 1523
Niah (Borneo)	Cranium	?40	Within modern human range
Europe			
Bacho Kiro (Bulgaria)	Fragments of bones	?43 +	Within modern human range
Hahnofersand (W. Germany)	Frontal	36	Within modern human range
Velika Pecina (Yugoslavia)	Frontal	34 +	Within modern human range
Predmosti (Czechoslovakia)	Many specimens	26	Within modern human range
Cro-Magnon (France)	Several specimens	20–30	Within modern human range

techniques, particularly carbon[14], are unreliable and because most of the supposedly old ancient specimens have one or more problems related to their geological or archaeological contexts. However, these caveats notwithstanding, it is quite likely that modern *sapiens* are much older than 40 thousand years in Africa at least, perhaps over 100 thousand years old, in contrast to the very restricted area of western Europe where no modern humans are known prior to about 35 thousand years. Not enough is known about Asia. It is unclear how localized was the area of origin of modern *H. sapiens* or where it occurred (other than in Europe), how time-transgressive it was, nor is the precise nature of the archaic–modern population interaction understood (which might have varied from place to place). Extremes of possibilities can be ruled out: that the transition to modern populations occurred everywhere, or that it occurred in a very restricted area, and that almost all other archaic populations were abruptly replaced.

A variety of genetic studies of living human populations show *H. sapiens* to be a very homogeneous species relative to other hominoids and non-primate mammals. Analyses of mitochondrial DNA, involving certain simplifying assumptions, suggest that all modern humans shared a common ancestor between 50 and 200 thousand years ago, perhaps before the appearance of anatomically modern *sapiens*. This implies that living humans descended mostly from one rather restricted geographical group, but perhaps with some admixture locally with archaic populations. One analysis involving mitochondrial DNA, based on the presence of assumed primitive traits in Asian populations, concluded that the earliest modern *sapiens* groups evolved in Asia. Another based on nuclear DNA points to Africa. The fossil record at the moment documents the oldest likely modern specimens as coming from Africa, but the record from Asia is too sparse and poorly dated for much to be made of that.

Anatomically, modern humans (Fig. 5.7) contrast with earlier archaic *Homo sapiens* in essentially all parts of the body. Skeleton and muscle were much less bulky, and body proportions fully modern. The skull was rounded, face flatter, front teeth smaller, and cranial base fully flexed. Although brain volume did not change with the appearance of modern humans, external brain shape and proportions did. It is not clear what, if any, internal changes this reflects. Tooth wear rates and features of bone organization suggest that early modern *sapiens* survived longer into post-reproductive life than archaic hominids did.

In only a few areas of the world do we have a reasonable anatomical record of the period during which the archaic–modern transition occurred. It is best documented in the north-western quadrant of the Old World. The rate of change of anatomy varies depending on the character. Some characters, brain volume for example, don't change. Others continue a gradual trend, cheek tooth size for example. Still others accelerate during the transition, like reduction in facial size and robusticity. A final group, such as skeletal robusticity and certain proportions — pubis length, hand and finger anatomy, and proportions — change abruptly in only hundreds or thousands of years. All this follows a long period of relative stability.

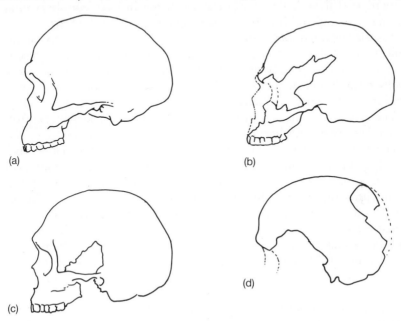

(a)

(b)

(c)

(d)

FIG. 5.7. Modern *Homo sapiens* crania, ages approximate. (a) Jebel Irhoud, Morocco, 0.06 Ma; (b) Skhul, Israel, 0.05 Ma; (c) Jebel Qafzeh, Israel, 0.06 Ma; (d) Border Cave, South Africa, 0.09 Ma.

If the archaeological records of 50 and 25 thousand years ago are compared they show very considerable differences, reflecting the emergence of fully modern behavioural and social systems. The period in between records the evolution, generally rapid (over thousands of years or even less), of the industries generally associated with modern humans (Upper Palaeolithic of Europe and west Asia, Late Stone Age of Africa and the rest of Asia), although the earliest of these industries 35–40 thousand years ago do not show in full form the attributes of later industries. As noted earlier, the precise pattern of the archaic to modern transition is unclear, but we can talk about the transition in terms of 'before' and 'after' states. The technology of stone tool making became more efficient and new methods of making blades produced significantly more cutting edge per unit of raw material. New tools and devices were invented: the bow and arrow, and spear thrower, complex bone tools, nets and snares, methods of food storage, improved hearths, clothing, more efficient and larger dwellings. Archaeological sites for the first time show the internal differentiation characteristic of modern humans into areas of living, tool and food preparation, sleeping, and so forth. The patterning of sites across the landscape implies a use of space like that of living foraging peoples. Faunal remains in the later archaeological sites come from animals that have been clearly hunted, often in large numbers implying planned or 'logistic' hunting by several co-operating hunters.

Burials became more elaborate. Grave goods imply not only changes in belief systems, but hint at variation in social status, a possibility also raised by sometimes extremely elaborate personal decoration. Some campsites imply quite large, dense, and sedentary populations of hunter–gatherers, similar to those of modern foraging peoples whose food supply is sufficiently abundant and predictable, and who have solved the problems of food storage. Earlier hominids had a modest aesthetic sense, but modern humans elaborated this to a tremendous extent. Rock art and cave paintings bloomed after 30 thousand years. We may never fully decipher their meaning, but at the least we can say that they encode information about hunting, critical life transitions such as puberty, and religion and belief systems. In short, the essential symbolically encoded information which must be transmitted between generations in a cultural-dependent species to ensure appropriate and adaptive behaviours. The degree of geographical differentiation of archaeological patterns of 'industries' and the rate at which they are replaced are much greater than for earlier periods.

The picture that develops is a very recognizable one. It reflects a behavioural system which is fully modern. Although the best, albeit still indirect evidence of behaviour comes from the archaeological record, morphology can also add something to the story. The reduction in skeletal and muscular robusticity points to an energetically less costly system. Longer legs, longer distal limb segments, and narrower pelves imply a more efficient pattern of bipedal walking and running. Possibly, the change in maternal pelvic size is associated with the birth of less mature infants after a shorter pregnancy. If so, this implies greater culturally based behavioural capabilities for the social group in maintaining such immature offspring. Front teeth were used less as a vice, and were smaller and wore less rapidly than in archaic *sapiens*. Cheek teeth were also smaller, and faces flatter and tucked in more under a shorter, higher braincase. This may have been related to more efficient food preparation. Skulls were more rounded, which goes with facial flattening and cranial base flexion as a developmental package. Which one or ones were directly influenced by selection is unclear. Cranial rounding could be produced by shifts in infant growth patterns in brain and skull, the brain growing more rapidly just after birth than in archaic hominids. Or rounding could be related to flexion of the cranial base consequent to the development of a fully modern vocal tract and phonetic capabilities.

All these changes can be explained as morphological expressions of behavioural and social systems that were flexible, efficient, and based on developed symbolling capacities. Most palaeoanthropologists would point to language as the critical ingredient. Fully modern language abilities may not have evolved until the appearance of modern *sapiens*: earlier hominid language was a less efficient method of information exchange and social control.

Why this change happened, and precisely when, where, and how cannot be answered at present. However, the shift reflected a change in the way behaviour could be organized, opening up the possibility of almost endlessly different patterns and making possible much more rapid behavioural shifts than would be the case for more biologically rooted patterns. After millions of years, quite suddenly and surprisingly recently, the modern system evolved. It is a distinct possibility that family organization and social patterns based on marriage and kinship, sharing, division of labour, hierarchic organization, social-cultural controls of aggression, and many other typically human characteristics evolved in their current form very late in human prehistory.

A major break in the human story is often drawn at the shift from food gathering (hunting) to food production (farming) which started 10 thousand years ago. This is certainly an important transition because it made possible, through control of plant and animal resources, the development of settled communities with much higher population densities, eventually to be linked by bureaucracies to form states.

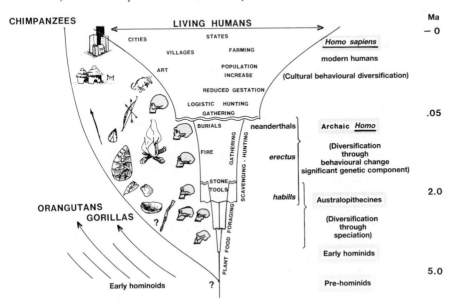

FIG. 5.8. Diagrammatic summary of major features of human evolution. (Courtesy of Jeanne Sept.)

'Civilization' is confined only to the very last few thousand years of human history, but there is behaviourally more to be said for dividing the hominid sequence at the evolution of modern humans 40 thousand years ago or more, with the appearance of the first hominid whose symbolling capacities made profound behavioural change possible. The shift from hunting to food production happened at slightly different times in several different parts of the world. Populations in those areas expanded, actually displacing hunting peoples most of whom also eventually changed their subsistence organization. Thus, a few geographical populations out of those originally many populations into which the human species was loosely divided, grew and ultimately came to include the bulk of humans. Fortuitous subsistence shifts thus led to a world population dominated by peoples originating in just a few places: west Africa, south-west Asia, and north-east Asia.

It is easy to be overimpressed by the superficial morphological differences between human populations, particularly the fortuitously expanded few, but the genetic differences are insubstantial. We already noted that *Homo sapiens* is a very homogeneous species. A recent study looked at 18 genetic loci in many individuals grouped in 180 populations sampled from six broad geographical areas (Europe, Africa, India, East Asia, Americas, Oceania). Considering the genetic difference between any pair of individuals, it was surprising to discover that 84 per cent of the total genetic variation was found between individuals within the same local group (tribes or nations), only 6 per cent between populations within broad geographical areas, and the remaining 10 per cent between those areas. We are truly one species.

Suggestions for further reading (Part I)

BEHRENSMEYER, A. K. (1982). The geological context of human evolution. *Ann. Rev. Earth Plan. Sci.* **10**, 39–60.

BENDALL, D. S., ed. (1983). *Evolution from molecules to men.* Cambridge University Press, Cambridge, UK.

BINFORD, L. R. (1983). *In pursuit of the past.* Thames and Hudson, London.

CARTMILL, M., PILBEAM, D., AND ISAAC, G. (1986). One hundred years of paleoanthropology. *Am. Sci.* **74**, 410–19.

CIOCHON, R. L. AND FLEAGLE, J. G. eds (1985). *Primate evolution and human origins.* Benjamin/Cummings, Menlo Park.

CRONIN, J. E. (1975). Molecular systematics of the Order Primates. PhD dissertation, University of California, Berkeley.

DAY, M. H. (1986). *Guide to fossil man,* 4th edn. University of Chicago Press, Chicago.

DEACON, T. (1984). Connections of the inferior periarcuate area in the brain of *Macaca fascicularis*: an experimental and comparative neuroanatomical investigation of language circuitry and its evolution. PhD dissertation. Harvard University.

DELSON, E., ed. (1985). *Ancestors*. Alan Liss, New York.

FOLEY, R., ed. (1984) *Hominid evolution and community ecology*. Academic, London.

GOODMAN, M. and TASHIAN, R. E., eds (1975) *Molecular anthropology*. Plenum, New York.

GOULD, S. J. (1985). A clock of evolution. *Nat. Hist.* April, 12–7.

—— and ELDREDGE, N. (1977). Punctuated equilibria: the tempo and mode of evolution reconsidered. *Paleobiol.* **3**, 115–51.

—— and VRBA, E. (1982). Exaptation — a missing term in the science of form. *Paleobiol.* **8**, 4–15.

GOWLETT, J. (1984). *Ascent to civilization*. Knopf, New York.

HAMBURG, D. A. and McCOWN, E. R., eds (1979). *The great apes*. Benjamin/Cummings, Menlo Park, California.

HINDE, R. A., eds (1983). *Primate social relationships*. Sinauer, Sunderland, Mass.

JOHANSON, D. and EDEY, M. (1981). *Lucy*. Simon and Schuster, New York.

JOLLY, A. (1985). *The evolution of primate behaviour*, 2nd edn. Macmillan, New York.

KREBS, J. R. and DAVIES, N. B., eds. (1984) *Behavioral ecology*, 2nd edn. Sinauer, Sunderland, Mass.

LEAKEY, R. E. (1981). *The making of mankind*. Michael Joseph, London.

LEE, R. B. (1979). *The !Kung San*. Cambridge University Press, Cambridge, UK.

LEWIN, R. (1984). *Human evolution*. Freeman, New York.

LIEBERMAN, P. (1984). *The biology and evolution of language*. Harvard University Press, Cambridge, Mass.

PFEIFFER, J. E. (1982). *The creative explosion*. Harper and Row, New York.

PILBEAM, D. (1984) The descent of hominoids and hominids. *Sci. Am.* **250**, 84–97.

PRICE, T. D. and BROWN, J. A. (1985). *Prehistoric hunters-gatherers*. Academic, New York.

READER, J. (1981). *Missing links*. Little Brown, Boston.

REYNOLDS, P. C. (1981). *On the evolution of human behaviour*. University of California Press, Berkeley.

RICHARD, A. F. (1985). *Primates in nature*. Freeman, New York.

SHIPMAN, P., WALKER, A., and BICHELL, D. (1985). *The human skeleton*. Harvard University Press, Cambridge, Mass.

SIBLEY, C. G. and AHLQUIST, J. E. (1984). The phylogeny of the hominoid primates, as indicated by DNA-DNA hybridization. *J. Mol. Evol.* **20**, 2–15.

—— and —— (1986). Reconstructing bird phylogeny by comparing DNA's. *Sci. Am.* **254**, 82–92.

SMITH, F. H. and SPENCER, F., eds (1984) *The origins of modern humans*. Alan Liss, New York.

SMUTS, B. B. (1985). *Sex and friendship in baboons*. Aldine, New York.

TRINKAUS, E. (1983). *The Shanidar neanderthals*. Academic, New York.

WAAL, F., DE (1982). *Chimpanzee politics*. Harper and Row, New York.

WOLPOFF, M. H. (1980) *Paleoanthropology*. Knopf, New York.

YOUNG, J. Z., JOPE, E. M., and OAKLEY, K. P. (eds) (1981). The emergence of Man. *Phil. Trans. R. Soc. Lond.* **B292**.

Part II

Human genetics and variation

G. A. HARRISON

6 Molecular and Mendelian genetics

General features of human variability

In this section of the book we examine the ways in which people differ from one another biologically. We shall be attempting to describe the variation as it occurs at every level of human bodily organization: cellular, tissue, and whole body, and answer questions concerning the origins, development, causes, and effects of the differences we observe.

As will be demonstrated in detail later the extent of human variability is enormous, so large that no two individuals who have ever lived or will ever live can ever be exactly the same. The fundamental causes for this variation lie in one or both of the two basic determinants of our being: the genes inherited from our parents, and the infinity of environments which act upon and within individuals from conception to death. In this section of the book we will tend to concentrate upon the heritable biological elements to variability (sometimes but misleadingly referred to as the elements of 'nature'), but not only is the environment (or 'nurture') equally important for life (since just as without genes so without environment there can be no life), it can also be the main if not the only cause for variability. More attention is devoted to the nature of environments in the last section of the book. Let it be sufficient to say here that the environments of no two individuals can ever be strictly identical and all environments are constantly changing so no single individual is likely to experience exactly the same environment twice. Yet notwithstanding the infinity of environments, we also tend to inherit our environments, through for example parental influences, as we do our genes and this can raise some extremely difficult problems in analysing variability.

The molecular basis of genetic variability

The genetic element to variability lies in the deoxyribonucleic acid code as it exists within the chromosomes of the cell nucleus and in certain plasmid elements, like mitochondria within cell cytoplasm. Knowledge of the nature, properties, and organization of DNA, and the way it governs biochemical processes within the cell, especially protein synthesis, is fundamental to understanding all life and a dramatic explosion in this knowledge has occurred in the last 30 years since Watson and Crick first suggested the way the DNA code was organized. All we shall do here is highlight some features of the organization of DNA which are crucial to understanding genetic variability.

The double helix of nuclear DNA is thought to exist as a single length (many *metres* long) in each chromosome, where it is folded and supported by proteins, particularly histones. It is made up of a linear sequence of nucleotides. Each nucleotide consists of the sugar deoxyribose, phosphoric acid, and a nitrogenous base. The nitrogenous base varies, and may be one or other of two purines adenine (A) and guanine (G) or one or other of two pyrimidines, thymine (T) and cytosine (C). To form the two stranded structure of DNA in all higher organisms the bases are bonded, with adenine always bonding with thymine and guanine always bonded with cytosine. The bonded bases form, as it were, the steps of a ladder and the deoxyribose, phosphoric acid molecules in sequence the sides. At one end of these 'sides' — the 5' end — a phosphoric acid molecule is combined with the fifth carbon of the deoxyribose. This is then linked to the second phosphoric acid through its third carbon atom and so on. The far end of the chain is therefore referred to as the 3' end. In simplified form, and without the curvature, the structure of DNA is shown in Fig. 6.1, which also shows how DNA has the capacity to replicate itself as is needed in the process of cell division. After the separation of the strands, two new strands are formed, each complementary to the original ones because one base precisely specifies the base with which it will pair. The genetic information in the DNA is conveyed from the cell nucleus to the cell cytoplasm by messenger ribonucleic acid (mRNA) (Fig. 6.2). RNA also consists of a series of nucleotides, but in it the sugar is ribose and the pyrimidine uracil replaces the thymidine of DNA. In the course of the synthesis of the mRNA the DNA message is said to be transcribed. Transcription is catalysed by polymerase enzymes; only one strand of the DNA is read, and the mRNA possesses the complementary arrangement of bases to this strand, but with the adenine of the DNA transcribed as uracil.

The mRNA becomes attached to the ribosomes in the cytoplasm, which themselves contain rRNA (ribosomal RNA and also transcribed from DNA by another group of polymerase enzymes), and from here directs the course of protein synthesis. The basic unit or 'word' in the message is a triplet of three nucleotides in which the three bases specify one of the 20 amino acids of which proteins are composed. These triplets, of which there are 64 possible combinations, are known as codons, and the relationship between the codons and the amino acids most of them specify is shown in Table 6.1. It will be noted that, whereas some amino acids are specified by only one codon (e.g. methionine by TAC in DNA, and AUG in RNA), others are specified by two or more codons (e.g. phenylalanine by both

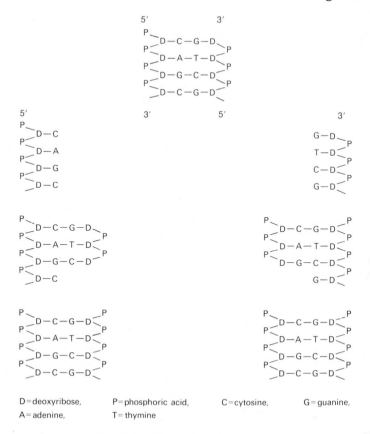

D = deoxyribose,   P = phosphoric acid,   C = cytosine,   G = guanine,
A = adenine,   T = thymine

FIG. 6.1.   The replication of DNA.

AAA and AAG in DNA). Because of the property of having more than one codon for a single amino acid the code is described as being degenerate or redundant. It will be noted that three codons ATT, ATC, and especially ACT in DNA (UAA, UAG, and UGA in RNA) do not code for any amino acid. They would appear to act as stops in the message, separating the synthesis of one polypeptide chain from that of another.

Within the cell cytoplasm there is a third type of RNA known as transfer RNA (tRNA) or soluble RNA and also transcribed from DNA by a third group of polymerases. Each molecule of this tRNA is differentiated for a particular amino acid and carries the anti-codon to that specifying the same amino acid in the mRNA. Thus, for instance, some molecules have the anti-codon AAA and these carry the amino acid phenylalanine to those parts of the mRNA on the ribosomes with the code UUU. Others have UGU and bring the amino acid threonine to the mRNA codon ACA, and so on. In this way amino acids are assembled and linked to form polypeptides in a sequence specified by the mRNA and hence the DNA.

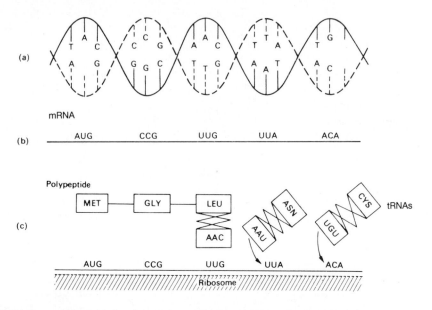

Fɪɢ. 6.2.   The process of protein synthesis. (a) The DNA code. (b) The mRNA produced by that strand of DNA. (c) the formation of a polypeptide on a ribosome.

TABLE 6.1
The DNA code

		Second base							
		A		G		T		C	
First base	A	AAA	Phe	AGA	Ser	ATA	Tyr	ACA	Cys	A
		AAG		AGG		ATG		ACG		G
		AAT	Leu	AGT		ATT	Stop	ACT	Stop	T
		AAC		AGC		ATC		ACC	Trp	C
	G	GAA	Leu	GGA	Pro	GTA	His	GCA	Arg	A
		GAG		GGG		GTG		GCG		G
		GAT		GGT		GTT	Gln	GCT		T
		GAC		GGC		GTC		GCC		C
	T	TAA	Ile	TGA	Thr	TTA	Asn	TCA	Ser	A
		TAG		TGG		TTG		TCG		G
		TAT	Met	TGT		TTT	Lys	TCT	Arg	T
		TAC		TGC		TTC		TCC		C
	C	CAA	Val	CGA	Ala	CTA	Asp	CCA	Gly	A
		CAG		CGG		CTG		CCG		G
		CAT		CGT		CTT		CCT		T
		CAC		CGC		CTC		CCC		C

(Third base column to the right: A, G, T, C)

Abbreviations for amino acids: Phe, phenylalanine; Leu, leucine; Ile, isoleucine; Met, Methionine; Val, valine; Ser, serine; Pro, proline; Thr, threonine; Ala, alanine; Tyr, tyrosine; His, histidine; Gln, glutamine; Asn, asparagine; Lys, lysine; Asp, aspartic acid; Glu, glutamic acid; Cys, cysteine; Trp, tryptophane; Arg, arginine; Gly, glycine.

Apart from having uracil instead of thymine the triplet bases in the tRNA, obviously, correspond to those on the DNA.

The first amino acid to be incorporated into a polypeptide appears always to be methionine. During the further synthesis of the polypeptide this terminal methionine may or may not be cleaved off, so many final proteins do not contain methionine in the first amino-acid position. Nevertheless, the code TAC in the DNA and AUG in the mRNA can be regarded as the 'start' codon to contrast with the 'stop' codons. It should, however, be noted that methionine can also occur in the middle of a polypeptide and there are two types of tRNA, one for terminal methionine and one for internal methionine.

The chromosome complement

The diploid number of chromosomes in man is 46 represented in the female as 23 homologous pairs. In males 22 of these pairs occur, constituting the autosomes and there are two chromosomes which are markedly different from one another, the X-chromosome and the Y-chromosome. These are the sex chromosomes and the X-chromosome is represented twice in females to make up the 23rd pair. The gametes, as a result of meiosis contain representatives of only one member of each homologous pair and therefore half the number of chromosomes of a somatic cell. This is the haploid number and the diploid number is restored at fertilization.

Human chromosomes have been classified in groups on the basis of their variations in size and position of centromeres. In Fig. 6.3 the karyotype of a normal human male is shown in which homologous chromosomes have been paired together and the autosomes arranged in seven classes A–G. Within each class it is often difficult in ordinary cytological preparation to distinguish some of the chromosome pairs from one another and, indeed, to be sure that homologous chromosomes have been accurately paired in drawing up the karyotype, but identification of most of the pairs is usually possible. Group A consists of three chromosome pairs which are relatively large, and the centromere is in a median position making the two arms of the chromosomes of approximately equal length. In Group B the two pairs are also large, but the centromere is placed much nearer one end than the other (referred to as being distal) so that the arms are of unequal length. The seven pairs of Group C are of medium size, and the centromere to varying degrees slightly off-centre (submedian). The X-chromosome is also assigned to this group. Group D consists of two pairs also of medium size, but with the centromere very close to one end. This situation is referred to as being 'acrocentric', though it should be noted that a short arm as well as a long arm exists. Three pairs of chromosomes are assigned to Group E, though in one pair the centromere

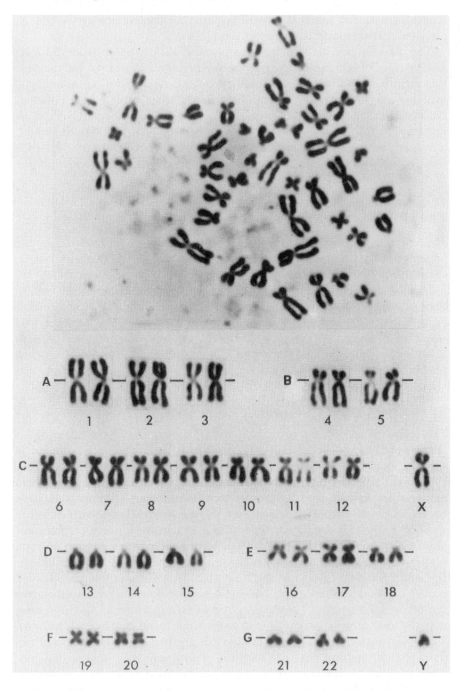

Fig. 6.3. Mitosis in a leucocyte of a human male after colchicine treatment, and karyotype.

is strictly median, while in the other two it is submedian. They are, however, all designated small. Even smaller with shorter arms are the two pairs in Group F and in them the centromere is in a median position. Finally, Group G consists of two autosomal pairs of very small size with the centromere acrocentrically placed though again short arms exist. To this group the Y-chromosome is also assigned. Sometimes all the chromosomes with median, submedian, and distal centromeres are referred to together as being 'metacentric' to distinguish them from the acrocentrics. Short arms are often designated p and long arms q.

The identification of particular chromosomes is facilitated by the fact that some show characteristic constrictions along their lengths or possess minute terminal structures known as satellites. However, the real breakthrough in chromosomal specification came when it was discovered that under certain conditions detailed banding patterns could be detected along the lengths of chromosomes and these bands characterize particular chromosomes. The dye quinacrine combines with AT-rich regions of DNA to produce fluorescent bands and these bands can also be detected by the long-used dye Giemsa after enzymic partial digestion of chromosomal protein. Because of the methods of detection the bands are known as Q or Q/G bands. Another form of banding occurs near the centromeres and is known as C-banding. Bands represent repetitive sequences of DNA.

Molecular and classical genetics

All genetic variation within a species arises because the base sequences in the DNA of homologous chromosomes are not exactly the same. At some site in the helix there will be one code in some chromosomes and another code in others. The difference can be, and often is, no more than in a single base. Such a base difference can lead to different amino acids being incorporated at a particular position in a polypeptide chain and from that a whole host of phenotypic differences at every level of organization may arise. Differences of this kind constitute the molecular basis for the allelic differences considered in classical genetics. When the base sequences in some region in the DNA are the same in homologous chromosomes one refers to the zygote as being homozygous at that locus, i.e. the genes are the same. When the sequences are differing one has the heterozygous condition.

Notwithstanding this very precise relationship between molecular, cyto-, and classical genetics there are a number of important features of the DNA code which could never have been predicted from classical analysis. One of these is the very high proportion of the DNA which consists of the same sequences of bases repeated many times over particularly in higher organisms. Cytogenetically, this comprises the heterochromatic regions of chromosomes. Perhaps even more surprising is the recent finding that within the DNA/base sequences for a particular polypeptide there are sections which are never represented in the polypeptide. These 'interons' are spliced out in the formation of the messenger RNA.

Although every nucleated cell of the adult organism typically possess the full chromosome complement and genetic code inherited by the zygote, the cells become differentiated in their form and activity to perform the functions of the various tissues. In one type of tissue some genes are active and others not, whilst in other tissues different sets of genes dictate what polypeptides will be synthesized. There is some evidence that in most tissues only about 5 per cent of the genes are active, though in the brain this seems to rise to about 12 per cent.

Following the pioneering work of Jacob and Monod we are just beginning to elucidate the mechanisms of gene control and the molecular basis of development. Which genes are active appears to be primarily a function of variable transcription of DNA rather than in the processing of RNA. With at least 50 000 genes in higher organisms it is unlikely that each is regulated separately. Recently, it has been shown that a number of genes concerned with developmental organization of particular structures in *Drosophilia* have a common segment of DNA. This segment has been termed the homeobox and governs the synthesis of a length of amino-acids which are thought to bind to nuclear DNA, and by so doing to switch genes 'on' or 'off'. Common DNA sequences have now been found in the genes of many organisms including man.

Simple versus complex inheritance

In examining the molecular basis of inheritance we only considered the action of genes to the stage of the production of a particular polypeptide. At this level the adage one gene/one polypeptide, though needing some qualification, especially for immunoglobulin synthesis, generally holds. However, whilst the effect of some allelic substitutions is only detectable in the chemical constitution of polypeptides, in other cases it can have ramifying effects throughout the phenotype.

Although all genetic variability resides within variations in the structure of DNA, the form in which it is expressed phenotypically is highly heterogeneous. One of the most significant distinctions at least from an analytical view is whether the variation is continuous or discontinuous. In the former, there are no clearly defined categories and the differences between people are quantitative as in human stature. When the variation is discontinuous or qualitative, clearly identifiable types are recognizable within a family or a population, and the number of these can be counted when questions of mode of inheritance, etc., arise. The essential distinction between these two situations is the complexity of the underlying genetic determination. If many gene loci are contributing to a particular character variation, the variation is continuous, if but one or a few, it is, typically, discontinuous. We may thus also regard the distinction as being 'complex' versus 'simple'.

Simple inheritance

The remarkable discoveries of Mendel were made possible because he directed his attention to qualitative differences in peas such as seed colour and shape. It is in comparable characters in man that one finds compliance with the Mendelian Laws of 'particulate inheritance' and 'independent segregation'.

The central feature of simple inheritance, however, is that the effect of a substitution of one allele by another is clearly recognizable in the phenotype at the level at which the observation is being made. At the molecular level all inheritance is simple in these terms, but much of the examination of organisms takes place at some developmental level removed from the actual DNA composition of the genes or the primary action of the genes in protein synthesis. Typically, it is in biochemical and serological variations that one finds the most numerous examples of simple inheritance since these variations tend to be close to if not the actual products of genes. Hence, the reason for so much attention to human blood, a relatively easily obtained tissue in which a lot of the observable variation is simply inherited. However, simple inheritance can be observed at more distant levels of organization as is well exemplified in many disease states. These, however, can usually be traced to some profound developmental alteration arising from an abnormality or deficiency in a particular protein.

Another feature of simple inheritance is that the environmental contribution to the variance is small; in particular, it does not mask the effects of gene substitution. This is a further reason why most of the best examples of simple inheritance are manifest at the biochemical or cellular level, since the nearer a character variation is to the primary action of genes the less 'room' there is in development for environmental factors to influence the gene expression.

In simple inheritance one finds Mendelian segregation ratios (as shown for the MN blood-group system in Table 6.2a, b) simply because the genes maintain their distinct identity from generation to generation. In other words the hereditary material derived from parents is not irrevocably mixed as was thought in the 'blending inheritance' theories before Mendel. This 'distinctness' is now recognized to be in the integrity of the DNA and so far as a particular gene is concerned, in that segment coding for a specific polypeptide.

TABLE 6.2a

Types of mating	Children		
M × M	M		
N × N			N
M × N		MN	
MN × M	$\frac{1}{2}$M	$\frac{1}{2}$MN	
MN × N		$\frac{1}{2}$MN	$\frac{1}{2}$N
MN × MN	$\frac{1}{4}$M	$\frac{1}{2}$MN	$\frac{1}{4}$N

TABLE 6.2b

Hereditary constitution of parents	Hereditary constitution of children		
$Ag^M Ag^M \times Ag^M Ag^M$	$Ag^M Ag^M$		
$Ag^N Ag^N \times Ag^N Ag^N$			$Ag^N Ag^N$
$Ag^M Ag^M \times Ag^N Ag^N$		$Ag^M Ag^N$	
$Ag^M Ag^N \times Ag^M Ag^M$	$\frac{1}{2}Ag^M Ag^M$	$\frac{1}{2}Ag^M Ag^N$	
$Ag^M Ag^N \times Ag^N Ag^N$		$\frac{1}{2}Ag^M Ag^N$	$\frac{1}{2}Ag^N Ag^N$
$Ag^M Ag^N \times Ag^M Ag^N$	$\frac{1}{4}Ag^M Ag^M$	$\frac{1}{2}Ag^M Ag^N$	$\frac{1}{4}Ag^N Ag^N$

Independent segregation versus linkage

Independent segregation occurs when the genes for two or more different character variations lie on different chromosomes. At meiosis the ways in which the products of one pair of chromosomes are distributed into gametes is quite independent of how others are distributed.

As was recognized soon after the re-discovery of Mendel's work there are many exceptions to his law of independent segregation because chromosomes carry many genes, and genes which are on the same chromosome, bound as it were in a common vehicle of hereditary transmission, tend to be inherited together. Such genes are said to be linked and the alleles of concern — typically the two recessive forms or the two dominant forms — may both be on one chromosome or they may be on different chromosomes of a homologous pair. The former is known as coupling or the *cis*-phase and the latter as repulsion or the *trans*-phase.

Linked genes are not, however, inseparable because, as is evidenced cytologically by chiasmata formation, exchanges between homologous chromosomes occur during meiosis. Genetically, this phenomenon is referred to as crossing over or recombination and the main factor determining whether or not two gene loci are separated by it is their distance apart on the chromosome. Genes lying very closely to one another are referred to as being tightly linked, those far apart as loosely linked. Between long chromosomes a number of chiasmata occur in every meiosis, in which case an odd number of cross-overs between two gene loci separates an original arrangement (*cis* or *trans*), whereas an even number keeps them together.

Apart from cases of very close linkage, such as occurs in the Rh blood group system and the genes controlling γ-δ-β polypeptide chain synthesis in haemoglobin, it is not easy to establish linkage groups in man by the standard methods used in experimental animals. Except in special situations to be discussed later (p. 174) linkage cannot be detected by looking for associations between two or more character variants in populations. At equilibrium, pairs of gene loci will be as often in repulsion as in coupling so no trait association exists, and when it comes to association studies within families, there are many difficulties arising from the limited proportion of marriages which can yield linkage information, small sibships, and the repulsion/coupling alternatives. Methods have been devised for attempting to circumvent these problems, but despite their analytical elegance they have not been very powerful in revealing anything but quite tight linkages such as those between the rhesus blood groups and some forms of elliptocytosis with a cross-over value of 3 per cent, the ABO blood groups and the nail–patella syndrome with a cross-over value of around 10 per cent, and Lutheran blood groups with ABH secretor status with about 15 per cent of crossing over.

However, other methods have now become available for detecting linkages, particularly the techniques for identifying chromosomal Q/G banding, and of preparing hybrids between human and other mammalian cells such as the mouse, which can contain various combinations of mouse and human chromosomes. These methods not only identify linkages, but also reveal the particular chromosomes on which the linked genes occur.

The first assignment through hybrid cell analysis was made when it was shown that a cultured cell line which lost its human thymidine kinase had also lost its human chromosomes 17. Now at least one gene locus has been assigned to every chromosome and a great many, over 20, to the larger chromosomes. These include Rhesus blood group and Phosphogluco-mutase 1 to Chromosome 1, PGM2 to Chromosome 4, HLA to 6, ABO to 9, and Haptoglobin α to 16.

Sex linkage

The one area where classical family studies have been revealing of linkages is where genes are on the sex chromosomes since genes which show the distinctive features of sex-linked inheritance must be on the same chromosome.

Fig. 6.4 shows some of the loci which have been identified by pedigree analysis as displaying X-linked inheritance and their relationships with one another. The gene assignments have been confirmed by hybrid cell analysis. The special features of recessive X-linked inheritance are

FIG. 6.4.   Gene symbols: Cd, deutan colour blindness; Cp, protan colour blindness; G6PD, glucose-6-phosphate dehydrogenase; Ha, haemophilia A; Ic, ichthyosis; Oa, ocular albinism; Xm, a serum protein antigen.

exemplified in Fig. 6.5 with respect to haemophilia, and arise from the facts that (1) every man receives his one X-chromosome from his mother and transmits it to all his daughters and none of his sons, and (2) every woman receives an X-chromosome from each of her parents and transmits either one or other of them to all her sons and daughters. The Y-chromosome which of course is not only confined to males, but also transmitted to all sons, has a testis determining factor (which acts also as a histocompatability antigen) linked to it.

Mutation

While 'crossing-over' releases genetic variability already in the genome the ultimate source of new variation is mutation. Mutation has been defined as 'the inception of a heritable variation', and in its widest sense can involve whole chromosomes as well as particular genes. The result of mutation is a mutant gene or chromosome, which can given rise to a mutant character. Gene mutation may occur at any time, but is more likely to occur while a cell is dividing than while it is 'resting' and in gametogenesis is most likely to occur during meiosis. From the genetic point of view, it is mutation during the formation of the gametes that is important, but it is of some medical consequence that somatic mutation occurs. A mutant gene will reproduce itself with self-copying precision until it is affected by the next mutation.

Although the cause for gene or 'point' mutation is still far from clear, the unravelling of the genetic code has clarified its nature. Typically, it involves the substitution of one base for another in the set of DNA triplets. Since the code is redundant some such substitutions will have no effect on protein synthesis, but often the change of one base will specify a different amino acid in a polypeptide sequence from that originally coded. Thus, for instance, if the cytosine in the triplet CTT is replaced by thymine to form TTT, the amino acid lysine instead of glutamic acid will be introduced at the corresponding place in the polypeptide being synthesized. This is the sort of change which has occurred in determining the β-chain of haemoglobin C as compared with the β-chain of normal adult haemoglobin (cf. p. 225). Sometimes, a base change will introduce a

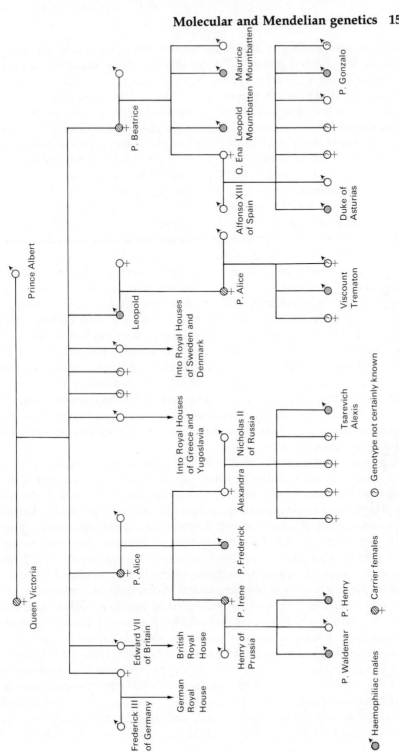

Fig. 6.5.   Pedigree of haemophilia in descendants of Queen Victoria.

'stop' in the reading of the code, as, for example, if adenine replaces cytosine in CTT to produce ATT. In this case there would be incomplete synthesis of the original polypeptide, only that portion prior to the stop being formed. There are also the possibilities of an 'insertion' of a totally new base, or the 'deletion' of an old one in the DNA sequence of triplets. Such events known as 'frameshifts' might cause changes of all the amino acids terminal to the mutation. For example, if there were an insertion of another cytosine prior to CTT (glutamic acid) AAA (phenylalanine) the code would read in triplets CCT (glycine) TAA (isoleucine) A--.

Gene mutation is a reversible process; thus, if gene A mutates to A′ it is equally possible for A′ to mutate back to A. Put in molecular terms, if one base is replaced by another this does not preclude the possibility at a later stage of return to the original base sequence. It is often said that mutation is a random process, but this does not mean that there are unlimited ways in which a gene can change, only that whatever changes occur are unrelated to any environmental demands that are being made on an organism or its environmental experiences. Quite clearly, when the concern is with base changes in the DNA, the possibilities, though numerous, are finite, and the effects of change very much dependent upon the code in which the mutation occurs. It follows that usually a mutant gene will affect the development of the same character system as does a gene from which it arose; but it will affect it in a different way or to a different degree. For instance, a gene which determines eye colour may mutate from a state in which it produces brown eyes to one in which it produces blue; it is highly unlikely that it will mutate to one which determines blood groups.

Chromosomal mutation may take two general forms; either there may be a change in the number of whole chromosomes, or a change within a single chromosome of the number and arrangement of genes. If the whole set of chromosomes is multiplied, a state of euploidy is produced. Euploidy arises as a result of failures in cell division during meiosis so that gametes with, for instance, the diploid number of chromosomes are produced. When such a gamete is fertilized by one carrying the normal haploid number a triploid zygote is formed; whilst if it is fertilized by another diploid gamete a tetraploid is produced. When only one or a few types of chromosomes are increased or decreased in number, rather than whole sets, the condition is known as aneuploidy. A now well-known example of a pathology arising from aneuploidy in man is Down's syndrome or mongolism. Individuals with this condition typically show retarded growth and development, abnormal dermatoglyphic patterns, the presence of a well-developed epicanthic fold over the eye-lid which somewhat resembles that found normally in Mongoloid peoples, various pathologies including increased susceptibility to leukaemia, and gross mental retardation. The probability of having an affected child increases strikingly with maternal age from about 0.04 per cent in women under 30 to 3.15 per cent for women over 45. The condition is usually due to one of the Group-G chromosomes being present three times rather than twice in all the body cells as a result of chromosomal non-disjunction during meiosis or some other abnormality of early cell division. For a number of years the extra chromosome was identified as chromosome 21, and the condition is still referred to as 'trisomy 21', though it has now been fairly certainly shown that the chromosome involved is number 22.

Gene number or arrangement can change in one of four ways. Part of a chromosome may be lost, a condition known as deletion. As an example, the condition known as 'cri du chat' can be given, which derives its name from the plaintive cry more or less continuously given by affected infants and involves severe mental and physical pathology. Affected individuals are heterozygous for a substantial loss of the short arm of chromosome 5.

The second type of change is in a sense the converse of deletion where a chromosome may be increased in length by the repetition of a segment, thereby forming a 'duplication'. Mechanisms are known whereby the full duplication and near complete duplication of at least a structural gene can occur. There is good evidence that the genes coding for the γ-δ-β haemoglobin polypeptides, which are closely linked to one another, arose by duplication of one original DNA sequence, and duplication has also been important in the evolution of the various immunoglobin molecules and some of their respective parts. Duplication is clearly a process whereby the sum total of DNA can be increased; and it produces new genes which can be subject to mutational change and 'offered' for selection while old ones continue to perform original and necessary functions. It has probably played an extremely important role in evolutionary differentiation.

A third type of change involves the transference of a piece of chromosome to a member of a non-homologous pair. This constitutes a 'translocation' and again a number of examples are now known in man. Thus, part of a 21 chromosome can become translated on to another autosome, e.g. number 14. An individual possessing this is normal since he possesses the usual DNA complement of one 14 chromosome, one 21 chromosome, and one 14.21 fusion chromosome, but, if during meiosis the 14.21 chromosome passes into the same gamete as the 21 chromosome and fertilization occurs with a normal gamete with a single 21 chromosome, the zygote effectively has the 21 chromosome represented three times and shows Down's syndrome. Thus, the basis for the condition can be transmitted to offspring in a Mendelian fashion and not all Down's children show a discrete extra chromosome.

Finally, a segment of a chromosome may become turned round to produce an 'inversion'. Thus, for instance, the order of loci may change from a, b, c, d, e, f, g, to a, b, c, f, e, d, g. This type of chromosomal change is frequent in *Drosophila*, where it can readily be detected and is of great importance in determining the evolutionary fitness of the organism. It has now been discovered in man through looking for changes in the position of the centromere and examining fluorescent banding patterns.

Rare versus common variation

A very striking distinction exists within what has here been called simple variation. In many examples, one form is extremely common and alternatives are very rare. Thus one has, for example, situations like that of albinism where only something of the order of 1 in 20 000 people in most populations are albino and the remainder have some capacity to synthesize melanin. By contrast, one has frequent situations where there are two or more variants all of which are quite common in a population. Such a situation is illustrated in the various blood groups like ABO. The latter condition is termed polymorphism which may be defined as the occurrence together in the same habitat of two more discontinuous forms of a species each with a population frequency of over 1 per cent.

The causes of polymorphisms will be discussed later, but it can be mentioned here that they are often seen as being of particular value in anthropological studies of the affinities of populations.

Gene expression

Even in 'simple' inheritance the relationship between a gene substitution and a change in the expression of a character or group of characters is subtle. Most striking is the contrast between characters which are expressed when only one gene is present (i.e. are dominant) and those which only appear in homozygotes (i.e. are recessive). Then there are co-dominant characters where heterozygotes have two characters in their phenotype (like the *AB* or *MN* blood-group heterozygotes) and intermediate inheritance where the character of a heterozygote lies somewhere between the expression of the two homozygotes. Biochemical explanations are usually available for these various situations, and often relate to the level at which an analysis is made. For example, the disease phenylketonuria is inherited as an autosomal recessive condition. It is primarily due to the absence of the enzyme phenylalanine hydroxylase which converts the amino-acid phenylalanine into tyrosine. Heterozygotes (or carriers) for the condition show none of the disease symptoms and the non-diseased state is therefore dominant, but such heterozygotes typically show less than normal levels of phenylalanine hydroxylase in their livers, so in terms of enzyme concentration the inheritance is intermediate. In the case of the co-dominant inheritance of the AB blood group the *A* and *B* genes of the heterozygote produce their respective glycosyl transferase enzyme, and so both acetylgalactosamine and galactose radicals (cf. p. 264) are added to carbohydrate moities of the H substance so conferring A and B antigenic specificities.

In many cases the substitution of one gene for another affects a number of characters. Thus, in the example of phenylketonuria, not only is a particular enzyme absent, but a whole host of other biochemical and developmental states are affected including defective growth, pigment formation and mental development. In the Laurence–Morn–Biedle syndrome, due essentially to a single gene in double dosage, obesity is associated with mental deficiency, polydactyly, and hypogenitalism. In arachnodactyly, which derives its name from the excessive length of the finger and toe bones, there are other skeletal abnormalities, and eye and heart defects. Such phenomena of multiple expressions of genes are known as pleiotropies. They can typically be traced to a change in the structure of a single polypeptide which plays a key role in a central or a number of developmental processes and do not invalidate the one gene/one polypeptide hypothesis. They forcefully demonstrate the fact that the genotype and the phenotype are often separated by a long and complex process of differentiation and development. They also demonstrate why it is desirable to restrict descriptions of the mode of inheritance to characters and not to genes, since some expressions of a gene are dominant and others recessive.

Another important factor in character expression is locus interaction in which genes at two or more loci interact in the determination of the phenotype. One example is 'complementarity' where two genes at different loci are required for the development of a character. Lewis β substance requires both a Secretor and a Lewis gene for its synthesis (cf. p. 264). Then there is 'epistasis' — a kind of dominance between loci — where the presence of a gene at one locus prevents the expression of a gene at another. Again in the ABH system, homozygotes *hh* are unable to synthesize A, B, and H substances whatever *ABO* genes they may have.

Another important type of locus interaction is known as 'position effect'. Here the expression of a gene is affected by its position on a chromosome, or by the nature of neighbouring genes. The Rhesus blood-group system affords a striking example, where individuals with the complexes *CDE/cDe* have the same genes as individuals who are *CDe/cDE*, but the former produce a large amount of E antigen and little C, and the latter produces a lot of c and little E. Thus, when *C* and *E* are on the same chromosome the synthesis of C is inhibited, and when *C* and *E* are on opposite chromosomes E synthesis is inhibited.

In more general ways there is also plenty of evidence for locus interaction in man. Quite often a gene which expresses itself in a definite way is not expressed at all in some individuals who possess it. Polydactyly is typically inherited as a dominant condition, but, occasionally, completely normal people have affected children. One of the explanations for this sort of situation is that the rest of the genotype acts in a way to suppress the action of the gene. Genes which may or may not find expression are said to show 'variable penetrance'. A similar phenomenon, known as 'variable expressivity', is the inconsistency in effect that a gene may have. In brachydactyly the fingers and toes are typically short and webbed, but one or more of these symptoms are frequently absent. This type of phenomenon indicates that the production of any character rests upon the action of many genes. The genotypes in a population have been conditioned by natural selection so that a particular gene replacement usually has a constant effect on the development of some normal character, but this does not mean that this gene is solely responsible for that character. It is therefore not surprising that deleterious genes, for which the genotype obviously has not been prepared, frequently have variable and multiple manifestation.

There is now every indication that what matters for the integrated development of the individual and, therefore, its evolutionary fitness is a co-adapted gene complex. It is easy to envisage how natural selection operates to produce this. Even when a gene whose overall effect is advantageous first arises, it is likely that some expressions will be deleterious and those individuals, whose genotype does not buffer development from these undesired expressions, will have fewer offspring than those in whose genotype there happen to be means of suppressing some of these effects. Such suppressing mechanisms will be further improved by natural selection and the undesired effects, by which one means those that reduce either viability or fertility or both, can be ultimately eliminated completely. The genes causing this suppression of effect are just one group of the so-called 'modifying genes' which have been built into the genotype by selection to control the expression of other genes. Some of them seem to have now no other effect. In addition to those that suppress deleterious effects, there are no doubt others that amplify advantageous expressions. Dominance itself can be evolved in this way. When a new gene arises and is therefore rare it will exist only in heterozygotes. If it is deleterious in effect, natural selection will operate in such a way as to suppress these effects through modifying genes. In other words, the gene will become recessive, and, in becoming so, the normal allele is obviously automatically made dominant. Conversely, if the new gene confers advantage on its possessor, natural selection will operate to confer these advantages on those individuals who only possess it in single dosage and thus make it dominant. The fact that when a deleterious gene arises by mutation it is often recessive from the outset is not inconsistent with this view, for mutation is a recurrent phenomenon, and a gene which arises now has probably arisen many times in the past, enabling a modifying system to be evolved which suppresses its effects in single dosage. Evidence in favour of this view has been obtained by hybridizing different populations of animals which possess the same dominant gene. When the dominance has been built up by different modifying systems in the two populations there is a complete breakdown of dominance relationships in the hybrid population.

7 Quantitative variation

By far the most important examples of many genes affecting the variation of a single character come when we turn to examine typical morphological, physiological, and psychological variation, such as stature and body weight, blood pressure, and many other features of the adaptability/homeostasis mechanism, and behavioural and ability traits such as intelligence. These characters are far separated from the immediate expression of genes and arise through a whole series of complex developmental processes. They do not show discrete variation within populations or even families, but continuous and quantitative differences with every degree of intermediacy between extremes.

The Gaussian curve

The actual form of the variation in a population is typically 'bell-shaped' as is shown in Fig. 7.1 where the frequency of men and various intervals of stature is shown in histogram form and a curve fitted to the distribution. The curve is the 'Normal' or Gaussian curve of variation and its properties afford the basis of much statistical science.

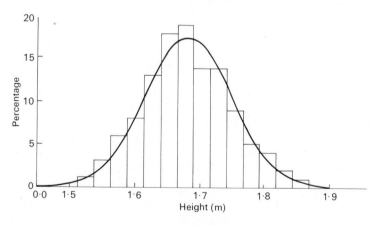

FIG. 7.1. Diagram showing the observed distribution of the heights of 117 males, exhibited in the form of a histogram together with a fitted 'normal' curve. (After Bailey, *Statistical methods in biology*, EUP, London, 1959.)

The point on the abscissa from which a perpendicular divides the curve into two halves of equal area is the average known as the mean (the sum of the measurements divided by the number of observations). Measures of the extent of the scatter on either side of this mean, are the variance; its square root, the standard deviation; and the coefficient of variation. In the Gaussian curve about two-thirds (68 per cent) of the observations lie within one standard deviation of the mean and practically 95 per cent of the observations within two standard deviations. The coefficient of variation is the variation expressed as a proportion of the magnitude of the dimension, i.e. the standard deviation divided by the mean × 100.

The distribution of some characters is not symmetrical and is said to be 'skewed'. In other instances it may be too flat (platycurty) or the tails too long (leptocurty) by comparison with the Gaussian curve. In skewed distributions other averages, the median, and the mode are different from the mean. The mode is the point of the abscissa from which a perpendicular bisects the apex of the curve, i.e. the most frequently occurring value. The median is the point which divides the horizontal extent of the scatter into two equal halves, i.e. the mid-point value. Skewed distributions can often be transformed to a more nearly Gaussian form by plotting some function of the original measurement such as logarithms. This is the case for measurements of subcutaneous fat with skin-fold calipers.

If samples of measurements of a character are taken from a population, and in biometry one is invariably dealing with samples, these will be distributed more or less normally also. The sample will have its own mean and standard deviation. This mean, however, is unlikely to be exactly the same as the true population mean; it will be affected by sampling error, and the smaller the sample and the larger the true standard deviation, the greater this error is likely to be. The degree of error is measured by the 'standard error of the mean'. Just how these various statistics are obtained can be found in any elementary statistical textbook.

For quite some time after the re-discovery of Mendel's laws it was generally felt that a particulate mode of inheritance could not explain the heritable basis of continuously varying characters and that some form of blending inheritance was involved. However, inseparable mixture of the hereditary material necessarily reduces the variation in a population by a half in every generation. This obviously does not happen and it is now known that the origin of new variation is quite inadequate to cope with the loss that would occur with blending. On the other hand, it can be shown that if a character is determined by a large number of genes acting together it will display the continuous variation that is observed in such characters as human stature.

The principle can be most simply explained by considering two unlinked loci whose genes, instead of producing different characters, are each responsible for a component of a single character. Suppose, for instance, that the effect of a gene A is to contribute a unit to some measurement and that its allele A' contributes two such units. If exactly the same effects are produced by the genes B and B', respectively, and the system is additive the double homozygote $AABB$ will produce *in toto* 4 units and the other double homozygote $A'A'B'B'$ eight units. If there is no dominance the double heterozygote $AA'BB'$ will be strictly intermediate phenotypically between the two homozygotes as it produces 6 units. Now consider an intercross of two heterozygotes, as in the following table, in which the number of units produced by the various genotypes are recorded. It will be seen that there are five phenotype classes and these occur in the ratio of 1 (4 units):4 (5 units):6 (6 units):4 (7 units):1 (8 units).

$$\frac{AA'BB'}{6} \times \frac{AA'BB'}{6}$$

Gametes		Gametes			
		AB	$A'B$	AB'	$A'B'$
			Zygotes		
	AB	$\dfrac{AABB}{4}$	$\dfrac{AA'BB}{5}$	$\dfrac{AABB'}{5}$	$\dfrac{AA'BB'}{6}$
	$A'B$	$\dfrac{AA'BB}{5}$	$\dfrac{A'A'BB}{6}$	$\dfrac{AA'BB'}{6}$	$\dfrac{A'A'BB'}{7}$
	AB'	$\dfrac{AA'BB'}{5}$	$\dfrac{AA'BB'}{6}$	$\dfrac{AAB'B}{6}$	$\dfrac{AA'B'B'}{7}$
	$A'B'$	$\dfrac{AA'BB'}{6}$	$\dfrac{A'A'BB'}{7}$	$\dfrac{AA'B'B'}{7}$	$\dfrac{A'A'B'B'}{8}$

In other words, the distribution of the variation of the measurement is symmetrical with the intermediate phenotypes more common than the extreme phenotypes. Something, in fact, resembling a normal curve is obtained. If the difference between the two extreme phenotypes, i.e. between 4 and 8 units of measurement, were determined by many pairs of genes instead of only two, any one particular gene contributing only some fraction of a unit, the number of classes of intermediate phenotypes would obviously be greatly increased. This itself means that the differences between any class and those phenotypically adjacent to it must be correspondingly decreased.

It is apparent, then, that as the number of genes responsible for a particular variation increases, it becomes correspondingly difficult to recognize distinct classes, and, if a very large number of genes are involved, the frequency distribution will be to all intents and purposes continuous. Furthermore, such discontinuity that might be recognizable genetically will invariably be marked by environmental contributions. Typically, characters whose variation is controlled by many genes are also affected by heterogeneity in environments, since the characters themselves are far removed from the primary site of gene action and this allows environmental factors to influence the various steps in the development of the phenotype.

Genes whose individual effects are small, but which act cumulatively to the variation in a single character are often referred to as polygenes, and the inheritance of quantitatively varying characters as polygenic or multifactorial. Such genes acting additively and independently produce a Gaussian or Normal curve. Environmental factors acting similarly do likewise.

A number of different genetic factors can affect the shape of the frequency distribution of a character. Firstly, there is the number of gene loci: the greater the number of genes responsible for a particular variation the more compressed will be the Gaussian curve, i.e. the smaller will be the variance or the coefficient of variation. Then the alleles at the individual loci can be at various frequencies in any population and this can cause skewness in distributions in a population. Inequality of gene effect may be present which in itself will act to produce discontinuity, i.e. qualitative effects in the curve. Just as with simple inheritance, dominance at the various loci can occur, so that the contribution of a heterozygote will be the same as that of a homozygote at some particular locus. However, so far as different loci are concerned, the dominance effects can be in opposite directions so that in the distribution they appear to cancel out each other's effects. Here it is necessary to distinguish between true dominance as it occurs from locus to locus and 'potence' which is the effect over all loci. Potence is the deviation of the multiple heterozygote from the midpoint of the distribution between the two multiple homozygotes: in classical genetic terms the deviation of the F_1 value from the mid-parent value. Situations are possible of no potence, but a great deal of dominance. Linkage between polygenes also occurs and tends to operate to reduce the apparent number of factors involved. It also acts to release variability slowly over generations. Then there are the potentially important complications of interlocus interaction and genotype environmental interaction. There are no *a priori* grounds for assuming that any polygene system is additive with a particular gene substitution having the same effect whatever other genes are present. Epistasis and complementarity can as well occur in quantitative inheritance as in simple inheritance. Likewise, a particular gene may have different effects depending upon the nature of the environment. A gene for high stature, for example, might contribute a great deal to human stature when nutrition is good, but be unable to express itself at all when nutrition is poor. Interactions of these kinds tend to lead to skewed frequency distributions.

Non-additiveness of gene or environmental effects often, however, depend upon the particular scale that is used for measuring a quantitative trait. If, for example, one uses the human eye for measuring skin colour, the addition of a fixed amount of melanin (either by a gene or tanning), to a black skin may be completely undetectable visually, but have a marked effect on the appearance of a mainly white skin. However, measuring the melanin addition biochemically would be totally independent of how much melanin was already there, and the system would be additive. Often enough some transformation of the scale of measurement to, for instance, logarithms will convert an interactive system to an additive one and remove the skewness from distributions.

Genetic analyses of characters whose determination is multi-factorial are much more complicated than those required for discovering the inheritance of differences due to one or a few major genes. Instead of being concerned with the frequency with which particular classes of phenotype occur, information has to be obtained from the parameters of distributions of variation and the nature and degree of resemblance between relatives. The ultimate aim of analyses, of course, is to determine the number of genes responsible for a particular trait variation and to locate the responsible loci on chromosomes, but usually one has to be satisfied with something much less than this, especially in man where large breeding programmes with known genotypes and selection experiments are impossible.

Heritability

One of the first questions which arise when analysing quantitative variation is how much of the observed phenotype variation is due to genetic causes and how much to environmental ones. This, of course, is the old nature versus nurture issue, but needs to be couched in terms of contribution to a character *variation*, since both genes and environment are equally necessary for the development of any character. It raises the concept of 'heritability' which can be defined as that proportion of the observed variance of a character in a population which can be attributed to genetic causes. There are two types of heritability estimate: broad heritability which represents the overall contribution of genetic factors, and narrow heritability which arises from only the additive components of the genetic variance. This additive component, as we have already seen, is that due solely to the average effects of individual genes on a character irrespective of what other genes are present. It is the part which allows the prediction of similarity between one generation and the next and determines what is presented to natural selection. Only when such additional components to the genetic variance as dominance and epitasis are trivial will narrow heritability approach the broad heritability. Generally speaking in man, where it is not possible to manipulate breeding and one doesn't have inbred lines, broad heritability is estimated from twin studies and narrow heritability from family studies.

Twin studies

Twins are essentially of two types: they arise either from the fertilization of two separate ova, in which case they are said to be 'dizygotic' or 'fraternal' and will be no more genetically alike than ordinary sibs, or from the division of a single fertilized egg, in which case they are referred to as 'monozygotic' and will be genetically identical except for somatic mutations which have occurred after the separation. Differences between monozygotic twins will, therefore, be mainly of environmental origin, whilst differences between dizygotic twins will be caused by both environmental and genetic factors. The contrast, therefore, in magnitude of these differences in the two types of twin will be a measure of the broad heritability of a trait.

Differences can be measured and expressed in a number of ways. For traits which can be categorized in discrete qualitative ways twins can be compared in terms of whether they both show the particular form of a character, in which case they are said to be 'concordant' or whether they possess alternative forms and are 'discordant'. Such comparisons are particularly useful in looking for heritable elements in disease in which one examines concordance rates for the disease in monozygotic twins and compares them with concordance rates in dizygotic twins. It is the magnitude of this difference which indicates the extent of genetic susceptibility. One finds high concordance rates for an infectious disease like measles in monozygotic twins, but practically as high ones for dizygotic twins. All concordance means here is that if one twin is infected the second is likely to be infected too. However, in the case of rickets or poliomyletis whilst the concordance rate of monozygotic twins is substantially lower than for measles it is very much higher than the concordance rate for the same diseases in dizygotic twins. This indicates a strong genetic susceptibility to these diseases even though one is essentially of nutritional origin and the other due to an infective agent.

For characters which are measured on a continuous quantitative scale the differences between pairs of twins can be expressed in such ways as the correlation coefficient (the magnitude of the correlation of the dimension of one twin with the other over all twin pairs of a particular type) or as the within-pair variance averaged over all twin pairs of a particular type.

The usual approach for estimating heritability is to express the difference in these measures between dizygotic and monozygotic twins, and express it as a proportion of the variation between dizygotic twins, i.e. $(V_f - V_i)/V_f$ where V_f is the within-pair variance of dizygotic (fraternal) twins and V_i is the within-pair variance of monozygotic (identical) twins. This is formally equivalent to the concordance rate comparison for discrete traits. Clearly, if V_i is low compared to V_f the ratio approaches 1 (or 100 per cent) and one would conclude that variation in the trait being considered is largely under genetic causation, i.e. is strongly heritable. Conversely, if V_i is practically as large as V_f the ratio approaches 0 and trait variation is mainly under environmental control, i.e. has low heritability.

More formally heritability

$$H = \frac{V_G}{V_T} = \frac{V_G}{V_G + V_E},$$

where V_G equals the genetic component to the variance; V_E the environmental component and V_T the total phenotypic variability as measured. V_i is the estimate of V_E and V_F of $V_G + V_E$.

The whole analysis is obviously a within-family one, comparing the magnitude of within-family genetic variation to the kinds of diversity of environmental factors that occur within the confines of a single family (such as heterogeneity within homes). The genetic variation between dizygotic twins (as sibs) also bears a relationship to the genetic variation in the population as a whole. Sibs share one-half of their genes in common (cf. p.189), but the remaining half are no more alike than unrelated people. Genetically therefore, one might consider dizygotic twins as half less variable than unrelated pairs in which case $V_f = \frac{1}{2}V_G + V_E$. With this viewpoint $V_f - V_i = (\frac{1}{2}V_G + V_E) - V_E = \frac{1}{2}V_G$. With heritability equalling V_G/V_T we need to double the difference in the variance of fraternal and identical twins to obtain our estimate, i.e. $2(V_f - V_i)/V_T$ and make a direct measure of the overall variability of the trait in the population V_T.

Occasionally, monozygotic twins are separated at birth or shortly afterwards, and are brought up in different homes and therefore often very different environments. Comparing the magnitude of the difference between such twins reared apart (A) and monozygotic twins reared together (T) can provide striking insight into the environmental lability of a character over a population. Thus, if using the formula $(V_iA - V_iT)/V_iA$ one obtains a proportion close to 1, the within-pair variance of monozygotic twins reared together is very small compared with that of those reared apart, and we conclude that between-family environmental factors play an important part in determining character variability. Conversely, a proportion approaching 0 signifies high heritability in the population. V_iA is, of course, also a useful parameter for indicating environmental contributions in the dizygotic twin/monozygotic twin formulae.

Approaches of this kind have been widely used in human genetics, and while greatly useful the results need to be considered with caution and are subject to many qualifications. Firstly and most importantly, it needs to be appreciated that a heritability estimate for a particular trait only relates to the particular population on which it was made and only to that population at the time it was made. In different populations and in the same population at different times a trait can have very different heritabilities. Clearly, the magnitude of environmental variation is never constant, nor need the genetic variation be. It is therefore very important to recognize the exact population on which the estimate is being made. Then, while the examination of monozygotic twins reared apart may cover a wide range of environmental heterogeneity, it is most unlikely that all the environmental influences that act on a whole population will be 'sampled' in this way, particularly with the practice of adoption agencies placing foster children in similar homes to those from which they come. Furthermore, none of the twin studies can tell us anything at all about the heritability of between-population differences. The variation of a character within any one population can be strongly heritable whilst the difference between two populations is mainly environmental.

Other major problems may arise from the phenomena of gene–environment interaction and gene–environment correlation. The approach to estimating heritability just discussed essentially assumes that the observed phenotypic variation arises from the *addition* of the environmental component of the variance to the genetic component, i.e. $V_T = V_G + V_E$. However, different genotypes may react differently to the same environment producing a non-additive or interactive situation. Likewise, different genotypes may not be randomly distributed over the environmental variation: monozygotic twins, for example, may seek out similar environments dependent upon their genotype. This is a case of gene–environmental correlation and tends to over-estimate heritability whereas interaction tends to under-estimate it. However, the magnitudes of these interrelationships are largely unassessed.

A matter which particularly affects family studies is that environments as well as genes can be transmitted and inherited between generations. The mode of transmission is very different, but some of the results can be similar. Books, for example, are handed down through families, thus generating a correlation between generations in the knowledge they contain.

It also needs to be borne in mind that twins are in many experiences different from singletons. Thus, they share a maternal uterus at exactly the same time and are inevitably competing for resources. Twins are also uncommon; only about 1 birth in 90 in the UK is a twin, and of these only about a quarter are monozygotic.

Difficulties of these kinds have bedevilled nature/nurture analyses and some human geneticists are completely opposed to them. This is especially so when the traits being considered are behavioural and ability ones, such as IQ where the nature of the interactions are likely to be particularly complex. However, this is an extreme view and some useful insight has been obtained about developmental flexibility from twin and family investigations of anthropometric and similar characters.

Genetic analyses

We can now examine the ways that have been employed for analysing the genetic components of quantitative variation. Clearly, this will be easier when environmental components are low, but even then presents many problems.

Occasionally, it is possible to identify a gene contributing to quantitative variation through its pleiotropic effects. Hair colour in man is a multifactorial trait which in most populations is distributed more or less normally. One gene which can contribute to this, though only to a small degree, is the phenylketonuria gene. Phenylketonurics tend to have less pigmented hair than non-affected individuals though there is great overlap in the distributions of the two types of people. The phenylketonuria gene, however, also has many other expressions in homozygous state. It tends to reduce growth, so measures of size, such as head size, are less. In this feature there is less overlap in the distributions between affected and normal people than for hair colour, but the phenylketonuria gene could still be regarded as one of a multiple series of factors affecting growth rates. Phenylketonurics also show retarded mental ability, but there is still a little overlap in the distributions of IQ. It is only when one examines the phenotypic expression at the biochemical level, in such terms as serum phenylalanine levels, that there can be no ambiguity between identifying affected and non-affected subjects, and inheritance can be judged as simple.

Linkage disequilibrium

A somewhat similar way of detecting genes contributing to quantitative variation is through linkage arrangements with genes of major effect. Normally, linked genes will be as often in repulsion as in coupling in a population so no association exists. However, genes which are very close to one another on the chromosome and therefore are rarely separated by crossing-over will be in linkage disequilibrium for a very long time. Whilst such disequilibrium exists, there is an association between characters in a population through an excess of either coupling or repulsion relationships. Thus, a polygene closely linked with a major gene is detectable by looking for associations between simply inherited traits and quantitative ones. this has proved a fruitful way of detecting polygenes in *Drosphila*, but has had limited success in man. The main problem is that other factors, especially heterogeneity in population structure, can lead to character associations within populations besides linkage disequilibrium (and pleiotrophy). Nevertheless, there is some quite strong evidence from population data that some genetic determinant of IQ is linked with the haptoglobin locus in man.

Biometrical techniques

A particular problem in analysing quantitative variation arises from not being able to identify genotypes. It is not possible to say that an individual of a particular stature has a particular set of genes; indeed as can be seen from the very simple example on p. 168 individuals of the same phenotype may be of a number of genotypes. Nor can it be said that subjects from the tails of distributions are homozygous for gene loci concerned — though they will tend to be more homozygous than individuals nearer the mean. Occasionally, however, two populations are so different in some trait that they can probably be regarded as each being homozygous for the genes which distinguish them. This has certainly been the assumption for skin colour differences between continental groups of human beings, particularly Africans and Europeans. Here there is not only no overlap in the distribution of character in the two groups but a very considerable separation of the distributions. If the assumption is valid it allows the procedures of biometrical genetics to be employed for analysing skin colour differences between 'blacks' and 'whites' through the study of hybridizing populations. In such populations, both 'parental' groups, F_1, F_2, and back-cross hybrids can be found in the earliest stages of the hybridizing processes. Comparison of their means and variances in skin colour enables some considerable dissection of the genetics of the trait to be done.

The problems of scaling are particularly well evidenced in this situation since it is possible to measure skin colour at a number of different wavelengths with a spectrophotometer (cf. p.309). It can be seen in Fig. 7.2 where reflectance is plotted by wavelength that the mean for F_1 hybrids is closer to the black parental value at the blue end of the spectrum, and to the white parental value in the red! Clearly one would come to very different conclusions about 'potence' depending upon the wavelength at which skin colour was measured. This is because the system is not additive at all these different wavelengths. To test for additiveness one needs to show that twice the mean of the backcross white hybrids equals the mean of the F_1 hybrids plus the mean of the whites, i.e. $2\bar{B}_W = \bar{F}_1 + \bar{W}$ and also $2\bar{B}_B = \bar{F}_1 + \bar{B}$ and $4\bar{F}_2 = 2\bar{F}_1 + \bar{W} + \bar{B}$. Such testing reveals that one needs some scale transformation of reflectance in the blue and red to achieve additiveness. Logarithmic transformation of the former and antilogarithmic transformation of the latter approach the requirement.

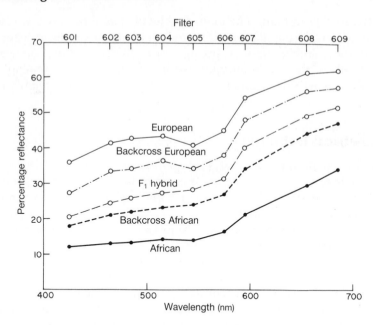

Fig. 7.2.   Mean reflectance curves of European, African, and various hybrid groups.

Once one has obtained the correct scales for analysis one can proceed to examine the variation of the different hybrids and partition this into three components: E, the environmental component; A, the additive genetic component; and D, the component due to dominance. By assumption, the two parental groups and the F_1 hybrid contain no within-group genetic variability (being either all homozygotes or heterozygotes). Such variance as is measurable must therefore be due to environmental factors and the average over all three groups provides the estimate of E. It has also been shown that the sum of the variances of the two backcross hybrids equals $\frac{1}{2}V_A + V_D + 2V_E$ and that the variance of the F_2 hybrid equals $\frac{1}{2}V_A + \frac{1}{4}V_D + V_E$. With 'E' known the equations can be solved for V_A and V_D. There is also a relationship between V_A and the number of genetic 'effective factors' responsible for the heritable element of the variation with the number of effective factors 'K' equalling the square of the difference between the means of the two parental groups divided by $4V_A$. 'Effective factors' are the minimal number of gene loci since without further generations of hybrids it is impossible to distinguish the effects of genes on the same chromosomes, i.e. linked genes. Using this kind of analysis, and with one or two other assumptions not mentioned here, it has been suggested that environmental factors contribute about 30 per cent of variance in skin colour in hybrids between Africans and Europeans, with the additive genetic contribution approximating 70 per cent and very little dominance. The same analysis indicates that between 3 and 4 effective factors are responsible for the parental difference.

Family structure

In the analysis of most other quantitative variation it is not possible to identify specific genotypes, such as multiple homozygotes or heterozygotes or groups of known genetic composition such as backcross hybrids. Under such circumstances the approach has to be based upon degrees of resemblance between relatives.

A parent passes on half its genes to any one offspring so offspring and parent bear half their genes in common. This 'half' is known as the 'coefficient of relationship' between parent and offspring. The coefficient of relationship between sibs is also a half since, if a particular offspring receives a gene from a parent, there is a half chance that another offspring will receive that same gene. If the likelihood of a gene being transmitted through one generation is a half then the likelihood of it being transmitted over two generations is a quarter ($\frac{1}{2} \times \frac{1}{2}$). This then is the coefficient of relationship between grandparent and grandoffspring as it also is between uncle/aunt and nephew/niece, and the coefficient between first cousins is one-eighth.

These varying relationships mean that in a totally additive genetic system an individual will be half like a parent (making due allowance for sex and age differences), a quarter like a grandparent or uncle/aunt, and an eighth like a first cousin.

To take a theoretical example, suppose human stature is entirely determined by heredity and the responsible genes are additive in effect and without dominance. If the mean height of a population is 1.70 m and mates are chosen at random so far as stature is concerned (and will therefore be on average 1.70 m) men who are 1.78 m tall will have sons whose mean height is 1.74 m, since the son will, on average receive from his father a half of the genes which made the father 0.08 m taller than the population mean and the other half of the son's genes, on average, will be like those of the population. Likewise, grandsons and nephews will average 1.72 m and great-grandsons and first cousins 1.715 m. It needs to be noted here that whilst there is a relationship (or regression) of sons on fathers there is equally a relationship of fathers on sons. Thus, men who are 1.78 m tall will have fathers whose mean height 1.74 m since on average a half of the genes which made the sons 0.08 m taller than the population mean will have come from the fathers. These relationships therefore in no way alter the distribution of stature in the population and 'regression of the mean' operates both forwards and backwards.

Actual similarity between relatives is expressed as the correlation coefficient 'r' which measures the degree of similarity or the regression coefficient 'b' which measures the form of the similarity. In the example just given where all the variance is genetic additive variance and V_A = the total genetic variance of V_G, the correlation coefficient between parent and offspring is 0.50 (i.e. the coefficient of relationship). Where there is dominance this will not hold, since parent and offspring will show no similarity when the heterozygous parent for the dominant gene is compared with homozygous recessive offspring. Generally, however it can be seen that $r_{(p/o)} = \frac{1}{2}(V_A/V_T)$ or $V_A/V_T = 2r_{(p/o)}$.

Although the coefficient of relationship between sibs is also a half, the similarity between them is less affected by dominance than that between parent and offspring. This is because if an offspring of two heterozygous parents happens to be a homozygote recessive there is a one in four chance that a sib will also be of this genotype. The sib–sib correlation coefficient therefore equals $\frac{1}{2}(V_A/V_T) + \frac{1}{4}(V_D/V_T)$. It follows that:

$$r_{(s/s)} - r_{(p/o)} = \frac{1}{4}(V_D/V_T) \text{ or } V_D/V_T = 4r_{(s/s)} - r_{(p/o)}.$$

Having obtained the proportions of the genetic variance which are due to additiveness and dominance, the proportion of the variance attributable to environmental factors can be obtained by subtraction $V_E/V_T = 1 - (V_D/V_T) - (V_A/V_T)$. We still need to remember the potential problems of epistasis and gene-environmental interaction. However, using such an approach it has been estimated that 16 per cent of blood pressure variation in one population is due to environmental factors, 48 per cent to additive genetic factors, and 36 per cent to dominance. Some levels of similarity between relatives for intelligence are shown in Table 7.1 in terms of the magnitude of correlation coefficients. Perhaps the most

TABLE 7.1

Summary of observed similarities in intelligence in a series of studies

Relationships	Genetic correlation	Observed correlation coefficients 0 0.1 0.2 0.3 0.4 0.5 0.6 0.7 0.8 0.9 1.0
Unrelated persons reared together	0.00	
Foster parent and child	0.00	
Parent and child	0.50	
Siblings reared apart	0.50	
Siblings reared together	0.50	
Dizygotic twins same sex	0.50	
Monozygotic twins reared apart	1.00	
Monozygotic twins reared together	1.00	

Horizontal lines represent the ranges in similarity in the different studies.
Vertical lines represent the mean values over the different studies.
Data based on the compilations made by Erlenmeyer-Kimling and Jarvik (1963).

striking feature here is the wide range of values that have been obtained in different studies, though, as has already been emphasized, the magnitude of environmental contributions is likely to be very different from case to case.

Between-population variation

As already explained, while twin and family studies can provide indications of the comparative role of hereditary and environmental factors in producing within-population variation, they are quite useless for analysing between-population differences. The only insight we can gain of this is through studies of migrants (and their offspring) from one population to another.

This approach was extensively applied by Boas in anthropometric studies of European migrants to North America and has recently been refined for the analysis of physiological variation. It is particularly useful in examining the physiological effects of living at different altitudes, since dramatic environmental changes occur in mountainous regions in relatively short geographical distances and often across populations between which there is so much exchange through migration that genetic differences could not arise. By comparing indigenous highlanders, recent lowland migrants to highlands, indigenous lowlanders, and recent highland migrants to lowlands it is possible to identify not only the immediate effects of altitude variation on development and physiology, but also any genetic components to differences between the highland and lowland indigines when these have been isolated from one another. The main difficulty in this approach is that migrants may often not be representative samples, at least in some traits, of the populations from which they come.

Environmental contributions to quantitative variation are often regarded by geneticists as nuisances which even further complicate the analysis of complex traits. Viewed in broader terms, however, environmental responses have important biological meaning in themselves as the last section of this book fully demonstrates. The genetic constitution of an organism sets many patterns and limits to the form of development; which particular pathway is followed depends for many traits on the environment in which the organism is living. The whole repertoire of potential pathways has been termed 'the norm of reaction' by Dobzhansky.

Broadly considered, a response or set of responses to an environment may be either disadvantageous or advantageous. As an example of disadvantageous responses one can cite the effects of large doses of ionizing radiation. Individuals who do not show these effects would be more fit than those that do in environments where there is much radiation. Whilst such responses can in the broadest sense be considered to be genetically determined they have not been evolved. Indeed, if an increased radiation load becomes a permanent factor in some environments natural selection will operate to increase an individual's resistance to it, provided, of course, that all individuals are not destroyed first! The capacity of the individual to buffer his existence from environmental factors which are constantly tending to disintegrate it is known as 'homeostasis'. Since an individual's environment is constantly changing, the states of the homeostatic mechanisms will themselves change correspondingly in providing the desired buffer effect. If, then, the differences between individuals are due to differences in their environment demanding different homeostatic states, these differences are adaptive. The ability of the individual to respond adaptively to environmental change is conveniently distinguished from adaptations involving genetic change by referring to it as his 'adaptability'. As examples of man's adaptability, one can mention the variation of erythrocyte count with altitude, basal metabolic rate with temperature, and state of skin colour with intensity of solar radiation. Most of this adaptability resides in reversible physiological and behavioural mechanisms, but it seems that irreversible growth and developmental processes can be adaptable also.

8 Population genetics — mating systems

Human biology as defined and conceptualized in this book could be more precisely termed 'human population biology' for here we are concerned with the nature, origin, development, causes, and effects of human variety, and variety can only exist in populations. The population concept is as focal to evolution and ecology as it is to genetics and ties together all of the many facets of human biology as they occur at the cellular, tissue, and whole body levels.

Notwithstanding this quite central importance of the population concept, the actual identification of populations in the real world is often difficult and frequently arbitrary. It may differ according to purpose so that a population recognized for demographic study may not be the same as one identified for genetic or ecological investigation. Number is, of course, of central concern as is sharing, but the features shared between collections of people may not be the same and thus the 'number' based upon what is shared will also vary. Populations defined for demographic purposes usually share 'space', those in ecology share environments and resources, and those in genetics and evolution share genes. Probably only in the case of small and long isolated island populations do the different kinds of sharing fully correspond.

Even a population considered for a single purpose can be difficult to define. Our concern in this section of the book is with genetic or Mendelian populations; groups of people who share in the same gene pool, but the extent of the sharing can vary greatly, is usually difficult to quantify, and is ever changing.

For some purposes the family can be envisaged as the smallest Mendelian population, but in population genetics one is dealing with units larger than this. Of these the 'isolate' is often recognized as basic and much attention in theoretical population genetics has been devoted to it. An isolate is a group of people, which is now and has been for some time past totally separated from other groups, and is reproductively unstratified so that all the individuals share equally in the same gene pool. Populations in the historic past may often have been sufficiently close to this definition for the theory of isolates to apply, but today groups of this form are rare and where they do occur they tend to be characterized also by the features of particularly small size which has its own genetic effects. More commonly today one finds groups of people, as for instance in villages or towns, separated by sparsely inhabited space, but over which there is gene flow through the movement of people between the groups. Such a situation has been characterized by Sewall Wright as the 'island-model', and its structure is primarily determined by the size of the groups, and the nature and extent of the gene flow between the groups. Most recently, and especially in developed societies where there has been dramatic urban growth and the appearance of the megalopolis, one finds people more or less evenly and densely spread over great areas. Clearly, not all the inhabitants share in the same gene pool since the likelihood of mating with someone nearby is substantially greater than with someone at the other side of the city, despite everyone being in the same demographically defined population. This situation has been termed by Wright the 'diffuse model' and the essential feature determining its genetic structure is the movement pattern as it affects the spatial distribution of mate choice.

This situation highlights another feature of Mendelian populations; the fact that they exist in a series of hierarchies depending upon the extent to which they share in common gene pools. The situation is analogous and could indeed correspond with demographic geographical populations in which the populations of towns and villages make up those of counties, and the counties those of states and countries. In the case of Mendelian populations the basic unit, as already identified, is the isolate or something like it such as the island model and the largest unit is the whole species, since all members of a species, potentially if not actually, share in the same gene pool. Different species, because of their reproductive isolation from one another which is recognized in their specific status, belong to different Mendelian populations.

Random mating

Clearly from what has just been said, the genetic definition of a population is primarily determined by the mating system. The mating system also controls the ways that genes are organized into genotypes within a population. One of the simplest types of mating system possible in populations of bisexual organisms is that in which any individual of one sex has an equal probability of mating with any individual of the opposite sex in the population. Such random mating is also referred to as panmixis. Strict panmixis is rare in man if only because it can only occur in geographical isolates. Then most forms of human population show some form of social or religious stratifications which precludes marriages being contracted at random. Added to this are the factors of personal mate choice and the fact that marriages between close relatives are forbidden. Nevertheless, it is usually true that genes, such as those which determine the blood groups, which are rarely taken into account in mate selection are distributed among the inhabitants of some fairly restricted locality as though mating were more or less at random.

The effect of random mating on the distribution of genes in a population is the foundation of population genetics. So that only this effect is considered, it will be assumed in the first place that there is no differential viability or fertility of genotypes, no mutation, and that the population is sufficiently large to obviate unrepresentative sampling of gametes. To understand the consequences of random mating it is convenient to envisage the formation of a new panmictic unit by the hybridization of a population which contains only one of a pair of genes, e.g. A with another population which contains only its allele, e.g. A'. All the individuals in these 'ancestral' populations are obviously homozygous at the locus considered for the two genes, respectively. In the hybrid population, however, there will not only be both homozygotes, but also heterozygotes, and the question of import is in what frequency will these three genotypes occur if mating is random? The answer obviously depends upon the proportionate representation of the two ancestral populations compounded in the mixed one. Suppose the population which is homozygous AA contributes a fraction p of the parents of the hybrid population and the remaining fraction q is contributed from the population which is homozygous $A'A'$. Then, if males and females from each ancestral population are equally represented, the frequencies of the different sorts of mating will be $p^2AA \times AA$, $2pqAA \times A'A'$, and $q^2A'A' \times A'A'$, as can readily be seen from the following table.

Frequency of matings		Frequency of two types of male	
		pAA	$qA'A'$
Frequency of two types of female	pAA	$p^2AA \times AA$	$pqA'A' \times A'A'$
	$qA'A'$	$pqAA \times A'A'$	$q^2A'A' \times A'A'$

Mating between homozygous parents of one type can obviously only produce children of the same type, whilst matings between homozygotes of different type will produce only heterozygous children, so the frequencies of children produced from these matings will be p^2AA, $2pqAA'$. $q^2A'A'$.

Frquency of types of mating	Frequency of types of children
$p^2AA \times AA$	p^2AA
$2pqAA \times A'A'$	$2pqAA'$
$q^2A'A' \times A'A'$	$q^2A'A'$

These are the parents of the next generation so, instead of only two types of parent as in the initial matings, there will be three types. The consequences of this, however, are determined in exactly the same way as in the preceding generation; first, the frequencies of the different types of mating are calculated and then the frequencies of the types of children from these matings.

Frequency of different types of mating		Frequency of three types of male		
		p^2AA	$2pqAA'$	$q^2A'A'$
	p^2AA	p^4 $AA \times AA$	$2p^3q$ $AA \times AA'$	p^2q^2 $AA \times A'A'$
Frequency of three types of female	$2pqAA'$	$2p^3q$ $AA' \times AA$	$4p^2q^2$ $AA' \times AA'$	$2pq^3$ $AA' \times A'A'$
	$q^2A'A'$	q^2q^2 $A'A' \times AA$	$2pq^3$ $A'A' \times AA'$	q^4 $A'A' \times A'A'$

In the above table three types of mating occur twice, so in considering the frequency with which children are produced reciprocal cross-matings are added together.

Types and frequency of mating	Type and frequency of children		
	AA	AA'	$A'A'$
$AA \times AA$ p^4	p^4		
$AA \times AA'$ $4p^3q$	$2p^3q$	$2p^3q$	
$AA \times A'A'$ $2p^2q^2$		$2q^2q^2$	
$AA' \times AA'$ $4p^2q^2$	p^2q^2	$2p^2q^2$	p^2q^2
$AA' \times A'A'$ $4pq^3$		$2pq^3$	$2pq^3$
$A'A' \times A'A'$ q^4			q^4
Total frequency of different types of children	$p^2(p^2 + 2pq + q^2) + 2pq(p^2 + 2pq + q^2) + q^2(p^2 + 2pq + q^2)$		

The total frequency of the children of the three types is obtained by adding together the contributions from the various matings to each of the columns. It will be seen that $(p^2 + 2pq + q^2)$ is common to each column and the genotypes are, therefore, produced in the frequencies p^2AA, $2pqAA'$, $q^2A'A'$. These are the frequencies of the different types of parents. Thus, if only the effects of random mating are considered, the frequency with which the three genotypes occur in a population is the same from generation to generation; the equilibrium frequency is p^2AA, $2pqAA'$, q^2AA'; and following the mixture of populations this equilibrium is established in a single generation. These conclusions are all aspects of what is known as the Hardy–Weinberg law.

The fact that the genotype frequencies are constant from generation to generation must, of course, also mean that the frequency of the two genes remains the same. It will be recalled that the symbols p and q were used initially to represent the frequencies of the homozygotes AA and $A'A'$ respectively among the parents of the new population. Since each of these individuals carries two genes of the same type, these symbols equally well represent the frequency of the genes A and A' in this population of parents. Now, whilst the frequencies of the genotypes changed in the first generation through the formation of a new genotype — the heterozygote AA' — the frequencies of the genes did not change, and they can therefore continue to be represented by p and q in every generation.

Gene frequencies are usually represented as decimals rather than as fractions or percentages. For example, if a quarter of the genes at a particular locus in a population are of the type A, this gene is said to have a frequency of 0.25. In this case, the allele A' must obviously have a frequency of 0.75, if it is the only other allele in the population, since $p + q = 1$.

It is often necessary to know the frequency of genes in a population. When the heterozygote is phenotypically distinguishable from both homozygotes, the frequency of two alleles can be counted directly. Thus, for instance, in the MN blood-group system every individual with only M-type blood possesses two Ag^M genes whilst MN individuals have only one Ag^M gene. The frequency of the Ag^M gene in a population is therefore

$$\frac{2M + MN}{2T}$$

where M and MN represent the number of M and MN individuals, respectively, in the population, and T the total number of individuals. The frequency of Ag^N can be calculated similarly or by subtracting the Ag^M frequency from 1.

When the heterozygote is not distinguishable from one homozygote this procedure is obviously not possible. In such cases the gene frequencies have to be estimated indirectly.

Consider the situation in PTC-tasting (cf. p.283) in which the ability to taste phenylthiocarbamide is dominant to the inability to taste it, with tasters being composed of the genotypes TT and Tt, and non-tasters all being tt. Under random mating the three genotypes will be in the frequency $p^2TT:2pqTt:q^2tt$. Now $p^2 + 2pq + q^2$ is the expansion of $(p + q)^2$ and the frequency of the gene T, i.e. p is the square root of the frequency of TT individuals and the frequency of t, i.e. q, is the square root of the frequency of tt individuals. The former are not distinguishable from the heterozygotes, but the latter are. The square root of the frequency of non-tasters in a population is therefore the frequency of the non-taster gene, and p can be obtained by subtracting q from 1.

In the case of a recessive sex X-linked trait like colour blindness or haemophilia, the position is even simpler. In females the situation is the same as for an autosomal condition, but in hemizygous males with only one X-chromosome, the frequency of a phenotype must also be the frequency of the responsible gene.

The principle of calculating gene frequencies from Hardy–Weinberg equilibrium expectations can easily be extended from a pair of alleles to three or more. In the ABO system A and B are dominant to O, but there is no dominance between A and B, i.e. the heterozygote AB is recognizable while the heterozygotes AO and BO are not. When other alleles such as A_2 do not occur there are therefore only four distinct phenotypes. If p represents the frequency of A, q of B, and r of O with $p + q + r = 1$, then, under a system of random mating, the frequency r is the square root of the frequency of individuals who are of O blood. q can be calculated in the following way:

$$p^2 + 2pr + r^2 = f_A + f_O,$$

where f_A and f_O represent the frequency of A and O individuals, respectively. That is,

$$(p + r)^2 = f_A + f_O$$

and so

$$p + r = \sqrt{(f_A + f_O)}.$$

But

$$p + q + r = 1,$$

therefore

$$1 - q = \sqrt{(f_A + f_O)}$$

and

$$q = 1 - \sqrt{(f_A + f_O)}.$$

p can similarly be shown to be equal to $1 - \sqrt{(f_B + f_O)}$. For use, these formulae need to be slightly modified, but this explanation covers the essential principle.

Gene frequencies rather than phenotype or even genotype frequencies are usually used for expressing the differences between populations and need to be known in many genetic problems. They are particularly necessary in calculating the frequencies of unidentifiable heterozygotes, which is a matter of some importance when these are carriers for deleterious genes. As has been seen for a two-allele system the frequency of heterozygotes is $2pq$ under random mating. This means that the highest frequency of heterozygotes occurs when the two genes have equal frequency, i.e. 0.5. At these frequencies 50 per cent of individuals will be heterozygotes. With a diminishing frequency of one of the genes there is a reduction not only in the homozygotes for this gene, but also in the heterozygotes.

A recessive condition may be very rare in the population and yet the frequency of carriers is comparatively high. Albinism, for instance, has an incidence of about 1/20 000. The frequency, therefore, of the albino gene q is $\sqrt{(1/20\ 000)} \simeq 0.00709$ and of the normal allele $p \simeq 0.99291$. The latter to all intents and purposes can be regarded as unity so $2pq = 2 \times 0.00709 = 0.01418$, which is approximately 1/70.

Different populations may have different gene frequencies even when they are quite close together and experiencing some gene flow between them, as in island-model type situations. We shall examine the causes for this later, but it may be pointed out now that in such cases the average frequency of heterozygotes over all the populations is always less than their frequency in a single population formed by coalescing all the separate groups into one randomly breeding group. Thus, if, for example, the frequency of one allele (of a two-allele system) in three populations were 0.20, 0.50, and 0.80, the frequency of heterozygotes ($2pq$) in each of these populations would be 32 per cent: 50 per cent: 32 per cent, respectively, with an average over all three groups (if they were of equal size) of 38 per cent. In a single population formed by their mixing and random mating the frequency of heterozygotes would be 50 per cent since the mean gene frequency is 0.5. This deviation was first noted by Wahlund and is often used for measuring the level of genetic diversity between a group of populations.

Inbreeding and outbreeding

Deviations from random mating can occur in two general directions. People who are related can either marry more frequently or less frequently than they would by chance. In the former case the mating system is one of inbreeding and in the latter one of outbreeding.

The closest form of inbreeding, and one possible in many plants, is self-fertilization. It demonstrates in the simplest way the consequences of inbreeding. When a heterozygote AA' fertilizes itself, a half of the progeny are heterozygotes, while the other half are homozygotes either AA or $A'A'$. The selfing of these homozygotes, of course, produces only more homozygotes, but again a half of the progeny of the heterozygotes are homozygotes. This means that under self-fertilization the frequency of the heterozygotes in a populations is constantly being reduced at the rate per generation of a half of the heterozygote frequency in the previous generation. Ultimately, to all intents and purposes, no heterozygotes will occur and the population will be composed exclusively of the two types of homozygote. In this model only a single locus has been considered, but of course, the trend to homozygosis will occur at the same rate at every locus and eventually, if no other factors come into play, every individual produces offspring which are identical to it in all their genes. Another consequence is that a series of quite distinct lines is established. Starting with a double heterozygote *AaBb*, for instance, there are four such lines possible, namely *AABB*, *AAbb*, *aaBB*, and *aabb*. The rate of reduction in heterozygosity as a whole is independent of the number of gene pairs concerned, but the probable number of homozygous individuals and the number of individuals heterozygous at 1, 2, 3, etc., loci in any generation depends on the initial number of gene pairs and can be calculated by expanding the binomial $1 + (2^r - 1)^n$, where r represents the generation concerned and n the number of gene pairs.

It must be noted that in this model of the reduction in heterozygosis it is being assumed that every member of a generation reproduces and all its offspring are reared. If there is some non-random selection of parents then the rate at which complete homozygosity is approached depends obviously upon the genotype of the individuals chosen. To take an extreme example, if only one individual were taken as the parent of the next generation and this happened to be a homozygote, then obviously all subsequent generations from that time on would be homozygotes of the type chosen. On the other hand, if the chosen individual were heterozygous, no reduction in the frequency of heterozygotes would have occurred. One can at least say, however, for self-fertilization that no individual can ever be more heterozygous than its parent; it may be the same or less, but whatever homozygosity is gained is always retained.

This is not true of the next closest forms of inbreeding, brother–sister and parent–offspring, which are, therefore, mating systems that require a greater number of generations to eliminate heterozygotes. Thus, for instance, six generations of self-fertilization are more effective than seventeen generations of brother–sister matings in bringing about homozygosis. Only occasionally, as among the Ptolemies and the royal house of the Incas, do human societies permit such close inbreeding as brother–sister unions. It is said that among the Ernadan of Malabar, a man takes his eldest daughter as his second wife, but usually the closest forms of inbreeding permitted by human societies are between niece and uncle (or nephew and aunt), and between first cousins. Even the former is forbidden in many societies, and a number of the States still discourage the latter. Continued cousin marriages will also tend to produce general homozygosis, but the rate at which heterozygotes are reduced in frequency is very considerably slower than with brother–sister or parent–offspring mating.

Marriages between related individuals, known as consanguineous marriages, are of considerable medical importance, since the likelihood of spouses having the same genes is obviously considerably greater if they are closely related than if they are unrelated. Relatedness, of course, is a state of infinite degree. In the evolutionary sense all people are related; at some stage in their history they had common ancestors. So far as some particular population is concerned, its past size, if all individuals were unrelated, would have to have been far greater than it actually could have been, since every individual has two parents, four grandparents, eight great-grandparents, and 2^n ancestors n generations ago. Assuming that on average there have been four generations per 100 years, any individual would have 2^{40} or approximately a million million ancestors a thousand years ago, if there had been no consanguinity. It seems probable that the total population of the world in the tenth century did not exceed 200 million and it was very much smaller in yet earlier times! In considering degrees of relatedness it is therefore usual to recognize some hypothetical basal population in which it is assumed all individuals are unrelated.

It is also necessary to distinguish between genes which are 'alike in state' and genes which are 'identical by descent'. Two Ag^M genes, for instance, are alike in state, they will replace one another with identity of genetic effect, but they may either have both originated from the replication of one gene in a previous generation, in which case they can be said to be identical by descent or they may have no common origin or a common origin so distant that it can be neglected, in which case they are non-identical in terms of descent, and can be referred to as 'independent'. Genes which are unlike in state must, of course, be independent in the absence of mutation.

The genetic relationship between two individuals is expressed as the probability that each possesses a gene identical by descent at a particular locus, i.e. as the coefficient of relationship (cf. p.172). This as we have already seen is one-half between parents and offspring, and between sibs; one-quarter between grandparents and grandoffspring, and between uncle/aunt: nephew/niece; and one-eighth between first cousins.

So far as some deleterious recessive gene is concerned this means that even in a family in which no-one is affected, the cousin of a carrier has, at least, a 1 in 8 chance of being a carrier also. The reason for giving a minimum probability is because no account has been taken of this cousin being a carrier by virtue of independent genes. If the gene is rare this is a remote possibility. Nevertheless, the situation illustrates strikingly the increased likelihood of having an affected child if a carrier marries a cousin rather than an unrelated person, for the likelihood of marrying a heterozygote from the general population is $2pq$. It may be added that the probability of a cousin of an affected person being a carrier falls to at least one-quarter, since one knows with certainty that both the normal parents of the affected individual must be heterozygotes and one of these is the uncle or aunt of the cousin in question.

The coefficient of relationship is a measure of genetic similarity between pairs of individuals. What one particularly needs to know in inbreeding is the degree of similarity or relationship between pairs of genes in the same individual. This is termed the coefficient of inbreeding and is defined as the probability that an individual will inherit at a given locus two genes which are identical by descent. It is usually represented as F and for the offspring of the two completely unrelated parents is O. In general form, the inbreeding coefficient for an individual is $F = \frac{1}{2}^{n+n+1}(1 + F_Z)$, where n and n' are the number of steps in the lines of descent from the common ancestor to the parents of the individual concerned and F_Z is the coefficient of inbreeding of the common ancestor. If this ancestor can be considered as not inbred, F_Z is 0 and the expression reduces to $F = \frac{1}{2}^{n+n'+1}$. The converse of the inbreeding coefficient, i.e $1 - F$, is, of course, the probability that an individual will have at a given locus two independent genes.

In the case of first cousins, these relatives have, in their grandparents, two common ancestors. The possible lines of descent connecting them are represented in Fig. 8.1.

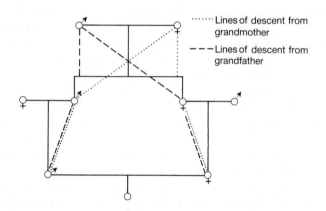

Fig. 8.1.   Pathways of common descent in first cousins.

Consider a gene present in the grandfather: the probability that it is transmitted to his son is a half, to his grandson a quarter, and thus by this route to the offspring of the cousins one-eighth. However, the same gene in the grandfather can also be transmitted to his daughter, and granddaughter to the cousins' offspring with the probability of one-eighth. Thus, the probability of it meeting itself is $\frac{1}{8} \times \frac{1}{8} = \frac{1}{64}$. There are, however, four genes that can be transmitted in this way: the one exemplified, its allele in the grandfather, and the corresponding two in the grandmother. There is thus a $4 \times \frac{1}{64} = \frac{1}{16}$ probability that a child of first-cousins will be homozygous for two genes identical by descent. Conversely, there is a $^{15}\!/_{16}$ chance that no gene meets itself. The same result is obtained by applying the above general formula. Two steps (in this case generations) separate cousins from their grandparents in both possible lines of descent, so if the grandparents themselves are not inbred, the coefficient of inbreeding of the offspring of cousins is

$$F = \tfrac{1}{2}^{2+2+1} + \tfrac{1}{2}^{2+2+1} = \tfrac{1}{16}.$$
(male common ancestor) (female common ancestor)

It may be noted that the coefficient of inbreeding for an individual is always one-half of the coefficient of relationship between his parents.

For a recessive gene whose frequency in the population is q the probability of a child with cousin parents obtaining two such genes which are identical by descent is $\tfrac{1}{16}\, q$. The probability of the gene not 'meeting itself' is $^{15}\!/_{16}\, q$, but in these cases it may come together with another of independent origin and the probability is $^{15}\!/_{16}\, q \times q$. The probability of the child being homozygous is the sum of these two probabilities, i.e. $\tfrac{1}{16}\, q +$ $^{15}\!/_{16}\, q^2$, which equals $\tfrac{1}{16}\, q(1 + 15q)$. These ideas have been explained in terms of cousin marriages and recessive genes, because cousin marriages are the most frequent types of close-relative matings in man, and because the paths of descent of recessive genes are particularly important in medical genetics, but the principles apply to all consanguineous matings and all genes. It is the laws of descent one is concerned with, not the nature or effect of genes.

The coefficient of inbreeding of a population can be represented as the mean of the coefficients of all the individuals who comprise it. Estimates of the level of inbreeding have been obtained for various groups, particularly small and isolated ones such as island populations and some religious sects. One of the highest levels found was among the Samaritans who have a mean coefficient of inbreeding of 0.0434, almost equivalent to the whole population being as inbred as the offspring of first cousins. Furthermore, it may be noted that coefficients of inbreeding tend to be minimal estimates, especially so when genealogical information is only available for a few generations, since often it is not possible to establish the extent to which common ancestors were themselves inbred.

While the offspring of consanguineous marriages are to some extent inbred, consanguineous marriages will, of course, occur in a randomly mating population. The frequency of cousin marriages in most Western European populations seems to be in the neighbourhood of 1 per cent or less. In certain populations, however, it is considerably higher as, for instance, in rural districts of northern Sweden, in Swiss alpine villages, and among Jewish communities in many German towns. Here it is not exceptional that 6 per cent of the marriages should be cousin marriages, and values as high as 12 per cent occur. It does not follow that in these populations cousin marriages are preferred, but rather that the populations are small. When the group from which a mate is chosen is small, the individuals composing the group will inevitably be more closely related than when the group is large. In particular, the frequency of first cousins is greater, and random mating will consequently lead to more consanguineous unions. In fact, when, as in some small Swiss communities, the frequency of cousin marriages is relatively low, it may be surmised that they are being deliberately avoided and that mating is, therefore, not at random.

The relationship between inbreeding and population size can be envisaged in the following way. Imagine a situation of N individuals all of whom have two alleles which are not only 'independent' of each other, but also independent of those of any other individual. The probability then of any zygote being formed containing two genes identical by descent at the locus is $\frac{1}{2}N$ (i.e. the probability of randomly picking the same gene twice). This then is the inbreeding coefficient for a single generation formed by the random union of gametes in population size N. The inbreeding coefficient increases progressively with every generation so that in generation n it equals $F_n = \frac{1}{2}N + (1 - \frac{1}{2}N)F_{n-1}$, i.e. the fixed level for a generation plus the accumulated level for all the preceding generations. When N is small the inbreeding coefficient becomes large very soon. Another way of looking at this is to consider the converse of the inbreeding coefficient: the panmixtic index. In the most outbred state where no two alleles of an individual are related $P = 1$ (and $F = 0$). As inbreeding necessarily increases in populations of limited size P declines ($P + F = 1$) and P becomes a measure of the relative amount of random mating heterozygosity which is reduced by the inbreeding causing homozygosis of genes identical by descent. Maximum heterozygosis ($2pq$) is when $P = 1$, equivalent to a generation of individuals who have no relationship with one another but happen to be under Hardy–Weinberg equilibrium. The rate of decline in P is a function of population size as is evident from the formula $P_n = P_0(1 - \frac{1}{2}N)^n$ [or in terms of F $P_n = F_n 1 - \frac{1}{2}N)^n$] and the effects of this are shown graphically in Fig. 8.2.

It has been calculated that in a population founded by 50 men and 50 women and composed of these numbers in every generation each individual in the eleventh generation would have 80 per cent of the founders in his ancestry under random mating. With 500 men and 500 women it would take only 15 generations to reach the same state of relatedness.

Assortative mating

Mating is described as assortative when there is a correlation between partners in some character. If the correlation is positive and individuals

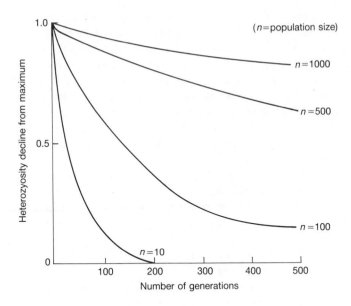

FIG. 8.2. Decline in levels of heterozygosity with time according to population size.

who resemble one another mate more frequently than would be expected by chance, the situation is one of positive assortative mating or 'homogamy'. If the correlation is negative then one speaks of negative assortative mating or disassortative mating.

There are many examples of similarities between spouses in man, but the characters involved tend to be quantitatively varying ones, typically with an environmental as well as a genetic component to their variance, such as anthropometrics like stature and weight, and psychometrics like IQ. This complicates analysis, especially if the similarity level is not established at the time of marriage. After marriage couples can and do become more similar in various characters as a result of sharing a common domestic environment including nutrition. Even phenotypic similarities at marriage may not represent any genetic correlation between the spouses though the characters involved have a heritable component to their variation. Stature typically has some genetic determination and in many societies there is a correlation between spouses of around + 0.2. However, it doesn't follow that stature is taken into account in mate selection or that the correlation relates to any aspect of the genetic component. We know that people of the same neighbourhood tend to marry one another and that neighbourhood in many human societies reflects similar socio-economic conditions. Thus, husband–wife correlations in stature could arise from nothing more than similar nutrition in childhood. It is problems like this which have inhibited research on assortative mating in man though preferential mate choice certainly occurs in our species.

The theoretical effects of assortative mating, however, are clear, and have been empirically documented in other animals and with respect to simply inherited traits with no environmental contribution. In many respects, the effects of positive assortative mating resemble those of inbreeding since the phenotypic similarity between mates tends to arise through possessing genes 'alike in state' if not identical by descent. Indeed, if the three genotypes in a diallelic system were each phenotypically distinct and there was strict assortative mating so that only identical phenotypes mated with one another, the result would be exactly the same as inbreeding under self-fertilization and a pair of homozygous lines would be established. However, whereas in inbreeding the whole genome is affected by the mating system, in the case of assortative mating only those loci controlling the characters involved in mate choice are affected. Inbreeding, particularly preferential inbreeding, has been termed 'assortative mating for ancestry'.

When the assortative marriage in man for quantitative traits is actually based upon character variation with a genetic component, the effect is to increase the variance of the trait in the population. This of course arises from the increased production of homozygotes at all the contributing loci as compared with the situation under random mating. The effect is also to increase the additive component of the total variance and to increase the correlation between parents and child (but not that between sibs because this depends only on the heterozygosity of parents which is not appreciably reduced by assortative mating if many gene loci control the character involved).

The biological effects of mating systems

The fact that continuously small populations, even when practising random mating, will tend to become inbred is of considerable importance, since it has frequently been shown that continued inbreeding usually, though not invariably, lowers the viability and fertility of many plants and animals. One possible explanation for this is that under inbreeding deleterious recessive genes, most of which are concealed in random or outbreeding populations, are brought into a state of homozygosity. It is theoretically possible that none of these genes will be fixed in some line, but if, as is usual, many loci are involved the likelihood is remote, particularly in a population which, by its very size, greatly limits the numbers of lines that can be established.

It has also been suggested that 'inbreeding depression' is directly due to the loss of heterozygosity as such and that heterozygotes, because they have two different genes at a locus, have a greater biochemical versatility than homozygotes. There is much evidence in favour of this view; nevertheless, it is apparent that organisms can become adapted to homozygosity, since many naturally reproduce by self-fertilization without any ill consequences. This suggests that the particular nature of the genes and the ways they are integrated into a balanced genotype is the fundamental basis to the fitness of organisms.

Over short periods there may be a considerable advantage in self-fertilization, particularly in immobile forms such as plants. The combined effect of inbreeding with natural selection will tend to produce eventually a population in which all individuals are homozygous for the same genes. This means that it is possible for every individual to have the most desired phenotype for prevailing environmental conditions, but it also means that the capacity of the population to meet changing environmental demands is low. On the other hand, a population with great genetic variability has a much better chance of evolutionary persistence for, although the inevitable phenotypic variability means that fewer individuals have the most desired phenotype, the population is, in a sense, pre-adapted to environmental change. One might, therefore, expect that under constant environmental conditions a self-fertilizing species would compete successfully with one which is more outbreeding, but natural environments are always changing and most organisms do not practise self-fertilization. The great evolutionary advantage of bisexuality and cross-mating is that it affords a means by which a variety of genes can be combined in a variety of ways. In these quite general terms one can see why populations are genetically heterogeneous, and more specific situations leading to the same result will be considered later. It may particularly be expected that in a population more than one gene will be available for most polygene loci, since it is the extent of heterozygosity at these loci which essentially determines the distribution of the variation of quantitative characters (p. 169). Incidentally, this distribution is the ideal compromise between the demands for a single phenotype and for variability, and it is of consequence that changing environments are particularly likely to demand changes in those characters that vary quantitatively. The real point at issue now, however, is that in outbreeding organisms the majority of individuals will be heterozygous at many of their polygene loci, and it is these individuals, rather than the rare homozygous ones, which will most closely approach the desired phenotype. The homozygotes can be considered as inevitably by-products of the way quantitative characters are determined. Furthermore, because heterozygotes are common, natural selection can operate to build up in them efficient homeostatic mechanisms. It is, therefore, not surprising that the inbreeding of organisms that are genetically constructed in this way will at first produce forms poor in vigour, for they are forms that only rarely occur naturally and cannot have the appropriate phenotype.

The general population structure of the human species throughout most of its evolution has been one of small isolates existing discretely for long periods, but intermittently broken down by population expansion, migration, invasion, and intermixture. The pattern of human evolution has to be seen in this context, with periods of differentiation in small populations interrupted by periods of gene flow. In recent times, however, the increased mobility that technological advance has given the individual has widened his mating circle and isolates are breaking down, probably forever, the world over. In many areas this has already progressed so far that anything like discrete units have vanished. However, in environments unsuitable for a high population density, and among closely knit social and religious groups, particularly when their technology is still primitive, the process has only just begun. Darlington believed that peoples of these small and long extant breeding isolates suffered no great disadvantage from the inbreeding that must have occurred, and thought that they must have become adapted, at least to some extent, to homozygosity. Consanguinity in such communities can however still be associated with reduced fitness. Sutter and George showed in two valleys in the Vosges that the offspring of first-cousin marriages showed a 23 per cent increase in mortality by comparison with the non-inbred offspring of the brothers and sisters of the cousins.

On the more positive side there is evidence that the widespread increase in stature that has occurred in Europe during the past 100 years is in part due to the breakdown of isolates. Hulse found that the offspring of marriages between Swiss cantons were taller than the offspring of endogamous marriages. It cannot be assumed that change in body dimensions are necessarily indicative of increasing fitness, but they often are. Among plants and animals that have suffered inbreeding depression, growth rate and adult body-size is typically diminished, and when such inbred forms are hybridized not only is vigour restored, but growth is also greater. The increased viability and fertility that follows hybridization of different homozygotes is known as 'hybrid vigour' or 'heterosis'. Heterosis not only occurs when closely inbred forms are hybridized, but also when some different subspecies of plants and animals are crossed. So far as viability alone is concerned it may even occur on the hybridization of good species. This phenomenon obviously supports the view that mere heterozygosity confers survival value. In these cases, however, the vigour of the second generation seems to be not only less than that of the first-generation hybrids, but is also less than the parent subspecies. Second-generation hybrids between inbred lines are less vigorous than first-generation hybrids because as a whole they are less heterozygous, but they are still more fit than their inbred parents. Just how far this sort of situation prevails after human race-crossing it is as yet impossible to say. Hybridization between the major races has been fairly common in many places. Despite various types of restriction there are large mixed groups in South Africa and the USA; and in South America and in some of the Polynesian islands, particularly Hawaii, race-crossing has been widespread. Most unprejudiced observers have not found any real evidence for breakdowns in fitness, and on Pitcairn Island, where the mutineers of the *Bounty* took their Polynesian wives, there is some evidence that heterosis occurred.

9

Population genetics — gene frequency changes

Whilst the breeding system practised within a population determines the frequency of the different genotypes, inbreeding tending to produce homozygosis, and outbreeding heterozygosis, it has no effect in itself on the frequency of genes. No genes are necessarily lost and none gained when a mating pattern changes. This is not to say that the breeding system is unimportant in evolution, but, since evolution means some degree of replacement of one set of genes by another, factors which change gene frequency are the basic evolutionary mechanisms. There are four such factors: hybridization, mutation, genetic drift, and natural selection. The first two introduce genes into a population; the last two determine what happens after they have been introduced.

Gene flow

Gene flow occurs when genes are transferred from one population to another. If we are considering spatially separated populations this clearly involves human movement.

The human species is an extremely mobile one and archaeological evidence and the occurrence of human beings over the whole land surface of the earth indicates that, although movement has increased dramatically in recent history, high levels of mobility have always characterized human kind.

Geographical mobility takes many forms, but that of genetic significance can be broadly categorized as local and long range. Local movement occurs in the daily habits of people and tends to be limited by regular return to a home base. It is typically the form which leads to choice of mate and much marital movement is especially important in determining gene flow, particularly in the past and in traditional societies. Marital movement is characterized by the extent of population exogamy, and the distances over which the exogamous unions are contracted. Figure 9.1 shows the changing pattern over historic time of the amount of village exogamy in a region of Oxfordshire and of the mean marriage distance of the exogamous marriages. Marital distance is in many societies the main component of parent/offspring distance (the distance between the birthplaces of parents and offspring) which is the most significant measure

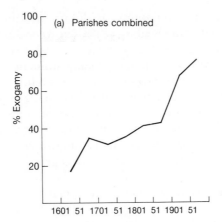

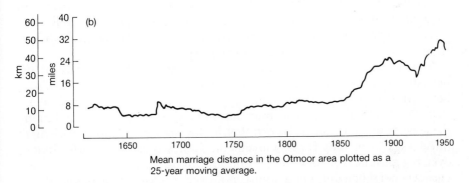

Mean marriage distance in the Otmoor area plotted as a
25-year moving average.

Fɪɢ. 9.1.   (a) Changing patterns of exogamy in 50-year periods in Otmoor parishes.
(b) Exogamy and marital distance in Oxfordshire villages.

of genetic movement. Where marriage is the sole vehicle for gene flow
parent/offspring distance equals a half of marital distance.

Long-range movement is mainly migration, where individuals or whole
groups of people move over considerable geographic distances making
new homes as they go. The movements may be progressive over a number
of years or generations, or largely take place at a single time as sea
migration has to be. In the past, long-range movement has often been into
uninhabited land, as in the various colonizations of the Americas, but it
can be at least as rapid when it is into already occupied land as, for
example, in the case of many of the westward migrations out of Asia. It is
this kind of long-range movement that one is essentially concerned with in
studying gene flow, because of the miscegenation of migrants and
sedentes.

Gene flow between populations, in all but one kind of situation, acts to make the populations involved more similar to one another. If it occurs in both directions it will eventually make populations genetically identical. Gene flow is thus the great homogenizing influence in human evolution acting to decrease between-population variability and increase within-population variability. This is why it holds such a central place in models of human population structure such as those of Sewall Wright (cf. p.182).

If two genetically different populations are geographically some distance apart, but interconnected by a series of intermediate populations through which genes can flow, there will tend to be a geographical gradient of gene frequency between the polar populations. This will only disappear when so much genetic interchange has taken place that they and all the intermediate populations are identical. Such gradients in gene frequency are known as 'genoclines', whilst the gradients in character variation they produce are known as 'phenoclines'. Clines are common in the pattern of human variation. One of the best known examples is the distribution of the blood group B gene in Europe, which progressively diminishes in frequency from east to west. This cline is usually interpreted as indicating a gene flow from Asiatic peoples, largely effected by the repeated invasions from the east to which Europe has been exposed.

Movement has been the main factor preserving the integrity of the human species. Although the geographical range increased strikingly between 40 000 and 20 000 years ago it was vast for a long time before that. While all the movement must have played some part in maintaining the species unity it can be shown that by far the most important element must have been long-range movement. Genes pass very slowly indeed over big distances if they have to depend upon local marital movement between neighbouring populations, however extensive that might be.

Exchanges between populations mean, of course, developing a common ancestry and becoming progressively more related. Relatedness between populations can be conveniently expressed in terms of the coefficient of kinship.

The coefficient of kinship of two individuals is the probability that a gene taken at random from one of the individuals is 'identical' (p.189) to a gene taken at random from the same locus in the other individual. The coefficient of kinship between sibs (and parent/child) is one-quarter since if a particular paternal gene is taken from one sib, there is a half chance that this gene was transmitted to the second sib and a half chance that of the second sib's two genes, a paternal gene is randomly taken. The coefficient of kinship between uncle/aunt: nephew/niece is one-eighth and between first cousins one-sixteenth. It will be noted that the coefficient of kinship is one-half of the coefficient of relationship. It is also the same as the coefficient of inbreeding of the offspring of the pair of individuals were they able to have children, i.e. be of opposite sex. Indeed, formation and combination of sperm and egg can be thought of as actually representing the random take.

For technical reasons the coefficient of kinship is a particularly good way to measure relationship and can clearly be used as an average over many pairs of individuals to represent levels of common ancestry between populations or such phenomena as declining relatedness with geographic distance.

Intermixture

The change in gene frequency in a population following intermixture with a second population clearly depends upon the differences in gene frequency between the two populations and the extent of the intermixture.

This has been used for estimating the amount of intermixture that has occurred in hybrid populations. The frequency of a gene in such a population will be intermediate between the frequencies in the parental groups and in proportion to the extent of the mixture. Thus, if q_1 is the frequency of a gene in one parental group, q_2 its frequency in the second parental group, and q_k its frequency in a hybrid group formed by the intermixture of the two parental ones the contribution of a parent group 1 is $(q_k - q_2)/(q_1 - q_2)$ and of parent group 2 is $(q_k - q_1)/(q_1 - q_2)$ or

$$1 - \frac{q_k - q_2}{q_1 - q_2}.$$

Using such an approach it has been estimated that the amount of white admixture in some American black populations is of the order of 25 per cent and in some Brazilian black groups about 40 per cent.

Very rarely do two human populations, on being brought into contact, form a single new panmictic unit immediately. This is particularly true if the populations differ greatly in their physical features and social customs, when some sort of restriction is invariably imposed against random intermarriage. For a time, then, different groups may live together in the same locality yet remain, in varying degrees, genetically isolated from one another. Inevitably, however, there is always some intermixture which can be viewed as a process of gene flow from one population to the other. The rate of the introgression can be calculated from the equation

$$(1 - m)^k = \frac{q_k - q_2}{q_1 - q_2}$$

where the right-hand side of the expression represents, as has been shown, the amount of admixture to date, and k is the number of generations over which gene flow has occurred. The modal rate of gene flow m is the fraction of genes in the mixed population with the gene frequency q^2.

Using this method, it has been calculated that the modal rate of gene flow from the white population to a black population in the USA is around 0.02 (i.e. 2 per cent) and in Brazil around 0.05 (5 per cent).

There are a number of points to be noted and problems to be overcome in making calculations of this kind. Clearly, the most reliable genetic systems to employ are those which show maximal contrast between the parental populations. In analysing hybrids of mixed European and African descent, systems like abnormal haemoglobin and rhesus are most useful. However, they can and do give rather different intermixture estimates. The most probable explanation is that the various systems have been exposed to varying régimes of selection in the hybrid group. It has been surmised that the gene for sickle-cell haemoglobin has been selected against in the non-malarious environments of North America thereby providing a lower estimate of the amount of African ancestry in American black populations than that offered by the Rhesus *cDe* haptotype which it is presumed has not been subject to selection since the intermixture occurred. It must also be remembered that considerable error can arise in determining gene frequencies for the parental group. Often it is difficult to establish what precise populations actually contributed to a hybrid group and how far their modern descendants may themselves have changed.

Mendelian populations are usually thought of in some sort of geographical context. However, more than one can occupy essentially the same space. This of course happens when there are social or religious barriers to marriage and these are widespread in many human societies. In Britain there are typically representatives of different social classes in any one place. Social class is mainly defined by occupation, with people from the professions in Class I, and unskilled manual workers in Classes IV and V. However, social class defines a whole life style as well as occupation and people do not marry randomly across the class divisions. The different classes need therefore to be viewed as different Mendelian populations. Marriage between spouses of different class is then to be seen as a vehicle of gene flow. There is in this case another vehicle of gene flow: social mobility. This is the phenomenon of an offspring moving from the social class of its parents into another class as a consequence of obtaining a different occupation through education or the like.

The effects of these two movements can be analysed separately or together in terms of their homogenizing effects over time as they act to disperse genes evenly through the sets of populations involved. Formally, this is done through constructing a migration matrix of observed levels of exchange in which all the possible channels of flow between every pair of populations are identified and magnitude of the flow multiplied up generation by generation. The same, of course, can be done in examining the exchanges between groups of spatial populations like villages, as an alternative to calculating coefficients of kinship.

Selective movement

Movement through social classes can be thought of as vertical flow since the classes form a hierarchy. This kind of flow highlights an important phenomena about the genetics of migration. As already pointed out we usually see movement as a factor acting to reduce genetic variation between populations. This, however, is only the case when the gene flow is random with respect to the genetic variety. If the probability of movement actually depends upon the genes an individual possesses then flow will tend to cause diversification rather than uniformity. Although there is some evidence that geographical migrants may not be fully representative of the populations from which they come in a number of morphological features, it is usual to assume that spatial migration is not associated with any particular genotypes. In the case of vertical movement, however, selectivity for a number of characters, particularly ones of behaviour, ability, and personality, is clearly very strong. Variation in these characters is likely to have some heritable basis and so we would expect social classes to become stratified for the relevant genes as a result of social mobility.

Mutation

Hybridization only forms a means by which genes already present in some populations within a species are introduced into other populations. It may result in genetic combinations which are entirely new, but the sole source of new genes is mutation (cf. p.158). Here we are mainly concerned with the rate at which mutation occurs and particularly the rate of point mutation.

Mutation at a particular locus is a rare phenomenon: of the general order of about 1 mutation per 100 000–1 000 000 or more germ cells. It varies from locus to locus, but is more or less constant for any one locus.

Determining the rate of mutation is in principle very easy for a condition which is recognizable in heterozygotes. Here one merely adds up the cases where an offspring displays a condition not present in the parents checking, of course, through other genetic systems, that illegitimacy is not likely to be responsible. A well-known case is the determination of the mutation rate for achondroplasic dwarfism (a dominant condition) in Copenhagen. Here, among 94 075 births 10 were suffering from achondroplasia. Two of these, however, were born of parents one of whom also had the condition, so the frequency of new cases was 8 in 94 075. This reduces approximately to 1:12 000 and since every individual possesses two alleles at the locus the mutation rate of a normal gene to an achondroplasic one is 1:24 000 — an unusually high rate.

This is the so-called direct method of establishing mutation rates. It obviously cannot be applied to a recessive character since, when mutation produces a gene which is recessive in effect, this gene cannot be recognized for counting. It will not have an opportunity of expressing itself until it comes into combination with another of the same type and that may be many generations after it first arose.

There is another way of determining mutation rates known as the 'indirect method' which can be applied to recessive as well as dominant conditions as long as these conditions are deleterious ones, and the responsible genes are selected against. Most fully recessive conditions are deleterious, since natural selection itself will tend to act to conceal disadvantages expressed in single dosages, i.e. in heterozygous states.

The indirect method depends upon the fact that if genes are constantly being eliminated by selection they are only present at all in a population as a result of recurrent mutation. Furthermore, given constancy of conditions, an equilibrium state is established in which the mutation rate equals the rate of loss through selection. This is, of course, because the rate of input determines the rate of output, rather like a bath in which loss of water through the plug hole is determined by the flow through the tap. In the equilibrium state then, the mutation rate μ for a recessive condition equals sq^2 (cf. p.210) where q^2 is the frequency of the condition and s is the selection coefficient acting upon it. If q^2 can be represented by the frequency of the condition at birth, b, then $sb = \mu$ and $b = \mu/s$. For a dominant character where the selection is essentially against the heterozygote, $u = ps$ [the loss of an Aa genotype is approximately $2ps$ since q is so near to 1 and in such an individual half the genes are A so $\frac{1}{2}(2ps) = ps$] and $b = 2\mu$. For a recessive sex-linked condition where almost all the selection is against hemizygous males $3u = sq$ since one-third of the population's X-chromosomes are in the males.

If the mutation rate is known then these formulae of course provide the equilibrium gene frequency so that for an autosomal recessive $q_e = \sqrt{(\mu/s)}$ and for an autosomal dominant $p_e = V/s$.

There are two main theoretical problems in the indirect method. Firstly, and particularly related to recessive conditions, is the assumption that the heterozygote has the same fitness as the normal homozygote. If the heterozygote is in any way favoured by natural selection then one has the situation of a balanced polymorphism (cf. p.212) and equilibrium is established through this, rather than through an inflow/outflow system.

Even more generally troublesome is the assumption in the method of an equilibrium having been established between mutation rate and selective loss. It takes many generations for such an equilibrium to be established and, whilst environmental changes have probably not much affected the mutation rate, they have clearly altered the fitnesses of genotypes, especially through recent medical advances.

Mutation alone will change the frequency of genes in populations very slowly. Although only strictly true for the generation following the one in which the mutant gene does not occur at all, the change in frequency from one generation to the next is essentially the mutation rate μ. In fact the frequency of a gene which is mutating to another will diminish according to the formula $p_n = p_0 (1 - \mu)^n$ where p_0 is the original frequency and p_n its frequency after n generations with a mutation rate μ. When the mutant gene is rare the number of generations (n) it needs to raise its frequency from q_0 to q_n equals

$$\frac{q_n - q_0}{\mu}.$$

This will generally be a very long time and does not take into account a number of other factors which can slow it down including the phenomenon of reverse or back mutation.

In the latter situation an equilibrium is eventually established which is set by the ratio of the two mutation rates v/μ where v is the back mutation rate. The frequency of the new mutant gene q is then $\mu/(\mu + v)$ and of the old ancestral gene $p = v/(\mu + v)$. The rate of increase in frequency of the mutant gene over n generations $= \log_e (q_0 - \hat{q})/(q_n - \hat{q})$ where q_0 and q_n represent the original and final frequencies and $\hat{q}$ the equilibrium frequency. For a gene to spread in any reasonable time other factors than mutation must operate on it.

Genetic drift

One way in which gene frequencies may change relatively quickly is by chance or genetic drift. The possibilities of chance acting as an evolutionary factor have been mathematically developed by Sewall Wright and drift is often referred to as the Wright effect.

As previously mentioned a species is usually broken up from the breeding point of view into many small and more or less isolated units such as in the 'island model'. The production of a generation of offspring by a generation of parents inevitably involves some sampling of the parents' gametes and the smaller the population the smaller the sample. In considering the Hardy–Weinberg Law it was assumed that the gametes carrying one or other of the two alleles A and A' were not only produced in the frequency p and q, but also combined in zygotes in this proportion too. The number of gametes produced is always large, especially in males, but when populations are small the number combined in zygotes is by definition small and substantial deviation from expected proportions can occur solely by chance. Thus, starting with two alleles at 50:50 frequency one is almost bound to get some deviation in even the first generation with small samples and, whilst in the second frequencies can move back towards the original state, they may well deviate even further. It is thus evident that a closed system which is small will eventually run to the fixation of one gene or the other. Once this has happened only mutation can restore the lost allele. It is, of course, also only a matter of chance which allele is fixed and which one is lost, so starting with a 50:50 situation and a large number of identical populations, approximately a half would eventually have one allele at 100 per cent and the other half the other allele. The length of time to fixation, however, would vary from one population to another.

The effects of varying population size on these processes is shown in Fig. 9.2 derived from Wright. The diagram indicates the proportion of populations which at equilibrium will have the specified frequency q of an allele, from a starting point in a two-allele system of $q = 0.5$. Because one is projecting to an equilibrium   state one needs to assume in the calculations a small amount of immigration in each generation. It is evident that when population size is large, most of the populations still have a gene frequency at or close to 0.5. With sizes of 5 000, however, all possible gene frequencies are equally likely and with sizes of 1 000 most of the populations are close to 100 per cent for one gene or the other.

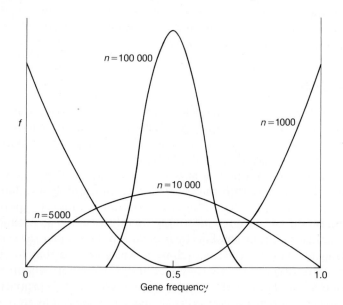

Fig. 9.2.   Effect of varying population size on genetic drift. Distribution of the frequencies of different populations at equilibrium by gene frequency and population size. Assumes very slight migration. After Sewall Wright (f = frequency of populations with a particular gene frequency starting from a situation where all populations had a gene frequency of 0.5).

For purposes of considering population size effects one needs to take account of the fact that real populations contain representatives of a number of generations, individuals who do not reproduce, and chance variability in family size, since if a few marriages produce many offspring and others only a few there is likely to be less genetic heterogeneity among the offspring than if the same total number is produced by all the different marriages making equal contributions. These factors are taken into account in determining what is called 'the effective population size', N_e. Where the average family size is two this can be expressed mathematically as $N_e = (4N' - 2)/(2\sigma_k^2)$, where N' is the number of individuals in a generation who are parents (and generations do not overlap) and σ_k^2 is the variance in the number of gametes contributed by the parents.

So far as rate of fixation is concerned Wright has shown that under drift this occurs in a two-allele system according to the formula $L_T = L_o e^{-T/2N_e}$ where L_o and L_T are the numbers of unfixed genes in the initial and T generations, respectively, and e is the base of natural logarithms. If, therefore, in a population whose effective size is 100 there are a number of loci occupied by pairs of alleles in equal frequency, chance will produce fixation at a half of these loci in 139 generations and at a half of the remainder in another 139 generations as can be seen by letting $L_T/L_o = 0.5$ in the above expression and solving for $T(L_T/L_o = 0.5 = e^{-T/200})$.

While the population structure has been ideal for genetic drift to operate throughout most of hominid evolution, it may still be questioned whether drift has contributed substantially to the processes of human differentiation or the differentiation of any species for that matter. Throughout the foregoing discussion it has been assumed that the two or more alleles at a locus were more or less of equal fitness. Genes will only drift when they are near enough neutral and may be replaced by one another with little or no effect on an organism's survival and reproduction. We will return to the important question of how often genes are neutral in a later section. What can be said here is that allowance for the chance or stochastic nature of many population phenomena has greatly increased the sophistication of our evolutionary thinking, especially in the development of mathematical models for understanding the genetic dynamics in small human groups. There can also be very little doubt that the kind of variation that is observed in blood groups and the like between groups of rather isolated and small communities, such as have been studied among South American Indians and in mountain villages in Europe, is mainly due to drift.

There are two other circumstances where chance is an important factor in the genetics of populations. In the first place it can profoundly affect the fate of new mutant genes. Even a highly advantageous gene may arise many times before it becomes established, because being at first so rare it is easily lost by chance. It is highly likely that it will not even find itself in a zygote. Conversely, quite disadvantageous mutant genes may for a time reach quite high frequencies in small populations if 'favoured' by chance.

Secondly, it is possible that a gene will by chance be absent in a small group migrant from it or in such low frequency that it is easily lost subsequently. Obviously, all the descendants of the group, which may be many, will then lack the gene, and the gene complex will have to become adapted to its absence, unless it is reintroduced by mutation. This phenomenon is known as the 'founder principle' and may have played a very significance role in determining geographical patterns of human variety, especially when small migrant groups have occupied new and uninhabited territory. Some of the variation between Polynesian islands is more than likely due to founder effects and it has been suggested, though it is rather unlikely, that the absence of the blood group B gene in most aboriginal American populations can also be explained in this way.

Natural selection

We now come to the fourth and last, and many believe the main, determinant of what genes are in a population and at what frequency, and, thus, of the evolutionary process itself. It is important to recognize at the outset, however, that natural selection can be as much responsible for the stability of the genetic composition of a population as for evolutionary change.

Natural selection is differential viability and or fertility, according to genetic constitution. In other words it is genetically determined variation in Darwinian fitness. It acts essentially on individuals and their gametes and on the totality of their phenotypes, since viability and fertility are properties of whole individuals. Only indirectly does it act on genes and genotypes. If, however, an individual possesses some character which confers a great viability or fertility on him compared with other individuals in the population, then the genes controlling this character will be preferentially represented in the next generation in the greater than average number of offspring of the individual concerned. We then speak of the individual having a high Darwinian fitness, of the character being adaptive, and of selection being in favour of the responsible genes or gene combinations. The gene can be thought of as the unit of evolution, but the unit of selection is the individual.

Individual fitness is a comparative phenomenon. While some genotypes produce phenotypes which are invariably lethal prior to the age of reproduction and Darwinian fitness is therefore always zero, the most important feature generally of the fitness concept is how it relates or ranks members of a population to one another. Thus, the absolute number of viable offspring an individual produces is only relevant in so far as it determines whether he leaves more or less of his genes in the next generation than other individuals.

One also needs to consider simultaneously the environment, for an individual who is relatively fit in one environment may be relatively unfit in another; a character which is adaptive in one environment may be maladaptive in another; and a gene which is selected for in one environment may be selected against in another. Environments exert the selective forces and pressures and through this organisms become adapted to the environments in which they live; in other words, to the environments that produced them. However, organisms also influence and change their own environments so we must expect to find co-evolution of genotypes and environments, and it must not be forgotten here that, in addition to the role of the environment as a selection agent, it also acts as an ontogenetic agent developmentally determining phenotypes which it may then select. Here, however, we shall focus on the effects of selection on gene frequency changes.

Normalizing selection

Selection can act in many different ways. One important way is to maintain the status quo and to remove unfavourable variety. As has been already discussed many mutant genes are disadvantageous and are selected against. Their presence in a population is solely due to recurrent mutation. Genes of medical importance tend to be in this category. Let us first consider such a situation where the mutant gene is dominant or at least expresses itself in heterozygotes in such a way that these never reproduce. Clearly, the frequency of such a condition in a generation depends solely on the mutation rate. By definition it is never transmitted. The situation is a little more complex for a recessive lethal condition, since selection here can only operate on the homozygote and a recessive mutant gene may exist in a population for many generations before it occurs in homozygous state. If such a gene a has a frequency q, and its normal allele A a frequency p, the three genotypes according to the Hardy–Weinberg Law will be produced in the frequency p^2AA, $2pqAa$, and q^2aa. After selection has been removed from the population all the homozygotes aa, the frequency of the homozygotes AA is $p^2/(p^2 + 2pq)$, i.e. $(p/p + 2q)$, and of the heterozygotes Aa $2pq(p^2 + 2pq)$, i.e. $2q/(p + 2q)$ (i.e. their original frequency divided by the total of the new population).

Substitute $1 - q$ for p in $p/(p + 2q)$ and the frequency can be expressed as $(1 - q)/(1 + q)$. Likewise, the frequency of Aa can be expressed as $2q/(1 + q)$. All the a genes remaining in the population occur in the heterozygotes, a half of whose genes are of this type. Consequently, the new frequency of the a gene is $q/1 + q)$. The frequency of the other allele A is $1 - q/(1 + q)$, which is $1/(1 + q)$. Under random mating, therefore, the frequency of the three genotypes in the next generation is $(1/1 + q)^2AA$, $2(1/1 + q)(q/1 + q)Aa$, and $(q/1 + q)^2aa$.

The relationship between the frequency of gene a from one generation to the next, i.e. q to $q/(1 + q)$, is a step in the harmonic series whose general term is $q_n = q_0/(1 + nq_0)$, where q_0 is the initial frequency of the gene, n is the number of generations of selection, and q_n is the frequency of the gene after n generations of selection.

Use of this formula clearly demonstrates the slow rate at which a recessive gene in low frequency is eliminated from a population even when its effects are as severe as they can be. It also demonstrates the inefficacy of the most severe eugenic measures to rid a population of some undesirable recessive trait. As an example, consider the effects of sterilizing before reproductive age all individuals who possess some rare recessive abnormality in each of 25 generations. If the initial frequency of the trait is 1/10 000, the frequency of the gene will be 1/100, i.e. 0.01. Substituting in the above formula one obtains

$$q_{25} = \frac{1/100}{1 + 25(1/100)} = \frac{1}{125}.$$

In other words, after 25 generations of such selection the frequency of the gene has been reduced to four-fifths of its original frequency!

Rarely does a genotype have zero fitness and partial selection usually operates. The relative fitness of genotypes can be expressed as the fraction of an offspring each produces for every one offspring produced by the genotype with the greatest fitness. The extent by which the less-fit genotypes deviate from unit fitness is known as the 'selection coefficient' and represented by the symbol s. In the case just considered, where the recessive homozygote has zero fitness $s = 1$. When character differences are of neutral survival value $s = 0$, and the gene frequencies remain the same from generation to generation. Sometimes in comparing genotypes (or genes) the complement of the selection coefficient, the fitness coefficient f, is used. $f = (1 - s)$ and represents the relative transmission rather than the relative elimination of genotypes.

The effects of partial selection can again be seen in terms of a dominant gene being favoured over its recessive allele, with the heterozygote having the same fitness as the homozygote AA. This situation is presented in Table 9.1.

TABLE 9.1

Genotypes	AA	Aa	aa		Total
Frequency before selection	p^2	$2pq$	q^2		1
Relative fitness	1	1	1	s	
Frequency after selection	p^2	$2pq$	$q^2(1$ | $s)$	1	sq^2

(The frequency of recessive homozygotes after selection can also, of course, be represented as q^2 sq^2, which shows why the total after selection is $1 - sq^2$ since $p^2 | 2pq | q^2 = 1$.).

After selection the gene a exists in the surviving homozygotes and in the heterozygotes, so its frequency is $pq + q^2(1 - s)/(1 - sq^2)$. If, again, $(1 - q)$ is substituted for p this expression becomes $(1 - q)q + q^2(1 - s)/(1 - sq^2)$ and this simplifies to $q(1 - sq)/(1 - sq^2)$. The change in the frequency of gene a from before to after selection Δq is therefore the difference $q - q(1 - sq/(1 - sq^2)$. It equals $q(1 - sq^2 - q(1 - sq)/(1 - sq^2)$, which simplifies to $aq^2(1 - q)/(1 - sq^2)$. This decrement in the frequency of a is, of course, an increment in the other allele A, so substituting $(1 - p)$ for q in the above expression one gets for the change in p, i.e. $\Delta p = s(1 - p)^2p/1 - s1 - p)^2$.

Use of the above equation further demonstrates how ineffective selection is at removing a rare recessive gene from a population. On the other hand, if the gene is common and homozygotes therefore frequent, even quite moderate selection coefficients can have marked effects. If, for example, q is 0.5 in a population and s only 0.1 the calculation shows that the decrement in the frequency of gene a in the population is 0.0128 in but a single generation.

Directional selection

This conclusion is of some importance when we come to consider the kind of selection which produces evolutionary change — directional selection. This is simply the opposite of normalizing selection with selection favouring mutant genes rather than eliminating them. The difference, of course, is that the mutant genes passes through every frequency from being extremely rare to becoming fixed as it replaces the previous allele, but what is seen from one viewpoint as selection in favour of a new gene can also be seen as selection against its allele. Thus, the equations provided for normalizing selection are equally applicable to directional selection.

Given a fixed level of selection, the length of time required for a dominant gene to replace completely a recessive one is, in essence, exactly the same as the time required for a recessive gene to replace a dominant one. However, the pattern of changing gene frequency is very different. Because it obtains phenotypic expression in heterozygotes a dominant gene can be selected for quickly when it is rare; as we have already seen the response to selection of a rare recessive gene is slow. However, as the dominant gene approaches fixation it is replacing a rare recessive so the final stages of change are very slow. Conversely, once a recessive has become common the selection is against a rare dominant and the final stages to fixation are very rapid.

We do not know with any certainty of genes that are currently spreading in human populations, except for some which are no longer disadvantageous as a result of medical treatment. A few of these may now have fitnesses greater than their normal alleles. Detecting progressive change in gene frequencies with time is a difficult business for a species with a long generation interval like man and this is the only unequivocal evidence for directional selection. Unfortunately, apart from abnormalities no major gene locus is recognizable in skeletal remains which would provide the best evidence. Skeletal variation does show change with time, including recent time, and some of this is almost certainly due to genetic change, such as increasing round-headedness, but all of this variation is multifactorial in origin. Indirect evidence also suggests that the genes for lactase tolerance have been changing in frequency in some populations since animal domestication and it could well be that the haptoglobin Hp^2 gene is also spreading. However, it is doubtful now if we shall ever witness again natural environments (i.e. non-man-made ones) causing directional natural selection.

Balancing selection and polymorphism

While a major gene is spreading through a population at the expense of an allele the character determined by the locus will exist in at least two and possibly three forms. This is an example of a polymorphic situation. Ford has defined polymorphism as 'the occurrence together in the same habitat of two or more discontinuous forms of a species in such proportions that the rarest of them cannot be maintained merely by recurrent mutation'. The definition is phrased to exclude: (1) differences between one population and another which is known as 'polytypism'; (2) continuous variation which, though it may often have a basically similar causation, is a directly contrasting form of phenotypic variation; (3) forms which are present solely as a consequence of mutation.

The polymorphism occurring while gene replacement is in process is transient, since as soon as the favoured allele is fixed the population becomes monomorphic for the new character. Many characters, however, in a population are more or less permanently polymorphic. One of the most obvious examples of such a 'stable polymorphism' is sex. Others that occur in many human populations, affect the blood-group systems, haemoglobins, red cell enzymes, and serum proteins. Unless affected by hybridization the frequencies of the genes determining the variation in these characters remain more or less constant over long periods of time: the antiquity of some blood-group polymorphisms being strongly indicated by their presence in apes.

Drift with low levels of gene flow can produce such stable polymorphisms, particularly if large increases in population size subsequently occur; these will hold gene frequencies constant deterministically. However, a stable polymorphism can also arise because of the opposition or balancing of selective forces, as was first recognized by Fisher, Ford, and Haldane. The polymorphism is then referred to as a balanced polymorphism. Balancing selective forces may act in a variety of ways.

The most important situation is that in which one allele of a pair is relatively advantageous in effect when in low frequency, but becomes relatively disadvantageous when in high frequency. This will happen automatically if the heterozygote is more fit than either homozygote. This is apparent from the fact that if selection is favouring heterozygotes it must maintain both alleles in the population, but can best be considered in detail by attributing the selection coefficients s_1 and s_2 to the two homozygotes in Table 9.2.

TABLE 9.2

Genotypes	AA	Aa	aa	Total		
Frequency before selection	p^2	$2pq$	q^2	1		
Relative fitness	$(1-s_1)$	1	$(1-s_2)$			
Frequency after selection	$p^2(1-s_1)$	$2pq$	$q^2(1-s_2)$	1	s_1p^2	s^2q^2

The frequency of gene a after selection is

$$\frac{pq + q^2(1 - s_2)}{1 - s_1p^2 - s_2q^2} \text{ , which equals } \frac{q(1 - s_2q)}{1 - s_1p^2 - s_2q^2} .$$

The change in its frequency as a result of selection is therefore $q - \{q(1 - s_2q)/ (1 - s_1p^2 - s_2q^2)\}$. The following steps in the simplification of this expression to

$$\Delta q = \frac{pq(s_2q - s_1p)}{1 - s_1p^2 - s_2q^2}$$

may prove helpful,

$$\Delta q = q - \frac{q(1 - s_2q)}{1 - s_1p^2 - s_2q^2} = \frac{q(1 - s_1p^2 - s_2q^2 - 1 + s^2q)}{1 - s_1p^2 - s_2q^2}$$

$$= \frac{q\{ - s_1p^2 + s_2q(1 - q)\}}{1 - s_1p^2 - s_2q^2}$$

$$[\text{substituting } p \text{ for } (1 - q)] = \frac{pq(s_2q - s_1p)}{1 - s_1p^2 - s_2q^2} .$$

It will be noted that if s_2q is larger than s_1p the expression is positive, and the gene a has increased in frequency relative to its allele, whereas if s_2q is smaller than s_1p the gene has decreased in frequency. In whichever direction the frequency has changed it will continue to change until $s_2q = s_1p$, after which no further change in frequency can occur. The system then is in balanced equilibrium. The frequency of the gene at equilibrium can be obtained thus: $(s_2q = s_1p) = \{s_2q = s_1 1(1 - q)\} = \{q = s_1/(s_1 + s_2)\}$ and the frequency of the other allele A at equilibrium is likewise $p = s_2/(s_1 + s_2)$. It will be seen, therefore, that when heterozygotes are at an advantage the frequency at equilibrium of the genes in a two-allele system will be determined by the relative fitness of the homozygotes.

The most outstanding example of a human polymorphism being maintained by heterozygote advantage is in the abnormal haemoglobins and especially in the sickle-cell case where the heterozygote is protected in early life against falciparum malaria.

Other situations, besides heterozygote advantage, can be envisaged in which there is an opposition of selective forces. A genotype, for instance, may be at a relative advantage when it is comparatively rare in a population but could come to have a disadvantage if it were common. In other words the nature of the selection on a gene may depend upon the frequency of that gene in the population. This is known as 'frequency-dependent selection'. Somewhat related to it is the situation in which the environment of a population is itself heterogeneous and different parts of it favour different genotypes. Genetic variability thus permits the more effective exploitation of the various ecological niches encompassed by the distribution of the population. This form of selection is known as diversifying selection or disruptive selection and has probably been very important in producing polymorphic variety in at least some organisms. How far it could operate in the case of a species like man with high individual mobility and behavioural plasticity is a matter of conjecture.

Drift versus selection

One of the most important and controversial issues in population genetics is concerned with the relative importance of genetic drift and natural selection in determining evolutionary change. The key question at stake is whether the immense genetic variety which is observable in populations of all species is inconsequential to survival and reproduction (i.e. is neutral), in which case drift will be the main determinant, or whether most gene substitutions do affect fitness, in which case natural selection is the main evolutionary force. The arguments over this issue have been intense during the past half-century and are little nearer resolution though some would say that the drift case has become progressively stronger. Drift by its very nature cannot be positively demonstrated. To do this it would be necessary to show that selection has definitely *not* operated, which is impossible. Much indirect evidence has been obtained, however, which purports to favour the drift position. This comes from a variety of sources.

Firstly, and in many ways most persuasively is the molecular and biochemical evidence. In summary, this indicates that for any one category of protein, evolutionary change appears to have occurred at a more or less constant rate over a great range of organisms and phyletic lineages. Thus, for example, it has been estimated by King and Jukes that in the evolution of all the groups of mammals whose amino acid sequences have been established the rate of substitutions per codon per year for cytochrome c was around 4.2×10^{-10}. The comparable figure for β-haemoglobin is 10.3×10^{-10} and for fibrinopeptide A 42.9×10^{-10}. Now whilst mutation and drift would be expected to produce more or less constant evolutionary rates, directional selection is much more likely to be very different between lineages and to operate in fits and starts within lineages. The environments of all organisms do not change at a constant rate!

So far as variety within species is concerned it has been shown that in many local situations, such as groups of Italian mountain villages, the distribution of blood groups and the like is completely compatible with a driftist explanation and therefore implies that the genes concerned are essentially neutral. Indeed one can 'explain' the whole of serological variety within the total human species on such a model including the definition of major geographical groups or races. Then biochemically it has been noted that many genic substitutions could not possibly have functional effects. After all, because the genetic code is redundant, some base substitutions in the DNA have no effect whatsoever on protein composition. In this connection it has been shown that the third base in the triplet code of DNA is substantially more variable than the first and second. It is mainly in this base that the redundancy lies. Furthermore, there appear to be sections of polypeptide chains in which amino-acid sequence is quite inconsequential to structure and function and, in general, functional differences between alternative enzymes have not been demonstrated. One might also mention here, although the evidence points both ways, that amino-acid substitutions do not occur equally frequently along the whole length of the polypeptides that occur in nature. Typically, there are regions with great variety and others which are more or less constant, at least within a single species. This is usually explained as selection holding the functionally important parts of the polypeptide constant, but having no effect in those regions whose only role is to hold the molecule together.

A totally different line of evidence for neutral genes comes from population genetics itself. As Haldane first recognized, assigning absolute or invariable fitnesses to alternative genotypes leads to an impossibly high genetic load. Load refers to the number of disadvantageous genes in a population. It can have two sources, mutational and segregational. Mutational load would be very high over all loci if it is assumed that most mutant genes reduce fitness in heterozygotes if dominant and homozygotes if recessive. Even more important is the segregational load if all balanced polymorphisms are maintained like the sickle-cell haemoglobin one and the model of balanced selection through heterozygote advantage. Clearly, acting over a large number of loci such a system would mean that essentially everybody in a population would be bound to be carrying disadvantageous gene combinations at a number of loci. This situation has been referred to as Haldane's dilemma, since Haldane was a strong believer in the evolutionary role of natural selection.

There is, of course, a selectionist answer to most of these points. First, constancy of molecular evolutionary rates is to some extent illusory and comes from averaging rates over long periods of evolutionary time. Different categories of protein clearly evolve at very different rates. Then failure to detect functional differences between proteins, certainly does not mean that they do not exist, especially *in vivo*. Furthermore, it may be necessary to view variety within the framework of whole body development and homeostasis, and the supply and costs of the ingredients of these processes. Two motor cars may appear identical in their structure and performance, but may have involved very different costs in their production. Haldane's dilemma particularly over segregational load arises in no small part from focusing directly on individual gene loci rather than on the real unit of selection — the individual. In the latter we are concerned with the comparative fitness of whole genomes and arrays of genes rather than single loci. However, perhaps even more important is the ascribing of absolute fitnesses. In the case of sickle-cell haemoglobin and malaria it is true that while we are concerned with how fit the different genotypes are with respect to one another, the fitness of one genotype does not affect that of another. In many situations, however, this is not the case and the selection operating on an individual may very well depend upon what other individuals are present. Comparative work capacity rather than absolute work capacity is likely to be the deteminant of socio-economic benefits and all the reproductive benefits or penalties that come from that. Under such circumstances the whole question of absolute load becomes meaningless.

The positive case for natural selection is more diffuse. One would not generally expect random changes through mutation to have no effect on highly co-adapted gene complexes. Clearly, most morphological and physiological evolution has been directed by natural selection. Why should it be different at the molecular and biochemical levels which, after all, determine the former? The same argument applies to within species. There is evidence that a number of morphometric traits are subject to natural selection in modern human populations and at the genetic level it is hard to see a fundamental distinction between such characters and biochemical ones.

What, of course, the selectionists need are many good empirical demonstrations of selection operating and these are embarrassingly few. There are, as we have noted, many examples of selection against mutant genes that cause easily recognizable morbidity. The evidence is also strong that polymorphic abnormal haemoglobin variation is due to falciparum malaria selection. Glucose-6-phosphate dehydrogenase deficiency is probably likewise related, as are features of the Duffy blood-group system to vivax malaria protection. However, apart from these systems, and possibly one or two others, evidence for polymorphisms being maintained by selection is largely non-existent. That is not to say that variety in some systems has not been associated with disease. In quite a number of cases it has. The ABO blood-group system especially has been shown to be significantly related with a large number of morbidities (p. 272) and until very recently strong selection is known to have been operating in the Rhesus blood-group system through maternal–foetal incompatibility. However, in these and other cases the selection is either operating too late in life to affect Darwinian fitness, or it is not operating with sufficient intensity or in the right direction to maintain the balanced polymorphisms.

The lack of success in detecting the functional significance of much human polymorphic variety, raises the question of what is needed to detect selection assuming it is operating. Evidence for natural selection has been sought in a number of ways and from a variety of sources. The best evidence is to show variations in fertility and/or mortality during the pre-reproductive and reproductive years which are unambiguously dependent upon genetic differences. This is extremely difficult to do when selection coefficients are moderate. It is only when they are very strong, as for many rare mutant conditions or cases like the sickle-cell haemoglobin system that it is possible. Here it must be remembered that most of the selection that produced any genetic variety would have done so in traditional societies mainly exposed to the rigours of truly natural environments. Such societies are rather rare nowadays, are invariably small and therefore incapable of demonstrating anything but the strongest selection, and anyway are very poorly studied from this point of view. The selection pressures, if any, exerted in large modern societies will be mainly new; they have usually been sought among hospital patients and people tend not to consult their doctors for high fertility!

Clearly, a prerequisite for natural selection is fertility and mortality variance, and the magnitudes of these variances have been used for calculating the so-called 'index of selection' which is a measure of the potential or opportunity for selection. The index for fertility is given by $I_f = V_f / \bar{x}_s^2$ where V_f is the variance in fertility and $\bar{x}_s^2$ is the mean number of offspring/parent surviving to reproductive age and the index for mortality is $I_m = V_m / \bar{x}^2$ where V_m is the variance in mortality and $\bar{x}_s^2$ the mean number of offspring/parent. Comparison of these indices for different societies shows that there is little 'room' for postnatal mortality selection to act in modern industrial societies, as compared with traditional ones, but the opportunities for selection through fertility variance is at least as high. Many people in modern societies have no offspring at all. However, it must be emphasized that because there is an opportunity for selection to act does not mean that selection actually is acting!

Another method which has been employed to detect selection is to search for systematic changes in gene or genotype frequencies, either from generation to generation, or more often, since it is more feasible, between different stages of the life-cycle. Were selection operating mainly in a particular age range, for example, one should find genotype frequency differences between the young and the old, with the latter out of Hardy–Weinberg equilibrium. Such approaches, however, have little resolving power, and enormous sample sizes are required to detect even quite large selection coefficients.

More penetrating is to look for functional differences between genotypes, *in vivo* or *in vitro*, which are likely to affect Darwinian fitness. Of course, this implies that one has some knowledge of the physiology of the systems being examined, which is far from being always the case. Then, as already mentioned, functional differences may be of no consequence to present environments. The approach has been valuable for examining the significance of some quantitative variations such as skin colour and body build, but has not added much to identifying selection for polymorphic traits.

Finally, evidence for natural selection has been sought by looking for concordance in the geographical or ecological distribution of a genetically determined character system and the distribution of some environmental factor that might be a selective force. Polymorphic haemoglobin variants, for example, typically come from regions which either now or in the recent past have had high falciparum malaria endemicity. The Gc polymorphism is related geographically to levels of incident ultraviolet radiation and nasal index distribution is highly correlated with atmospheric humidity. Such relationships can provide important first clues, and we will systematically consider geographical distributions of characters later in this book. Of course, they need to be followed up by other methods before selection can be confirmed.

This review leads one to conclude that at least one important reason why we have so few unambiguous cases of natural selection is because it is so hard to detect even when selection coefficients must be of evolutionary significance.

The critical parameter as to whether selection is an important determinant of the genetic composition of a population is, according to

Kimura $N_e s$ where N_e is the effective population size (see p.207) and s the selection coefficient. If $4 N_e s$ is less than unity then the allele can be considered effectively neutral and drift will be the critical determinant of gene frequencies. If greater than unity then selection is likely to be the important factor. What is needed are good empirical studies in such appropriate populations as still exist to try and establish selection coefficients and effective population sizes.

Polytypism

Drift and natural selection are forces that operate within populations, but the consequence of their action is that populations tend to become different from one another. In general, we can say that no two human populations are ever genetically the same. Only if reciprocal gene flow were massive would identity be expected and then it would be debatable whether one was dealing with two Mendelian populations or one. Neighbouring populations will, of course, commonly be very similar, partly because of gene flow and partly because they probably recently shared a common ancestry. Conversely, excepting the results of recent long-range migration, populations which are far apart are most different from one another, because of remote ancestry, little gene exchange, and the long action of selection and drift. When the populations within a species are clearly genetically differentiated from one another the species is said to be a polytypic one. *Homo sapiens* is highly polytypic and the patterns of genetical differentiation tend to follow broad geographical regions.

Race

Races have been defined as 'populations which differ in gene frequencies' so with this definition races must exist in the human species as they do in most organisms. The question of importance, however, is whether the geographical variety that can be observed can be meaningfully and usefully categorized into named racial groups. This is a matter on which there has long been great controversy and to which we will return later.

10 Population variation in qualitative traits (proteins)

In this and following chapters we shall consider the kind of genetic variation which characterizes human populations. Genes which vary within populations usually vary between them so this is also the variation of anthropological significance.

We shall first devote attention to simple inherited variation. As will have become clear from the previous sections, such variation is invaluable in population genetics since populations can be described in terms of the frequencies of the genes they possess. For anthropological purposes it is also ideal since populations can be compared in terms of their differences in gene frequencies.

Much of the simple polymorphic variation occurs, not surprisingly, in proteins, since their polypeptide chains are the immediate product of gene action. The primary structure of a protein is the peptide-bonded amino-acid sequence of the polypeptide chains. These chains may be thrown into spirals at some parts of their length to form the secondary structure and then further folded in a specific pattern to form a globular unit. This 'tertiary structure' of the molecule is vital to its function. In many cases, the functional molecule consists of two or more identical or different subunits bonded together to form a larger structure (quaternary structure). The conformation of the molecule is often maintained by relatively weak bonds (hydrogen bonds and hydrophobic bonds), but in many cases firmer (covalent) bonds formed by sulphydryl (SH) groups contribute to stability. These various bonds are formed between the side chains of the amino-acid residues which project at regular intervals from the main (backbone) chain. The side chains vary in size and chemical properties. Some are water-repellent (hydrophobic) and, in the case of soluble proteins, are therefore directed towards the interior of the molecule; others are water-attracting (hydrophilic) and make contact with the aqueous medium at the surface. Some of these hydrophilic side chains are able to ionize and become positively or negatively charged. Changes of surface charge, due to substitution of a different amino acid as a result of a mutation, have made it possible to detect genetical variants by fairly simple methods (electrophoresis).

Haemoglobin (Hb) and its variants

Of the many proteins which have been studied, none is better known than haemoglobin, the pigment of the red-blood cell which is responsible for the transportation of oxygen in the blood. A haemoglobin molecule has a molecular weight of 64 000 and consists of four prosthetic groups of haem, each containing an atom of ferrous iron, and four polypeptide chains. The latter typically occur as two identical pairs and in normal adult human haemoglobin (HbA) these are referred to as two α-chains and two β-chains. The general structure of the molecule is shown in Fig. 10.1. The α-chains contain 141 amino-acids and the β-chains 146. Not all adult haemoglobin is $\alpha_2 \beta_2$; a small proportion, usually about 2 per cent, is made up of a combination of two α-polypeptide chains and two δ-chains (HbA$_2$). The latter are very similar to β-chains, not only in having 146 amino-acids, but also in the fact that the amino-acid sequence is the same at all but 10 positions.

The haemoglobin of the foetus differs from that of the adult and has different physiological properties associated with its need to take up oxygen from maternal tissue. In foetal haemoglobin the two α-chains are combined with two γ-chains. Like β and δ these also contain 146 amino-acids and β/γ have the same amino-acids at 107 positions; δ/γ share 105 positions in common. An interesting feature of γ-polypeptides is that two different γ-chains occur in more or less every individual; one of these has the amino-acid glycine at position 136 and the other alanine in this position. At an earlier stage of development there are yet other kinds of haemoglobin characterized by ζ and e chains (e.g. $\zeta_2 e_2$).

The essential similarity between the various haemoglobin polypeptide chains arises from the common origin of their determining genes. Those for the β, δ, γ, and e chains are closely linked on the same chromosome — chromosome 11 — and undoubtedly arose as a series of gene duplications. We now know the order to be $e - \gamma^G - \gamma^A - \psi\beta_1 - \delta - \beta$ (Fig. 10.2). It will be noted that two γ loci are recognized corresponding to the polypeptide with glycine at position 136, γ^G, and the polypeptide with alanine in this position γ^A. The DNA sequence also indicates the presence of another haemoglobin gene $\psi\beta_1$ but this apparently produces no polypeptide product and is known as a pseudogene.

The DNA coding for the α-polypeptide is on a different chromosome — chromosome 16 — but is also thought to have a common origin with $e - \beta$ genes. Even though the α-polypeptide is shorter, there is a clear similarity in amino-acid sequences with the other haemoglobin chains. We also know that there are two α gene loci closely linked with one another, rather like the two γ loci, but without any difference in the polypeptides they code for. A pseudo-haemoglobin gene also occurs linked to the α locus and on

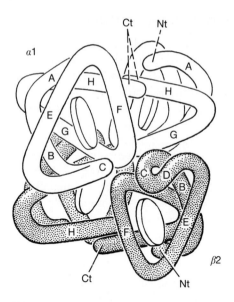

Fig. 10.1. Much-simplifed model of a haemoglobin molecule. The two α-subunits are light and the two β-subunits dark. A disc-shaped haeme group is contained in a pocket of each subunit. The helical regions, A–H, are labelled completely for one subunit only. Nt = N-terminal end, Ct = C-terminal end. The view is such as to show the proximity of the F-G and C regions where important $\alpha_1 \beta_2$ contacts occur.

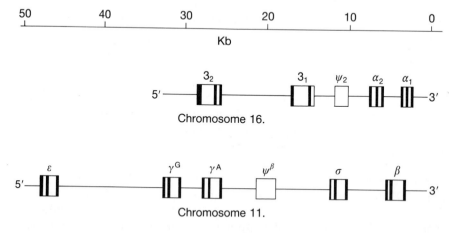

Fig. 10.2. Disposition of haemoglobin genes along chromosomes 11 and 16. The boxes represent the transcribed genes and the connecting lines the intervening untranscribed spacer DNA. The exons are shown in black and the introns in white. (After Dickerson and Geis, *Haemoglobin*, 1983.)

the chromosome there are two loci coding for ζ chains. It will be noted (Fig. 10.2) that on both chromosome 11 and 16 the genes coding for the developmentally early haemoglobin are separated from genes controlling adult chains by pseudogenes. This is a common condition in many species, but is not invariable. The figure also shows the characteristic distribution of the exons (base-pair sequences which actually code for amino-acids in the polypeptide chains) and introns (base-pairs which are translated into the messenger RNA, but then excised before protein synthesis) in the haemoglobin genes.

Genetic variation occurs in all the various polypeptide chains of different human haemoglobin molecules. This variation affects amino-acid sequences or the processes by which the polypeptides are synthesized.

The simplest form of variation arises from point mutation in which as a result of a base change a different amino-acid is substituted at a particular position in the polypeptide chain. Over 150 such variants affecting 35 positions are known in the β-chain; and over 70 are known in the α-chain. Most are rare, but provide important insight into the way the haemoglobin molecules functions (see Table 10.1). Usually, the heterozygote shows no pathological abnormality, but the homozygote is frequently anaemic to at least some degree.

TABLE 10.1

Haemoglobin variant	Site	Change	Effects
M Boston	α58 (E7)	His→Tyr	Haeme iron of the abnormal chain is stabilized in the ferric state preventing its combination with O_2.
M Iwate	α87 (F8)	His→Tyr	
M Saskatoon	β63 (E7)	His→Tyr	
M Hyde Park	β92 (F8)	His→Tyr	
Torino	α43 (CD1)	Phe→Val	Destroys a contact with haeme group. Unstable at 50°C. Inclusion body anaemia.
Hammersmith	β42 (CD1)	Phe→Ser	Haeme pocket accessible to water. Unstable at 50°C. Inclusion body anaemia.
Sydney	β67 (E11)	Val→Ala	Destroys a haeme contact. Unstable. Haemolytic anaemia.
E	β26 (B8)	Glu→Lys	Weakens $\alpha_1\beta_1$ contact; lowered O_2 affinity.
Chesapeake	α92 (FG4)	Arg→Leu	Affects $\alpha_1\beta_2$ contact; raised O_2 affinity.
J Capetown	α92 (FG4)	Arg→Gln	
Yakima	β99 (G1)	Asp→His	
Kempsey	β99 (G1)	Asp→Asn	
Wien	β130 (H8)	Tyr→Asp	Polar group replaces internal non-polar group. Unstable. Haemolytic anaemia.
Genova	β28 (B10)	Leu→Pro	Proline disrupts β-helix. Haemolytic anaemia.
Gun Hill	B92–96 deleted		Affects haeme contacts and contacts between subunits. Loss of haeme. Haemolytic anaemia.

A few of the variants reach polymorphic proportions in some populations. These include HbS, HbC, HbE, and HbD Punjab which are β-chain variants, and HbJ Tongariki which is an α-chain variant.

Sickle-cell haemoglobin (Hb-S)

The beginning of the story was the recognition of a severe type of congenital anaemia in a West Indian in 1910. It was called sickle-cell anaemia (SCA) because blood-films showed that many of the red cells were distorted into an elongated, curved shape. Work on American black and African families established that this disease is simply inherited, the SCA cases being homozygous for an allele Hb^S. Most of these homozygotes die prematurely unless advanced medical care is available, so that most cases arise from a mating of two heterozygotes who both carry a sickle cell and a normal gene (genotype Hb^A/Hb^S). The blood of these heterozygotes looks normal when it is well oxygenated, but if oxygen is removed the cells assume the sickle form, reverting to the normal shape when oxygen is readmitted. This heterozygous phenotype, known as 'sickle-cell trait' (SCT) is not usually harmful though it can cause trouble at high altitudes when there is oxygen deficiency. The kidney is also mildly deficient in ability to concentrate the urine, and this could be a disadvantage in dehydrating conditions.

An important advance was made by Pauling, Itano, Singer, and Wells in 1949 when they showed the SCT cells contain two types of haemoglobin that can be separated by electrophoresis. We have already mentioned that proteins have a surface charge; the magnitude and sign of this charge depends on the acidity (pH) of the medium. If an alkaline (pH 8.6) solution of normal haemoglobin is placed between electrodes and a voltage is applied, the red protein is seen to migrate towards the positive pole (anode). If this is done with SCT material a more slowly moving type, Hb-S, gradually separates from the normal haemoglobin, Hb-A. Solutions from SCA cases show no Hb-A, but only Hb-S, together with variable amounts of foetal haemoglobin, Hb-F.

The sickling phenomenon is evidently due to a mutation that makes the Hb much less soluble than normal when in the de-oxygenated state. It therefore precipitates in semi-crystalline sheaves that distort the red cell. Ingram went on to analyse Hb-S in more detail. He did this by first breaking the molecule into smaller fragments (peptides) by digestion with the enzyme trypsin. The mixture of peptides can be separated on filter-paper by applying electrophoresis in one direction followed by chromotography at right angles to this. Chromotography has been an immensely useful method for separating small amounts of biological substances. In one version of the method a strip of filter-paper bearing a spot of the material to be analysed is dipped into a suitable mixture of organic solvents. As the solvent creeps along the paper the more hydrophilic components adhere to the thin water layer on the cellulose and are retarded in relation to more hydrophobic constituents. Ingram was thus able to spread the mixture of Hb-S peptides out on paper and, after staining to show their positions, to compare the pattern of spots (fingerprint) with that of Hb-A (Fig. 10.3). This revealed a change of position of one peptide and further analysis showed that it differed from normal in having valine (Val) instead of glutamic acid (Glu) at position 6 in the β-chain. This is the only difference between Hb-A and Hb-S, and it can be accounted for by a change of one base in the corresponding base triplet of the gene (CTT or CTC to CAT or CAC). The amino-acid substitution explains the electrophoretic difference, since Glu is negatively charged while Val is neutral.

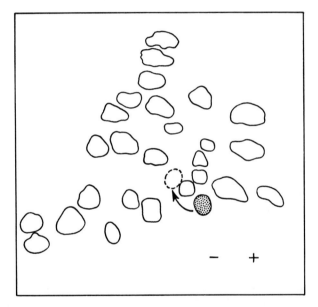

FIG. 10.3.   Map of tryptic peptides of whole haemoglobin A. In the case of a digest of haemoglobin S the stippled dot is found to have moved to the position shown by the arrow. The direction of electrophoresis is horizontal and that of chromatography vertical. (Based on Baglioni 1961.)

Hb-C is a variant that migrates even more slowly than Hb-S at pH 8.6 (Fig. 10.4) because lysine (Lys) with a positive charge is substituted for glutamic acid at the same position 6 in the β-chain. The homozygote Hb^C/Hb^C shows mild abnormalities of the blood film, but suffers no great disadvantage.

Hb-E is another slowly migrating variant in which the substitution is $\beta26$ Glu-> Lys. Again the homozygote is only mildly affected despite the

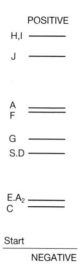

FIG. 10.4.   The approximate relative positions of a number of different haemoglobins when separated by electrophoresis on paper. (Veronal buffer, pH 8.6.)

fact that the substitution is in a subunit contact area and the O_2 affinity is somewhat lowered as a result.

Hb-D has been found in various parts of the world and reaches its highest frequencies (1–5 per cent) in Sikhs and Gujeratis. It has the same mobility as Hb-S, but does not cause sickling. Later studies on Hb-D from various regions showed that it is heterogeneous, thus illustrating the limitations of simple electrophoretic screening of whole Hb for the identification of mutants. Hb-D Punjab (β121 Glu–>Gln) proved to be the same as half a dozen variants that were at first given different names. On the other hand, three other Hb-Ds have different compositions: Hb-D Ibadan is β87 Thr–>Lys; Hb-D Bushman is β16 Gly–>Arg; and Hb-D Baltimore is α68 Asn–>Lys. Haemoglobin J Tongariki is characterized by the substitution of aspartic acid for alanine at position 115 in the α chain. The characteristics of some other variants are presented in Table 10.1, but the genes responsible for these are all rare.

Most variants are in the form of single amino-acid substitution, but more complex situations are found. In haemoglobin Harlem, the glutamic acid normally present at position 6 in the β chain is replaced by valine (as in sickle haemoglobin), but in addition the usual aspartic acid at position 73 is replaced by asparagine. The latter substitution, on its own, causes haemoglobin Korle-Bu. Such a situation may arise as a second mutation in the sickle-cell gene or through intragenic crossing-over between a chromosome carrying Hb^S and one carrying Korle-Bu.

Individuals are known who carry both a β-chain and a α-chain variant.

There is an α variant Hb-G which has been found in heterozygous state with a carrier for Hb-C. The genotype is $\alpha/\alpha^G:\beta/\beta^C$ and leads to the production of four types of adult haemoglobin: $\alpha_2\beta_2$, $\alpha_2\beta_2{}^C$, $\alpha_2{}^G\beta_2$, and the hybrid $\alpha_2{}^G\beta_2{}^C$.

Other variations arise by deletion and addition. In haemoglobin Freiburg position 23 in the β-chain is completely missing and in Gun Hill five amino-acids from positions 92–96 are lost. As we shall see later, deletion plays a very important role in the causation of thalassaemias. In the case of haemoglobin Constant Spring the α-chain possess an additional 31 amino-acids; this is due to a mutation in a normal stop codon.

The Le Pore haemoglobins are rare variants in which the polypeptide chains are of mixed type. For example, in Le Pore Washington the beginning of the chain has a δ sequence of amino-acids, but the end is of β sequence. Such situations arise from unequal crossing-over at meiosis. When one has a linear series of closely linked genes arising by duplication, the DNA sequences adjacent to one another are very similar. Under such circumstances mismatching during meiosis is likely and pairing with not fully homologous segments occurs. Thus, instead of $(\delta-\beta)/(\delta-\beta)$ pairing so, it can easily happen that the arrangement is $(\delta-\beta)/(\varsigma.\delta-\beta)$. Crossing-over in the region of δ/β pairing will lead to a DNA code starting with the δ sequence and finishing with the β one, and this will produce a Le Pore haemoglobin. These vary according to where the cross-over occurs and to produce Washington this must occur between positions 87 and 116 whilst in Hollandia it must be between 22 and 50. Le Pore haemoglobins also occur through mismatching in the β and γ genes. This is the cause for the so-called Kenya haemoglobins. Such unequal crossing-over, of course, also produces chromosomes which contain the normal genes as well as a mixed one, e.g. in the case just considered $\delta-\beta\delta-\beta$. This is the so-called anti-Le Pore situation.

Molecular genetics

Recently, attention has been moving in studies of haemoglobin variation from analysing the polypeptide chains themselves to analysing the DNA sequences in the responsible genes using the techniques of molecular genetics. Here the existence of a whole host of endonuclear restriction enzymes, obtainable particularly from bacterial cells, is of paramount importance. These enzymes recognize particular sequences of DNA base-pairs and 'cut' the DNA into pieces where the specific recognition sites occur. For example, there is a restriction enzyme Hpal which recognizes the sequence of bases GTTAAC. Such sites occur in the region of the β-haemoglobin gene and the enzyme typically produces a fragment of DNA in this region 7.6 kb long. A point mutation can, however, obliterate the recognition site, e.g. CTTAAC—>GCTAAC and when this occurs 'downstream' (i.e. towards the 3' end of the DNA) of the β gene a fragment of 13.0 kb length is produced as the enzyme 'misses' the original site and picks up the next one along. Conversely, a point mutation can introduce the specific recognition site where one does not normally exist. Such a situation leads to the production of a 7 kb fragment in the β-haemoglobin gene area.

Interestingly, the 7.6 kb fragment is predominantly associated with the sickle-cell gene in Kenya, Saudi Arabia, and India, but in Ghana, Nigeria, and around the Mediterranean the sickle association is mainly with the 13 kb fragment. In the Ivory Coast and Sierra Leone both associations are common. HbC tends to be associated with the 13 kb fragment.

Another DNA variation in the β-globin gene is detected by the restriction enzyme βamH and these variants like the Hpal ones are polymorphic.

Restriction enzymes have been useful in prenatal diagnosis of sicklaemia. The base sequence of the amino-acids 5:6:7, i.e. proline: glutamic acid: glutamic acid in the β-chain is CCT-GAG-GAG. This sequence contains recognition sites for two restriction enzymes Mn/e recognizing GAGG and D-del recognizing CT(X)AG. The first generates too many fragments, but D-del normally produces two fragments 180 and 201 base pairs (bp) long in the β region. The sickle mutation GAG-GTG abolishes the recognition site and in sicklaemics only one fragment 381 bp long is generated.

Thalassaemia

So far, we have considered genetic mutations which affect the structure of the haemoglobin molecule. Others occur which determine the synthesis and, in particular, the extent to which the various polypeptide chains are formed. In normal development a succession of synthetic events occurs with the production of embryonic haemoglobin being followed by that of foetal haemoglobin and the ultimate replacement of this by adult haemoglobin (see Fig. 10.5). Variants occur which disrupt this sequence, and those which lead to an interference with β-and α-polypeptide production cause, respectively, the β-and α-thalassaemias.

β-thalassaemia

The name 'thalassaemia major' (thalassa = Greek, forged) was given to a foetal type of congenital anaemia because it was commonly found in children of Mediterranean origin; it is also known as Cooley's anaemia. The blood-film shows many red cells of grossly abnormal size and shape and with a low mean Hb content. Some of the pathology is due to unused

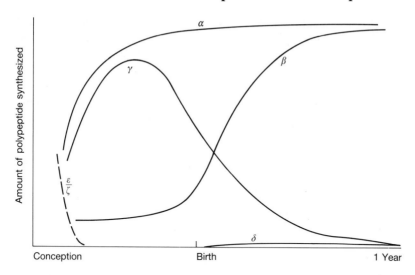

Fig. 10.5.   Relative amounts of the various haemoglobin polypeptides at different times during prenatal and early postnatal development.

α-chains which denature, are precipitated on to the cell membrane, and cause increased red-cell destruction. Compensatory growth of the bone marrow causes thickening of the skull vault and facial bones, and growth is retarded. Electrophoresis shows Hb-A together with a variable amount (10–90 per cent) of Hb-F.

Cases of Cooley's anaemia are typically homozygous for genes which control β-polypeptide production. It is now known that there are many such genes, which is why heterozygotes in particular show an enormous amount of variation and are not always easy to detect in population surveys. Usually, they show a milder version of the disorder — thalassaemia minor — but they can be practically normal or almost as badly affected as the homozygotes. An increase in the relative concentration of Hb-A_2 to about 5 per cent on average and in many cases a slight rise of Hb-F level are useful diagnostic signs.

β-thalassaemias can be broadly divided into two categories $β^+$, in which at least some β-chains are synthesized, and $β^0$ in which there is no production at all. $β^+$ thalassaemias characteristically arise from some defect in the mRNA stage with the messenger being formed in reduced and inadequate amounts. In some cases at least this appears to be due to mutation in the introns. In the $β^0$ thalassaemias, mRNA may or may not be produced but there is no β-chain synthesis. One cause can be mutation to a stop codon within the normal full sequence, in which case RNA is formed, but cannot complete the synthesis. In other cases there can be actual deletion of the β-gene.

Two related conditions which are caused by gene deletion are δβ-thalassaemia and hereditary persistence of foetal haemoglobin (HPFH). Such deletions are typically caused through unequal crossing-over and the Le Pore haemoglobins can be considered thalassaemic (cf. p.227). The extent of the deletion not surprisingly affects what the affected chromosome can synthesize, but it does not follow that the greater deletion the greater the severity of the pathology. Deletion of the β gene and part of the δ gene is much more severe than a somewhat longer deletion embracing the whole of the δ as well as the β gene and some upstream spacing sequence. In the latter, foetal haemoglobin can continue to be produced in large amounts in the adult, probably because the gene which normally turns off γ-chain production is also deleted. This is the basis for HPFH and as long as both γ genes are present even the homozygote for the deletion is practically normal. Interestingly, a large deletion which embraces both γ genes and the δ gene but not the β gene leads to a situation in which no β-chains are produced.

α-thalassaemias

The main cause for α-thalassaemia appears to be gene deletion and, with the possibility of losing from 1–4 α genes, four abnormal conditions are recognized. The loss of a single α gene causes no pathology: the condition is only recognized by the presence of tetramers of γ-chains($-\gamma_4-$) or Haemoglobin Barts in the neonate. Around 2–3 per cent of the haemoglobin is of this form; the condition is known as α-thalassaemia 2. When two α genes are missing one has α-thalassaemia 1, and at birth 5 per cent of the haemoglobin is the Hb Barts. The situation is associated with minor blood abnormalities in later life. When only two α loci are present they may be on the same chromosome, i.e. the homologe has lost both α loci, or they may be on different chromosomes — the so-called homozygous situation — in which a single gene has been deleted from each chromosome. This 'homozygous state' tends to characterize the situation in African and Mediterranean countries, and the heterozygous one in Asia. The loss of a third α gene leads to HbH disease which is an anaemia with severity intermediate between β-thalassaemia minor and major. Biochemically, it is characterized by the occurrence not only of γ_4 chains in the infant, but also of HbH-β_4 in the adult. Thus, β-chains like γ-chains can form into homotetramers and will do so when in relative excess of α. Incidentally, tetramers of α-chains cannot occur so there is no α_4 equivalent to γ_4 and β_4 in the β-thalassaemias. Complete absence of α-loci leads to a condition — hydrops foetalis — which is invariably fatal in pre-natal development. No α-chains are produced at all but production of ζ- chain continues for longer than normal. This, however, is insufficient for postnatal existence. α-thalassaemias not only arise by gene loss, since the presence of the variant Constant Spring, in which α-chains are extended by an extra 31 amino-acids, has very similar effects to a gene deletion.

Population genetics and geographical distribution of Hb variants

HbS, HbC, and HbE are found in fairly large areas of the world at heterozygote frequencies of 10 per cent or higher. Many populations in the wet tropical belt of Africa have 20–30 per cent of the sickle-cell trait and comparable high values occur in a few Mediterranean localities and in various peoples of India (Fig. 10.6). In northern Europe and eastwards from India the gene is rare, and it was almost certainly absent in the New World before the advent of African slaves. HbC is frequent in a somewhat restricted area of West Africa centring on Upper Volta and Northern Ghana, while HbE is remarkably common in Thailand and immediately adjacent countries.

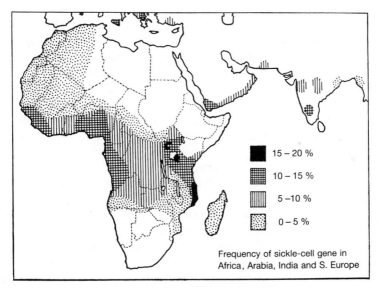

15 – 20 %

10 – 15 %

5 – 10 %

0 – 5 %

Frequency of sickle-cell gene in
Africa, Arabia, India and S. Europe

Fig. 10.6.   Frequency of the sickle-cell gene HB^S in various parts of the Old World. (After Allison in *Genetical variation in human populations*, 1961.)

A few other variants attain heterozygote frequencies around 5 per cent locally, e.g. HbD in the north-west of the Indian subcontinent, HbK in parts of west and north Africa, and HbO in parts of Indonesia. Haemoglobin J. Tongariki is the only α-chain variant known to reach polymorphic proportions and this it does in various islands of the South Pacific including parts of coastal New Guinea.

It is worth re-noting here that the incidence of genetical variants is sometimes given in terms of the frequency of the gene and sometimes of the heterozygote (trait frequency) and it is important not to confuse the two. The gene frequency is the more useful for theoretical work, but it is an inference from the phenotype and may therefore be uncertain unless the relationship between genotype and phenotype are entirely clear. It is worth noting that in simple cases, where one of a pair of alleles has a frequency of 10 per cent or less, the gene frequency is roughly half the heterozygote frequency since most of the variant genes are in the heterozygous state; However, the error of neglecting the homozygotes increases as the gene frequency rises.

As we have already noted the β-thalassaemias constitute a variety of different genetical conditions, but the disease is quite common in many tropical and subtropical countries of the Old World including areas of West Africa, around the Mediterranean, through the Middle East to India and South-east Asia, and extending into Indonesia, New Guinea, and some Pacific Islands. Since β-thalassaemia is quite common in certain regions where the structural variants Hb-S, Hb-C, and Hb-E are also frequent we can find heterozygotes carrying both types of mutation. In the case of sickle-cell thalassaemia (β/β'thal') the clinical and biochemical manifestations are extremely variable. Some cases are only mildly anaemic, others are as severely affected as sickle-cell anaemics. Depending upon the degree of suppression of β-chain synthesis and upon whether the sickle and thalassaemia mutations are on the same chromosome or are essentially allelic the amount of HbS can vary from almost 100 per cent to none.

Hereditary persistence of foetal haemoglobin reaches polymorphic proportions in west Africa and α thalassaemia has been found in surveys of populations in west Africa, around the Mediterranean, and South-east Asia. In parts of Thailand up to 30 per cent of infants have the Hb Barts, but the concentration varies from 1 to 90 per cent of the total haemoglobin and not surprisingly falls into several distinct ranges according, one presumes, to the number of deleted α genes.

The presence of so many different genes affecting haemoglobin structure and synthesis at polymorphic levels in many populations raises important questions about causation when homozygotes at least tend to suffer to at least some degree from anaemia and are the cause for so many haemoglobinopathies. The position is highlighted and best understood for Hb^S. Since sickle-cell homozygotes die prematurely, especially where medical facilities are lacking Hb^S genes are lost from a population every generation. Thus, even were the frequency strikingly raised in a population by drift one would expect it to fall to quite low levels in a few generations with such strong negative selection. Clearly, this has not happened and some alternative explanation is required. As we have already seen (p.212) the loss of genes through an unfit genotype can be compensated if the heterozygote is fitter than the other homozygote. This is the case of balanced polymorphism maintained by heterozygote advantage and a stable situation is reached determined by the relative fitnesses of the two homozygotes.

In the case of the sickle haemoglobin situation there is impressive evidence now that the advantage of the heterozygote is conferred by the protection it affords to malaria: specifically the malaria caused by the protozoan *Plasmodium falciparum*. This parasite, which is transmitted by anopheline mosquitoes, produces the malignant tertian form of the disease and is the malaria parasite most likely to cause death. In equatorial zones falciparum malaria is typically holendemic (i.e. transmitted throughout the year); in places with more seasonal rainfall it is mainly transmitted in the wet periods, but in most parts of the tropics it is not only the commonest form of malaria, but also the commonest important infectious disease. In holendemic zones nearly all children are infected by 2 years of age. The number of parasites in the blood declines with age as the acquired immunological defences of the body become more effective. In impoverished and disease-ridden countries it is often difficult to assign a precise cause of death in early childhood, but the sharp decline in deaths of this age group that often follows malaria eradication leaves little doubt the disease is a major killer. There is also a particularly severe form of the disease, when the parasite invades the brain. This is cerebral malaria which can be diagnosed with confidence.

In 1954, A. C. Allinson working on East African children, found that individuals with sickle-cell trait were less frequently infected with *P. falciparum* than were individuals with normal haemoglobin and that the number of parasites in a unit volume of blood was also lower on average. This observation has been confirmed a number of times and suggests that the sickle-cell trait confers some resistance to malaria in young children before full acquired immunity has been achieved. Assuming that the position is in equilibrium and the gene frequencies stable one can estimate the relative fitness of the heterozygote and the normal homozygote from the gene frequencies on the basis that the homozygote sicklaemic has zero fitness ($s = 1$). For example, if Hb^S has a frequency of 15 per cent — a figure which is reached in a number of places — those with only normal adult haemoglobin should be about 17 per cent less fit than sicklers. In stable situations the equilibrium gene frequency is determined by the comparative fitness of the homozygotes; with one of these always fixed at 0 it is the extent to which the other is less fit than the heterozygote which matters and this, of course, is dependent upon the amount of malaria in the population. The conclusion is strongly supported by the fact that, among deaths from cerebral malaria, the frequency of sicklers is strikingly low.

There is also some evidence that sickler women may be more fertile than non-sickler women, so the fitness differential may not be wholly due to mortality in childhood. It is thought that the invasion of the placenta by malaria parasites may sometimes cause stillbirth or neonatal death. These deaths would presumably be less frequent in births to sickler mothers if the trait confers some immunity during pregnancy.

It is far from easy to measure the exact fitness of the different genotypes in real conditions of field situations and no worthwhile figures are available. In other words we expect the Hb^S gene frequency to vary according to the degree of falciparum malaria selection. Just how Hb^S acts to control parasitaemia is not properly understood, but the single amino-acid substitution has a profound effect on the tertiary structure of the haemoglobin molecule through the introduction of a hydrophobic residue on the surface, and this increases the tendency for the haemoglobin to crystallize. Presumably, such crystallization and consequent cellular collapse on infection prohibits the proliferation of invading parasites.

The geographical distribution of Hb^S is consistent with the malaria hypothesis in that the highest gene frequencies occur in regions where malignant tertian malaria is highly endemic, or has been so until quite recently. An exact correspondence is not to be expected because population movements may have changed the local gene frequency pattern and there has been too little time for equilibrium levels to be re-established. In East Africa the gene is infrequent or absent in Nilo-Hamitic pastoralists. This may be because they came from, and still inhabit, less malarious regions. It has also been suggested that the presence of cattle may divert the mosquitoes from biting humans. It may be that malaria only became a serious disease in the wet tropics when the rainforest was opened to agriculture and population densities increased. Possibly certain Liberian tribes have low sickling frequencies because they represent such an ancient forest population, but it is worth noting that sickling is frequent among the Efe Pygmies of the Ituri forest. In India the highest frequencies of Hb^S are found in some of the tribal peoples, who are hunters and slash-and-burn agriculturalists, and also in certain lower Hindu castes that may be partly derived from them. Patches of high sickling, as in parts of Greece and certain Southern Arabian oases, may be correlated with locally intense malaria.

In general the Hb^S frequencies in American blacks are around 5 per cent which is about half that in many West African populations, but there are some more or less isolated groups in which the gene frequency reaches 10 per cent. When slaves were taken from West Africa to the generally less malarial environment of the New World the selective advantage of the heterozygotes presumably declined or disappeared and the gene frequency would then have fallen due to selection against the homozygotes, but there are several uncertainties that complicate the interpretation of this situation. The initial gene frequencies are open to some doubt because the levels in West Africa are not uniform and slaves were exported from different regions to various parts of the Americas. In addition there has obviously been a good deal of intermixture with Europeans or American Indians and this would also lower Hb^S frequencies. The extent of this intermixture no doubt varied in different places. In principle it can be estimated by studying a number of genes that have widely different frequencies in the parental populations, but are not considered to be so strongly influenced by selection as Hb^S. The frequencies of the blood-group genes R_0 and Fy^A, for example, are very different in Africans and Europeans, but again the data are limited and there are some doubts about the relevant parental frequencies. It is also an oversimplification to suppose that the selective advantage of the Hb^S heterozygotes was entirely abolished in the New World. Whether malaria of any kind was present in pre-Columbian times is disputed, but *P. falciparum* is now well established in parts of Central and South America and it was present in the southern USA until quite recently.

Whether the other haemoglobinopathies are related to falciparum malaria is far less certain, but the evidence is suggestive. Certainly, they only occur at polymorphic proportions in regions where the disease is endemic.

The gene Hb^C has a focus of high frequency (10–15 per cent) in a limited area of West Africa. Since the homozygous state for this gene is only mildly deleterious a small advantage of the heterozygote would suffice to maintain quite high equilibrium levels, but would be difficult to detect. However, the situation is more complicated because Hb^C occurs in regions where its allele Hb^S is also frequent. The heterozygote Hb^S/Hb^C is clearly at some disadvantage though it is not so severely affected as Hb^S homozygotes. Whether the three alleles, Hb^A, Hb^S, and Hb^C can co-exist at equilibrium or whether either Hb^S or Hb^C will be excluded depends on the relative fitnesses of the various genotypes and it is difficult to measure these fitnesses with the required accuracy. In those parts of West Africa where both Hb^S and Hb^C occur it is found that there is a negative correlation between their frequencies, Hb^S being low where Hb^C is high. It is possible that Hb^C is spreading and ousting Hb^S, but this is by no means certain. We have already mentioned the diseases due to interaction between β-thalassaemia and β-chain structural variants. Since β-thalassaemia is known to occur in West Africa this is an additional, though perhaps minor, complication. β-thalassaemia would itself appear to be related to malaria protection. Its distribution in Sardinia, for example, closely follows that of anopheline mosquitoes and the disease.

Isoenzymes

An enormous amount of genetic variation has been detected in the structure of enzymes of the human body. Enzymes which have the same function, but different structures are known as isoenzymes or isozymes. Many, but by no means all the variants have been discovered by electrophoresis and this has implications when considering the extent of the variation. For obvious reasons a considerable number of the variant systems were first found in red-blood cells, but they also often occur in other tissues and some have been found in body fluids such as saliva, e.g. salivary amylase. Table 10.2 provides information on a number of examples, all of which display polymorphism in at least some populations. As can be seen variants also tend to vary in frequency between populations. Typically variants have somewhat different physical and chemical properties but in most cases there is no known functional explanation for the existence of the variants. Amino-acid sequence analysis has not been widely undertaken, but where it has variants appear to have arisen as a result of point mutation and single-base pair changes.

Whilst in general the causes for the isozyme polymorphisms are unclear, there is one case of great anthropological interest.

Glucose-6-phosphate dehydrogenase G6PD

Glucose-6-phosphate dehydrogenase occurs in various tissues including red-blood cells and catalyses the oxidation of glucose-6-phosphate to 6-phosphogluconate with nicotinamide adenine dinucleotide phosphate (NADP) as a co-enzyme in the so-called 'pentose shunt' pathway of the catabolism of glucose. This shunt leads to the formation of reduced glutathione.

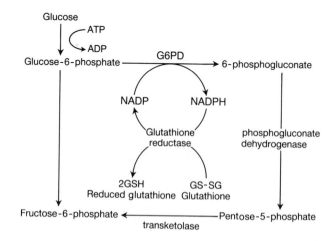

TABLE 10.2
Some polymorphic enzymes of anthropological significance

Enzyme	Abbreviation	Action	Detection	Frequent alleles	Distribution
Red cell acid phosphatase	AP	Splits phosphate from various organic phosphates	Electrophoresis	ACP_1^A, ACP_1^B ACP_1^C (at one of three loci)	ACP_1^B commonest ranging 60-80% ACP_1^C least frequent, but reaches 4-9% in Europe.
Glucose-6-phosphate dehydrogenase	G6PD	Conversion of glucose-6-phosphate to 6-phosphogluconate	Electrophoresis & quantitative assay	Gd^{B+}, Gd^{A+} & Gd^{B-} Gd^{A-} (X-linked)	Gd^{B+} common everywhere Gd^{A+} & GD^{A-} in Africa Gd^{B-} in some Mediterranean countries.
6-phosphogluconate dehydrogenase	6PGD	Follows G6PD and oxidizes 6-phosphogluconate to ribalose-5-phosphate.	Electrophoresis	PGD^A, PGD^C	PGD^A common everywhere. PGD^C around 3% in Europeans. Higher in African & Far East populations, e.g. 15% in South African Bantu.
Phosphoglucomutase	PGM	Conversion of glucose-1-phosphate to glucose-6-phosphate	Electrophoresis	PGM_1^1 & PGM_1^2 (at one of three loci)	PGM_1^1 typically has a frequency of between 65-85% with a tendency for high values in Africa and Amerindians.
Phosphoglycerate kinase	PGK	Conversion of 1:3 diphosphoglycerate to 3-phosphoglycerate in glycolisis	Electrophoresis	PGK^1, PGK^2 & PGK^3 (X-linked)	PGK^2 at 7% in Micronesians PGK^4 at 5% in Papua New Guinea.

TABLE 10.2 (continued)

Enzyme	Abbreviation	Action	Detection	Frequent alleles	Distribution
Adenylate kinase	AK	Conversion of adenosine-diphospate into ATP and AMP	Electrophoresis	AK^1 AK^2	AK^2 rare in Africa and Central America; reaches frequencies of up to 5% in Europe and is highest in the Indian subcontinent at 13%.
Adenosine deaminase	ADA	Deaminates adenosine	Electrophoresis	ADA^1 ADA^2	ADA^2 rare in Africa, Eskimo and Japanese (0–3%); around 5% in most of Europe, but higher in S. Europe, Middle East, and India. Up to 16% in Lapps.
Peptidases	PEP	Hydrolizes dipeptides and polypeptides	Electrophoresis	Four loci. $PEPA^1$ $PEPA^2$ $PEPD^1$ $PEPD^2$	$PEPA^2$ and $PEPD^2$ commoner in Africans than Europeans.
Alcohol dehydrogenase	ADH	Reduces ethyl alcohol	Electrophoresis Activity Ratios	ADH_2 typical ADH_2 atypical	6% of Europeans have the atypical phenotype whilst 90% of Japanese are 'atypical'.
Placental alkaline Phosphatase	PC	Splits organic phosphates	Electrophoresis	3 loci PC^1 PC^2 PC^3	PC^2 has frequency of 9–26% in Europeans; much less in Africa and the Orient, but PC^3 common in the latter.

It has long been known that some people of Mediterranean descent could suffer a severe haemolytic crisis on ingestion of the bean *Vicia fava* — a condition known as 'favism'. The same crisis is initiated by some drugs such as the anti-malarials primaquine and pentaquine, and sulphonamides. It is caused by a deficiency of the enzyme G6PD in the red blood cells and this deficiency is simply inherited on the X-chromosome. Thus, while in females there are three genotypes, in males there are only two and these correspond to those with normal G6PD and those who are deficient in G6PD. The frequency of affected males is the gene frequency in a population.

Electrophoretic screening of haemolysates, using specific stains to reveal the position of G6PD has shown that there are many variants of the normal enzyme [Gd(B +)] and that a number reach polymorphic proportions; more than one is associated with enzyme deficiency. In Africa two variants of the normal form are remarkably frequent — about 20 per cent each. At pH 8.6 both migrate more rapidly and to about the same degree towards the anode than the normal form. One of them [Gd(A +)] has only slightly reduced activity in laboratory tests and only differs from Gd(B +) by a single amino-acid substitute — aspartic acid in place of asparagine. The other variant Gd(A –) has much lower activity, to 8–20 per cent of normal levels. It is associated with the drug sensitivity. The Mediterranean form of the deficiency is more severe with haemolysate activity less than 7 per cent of normal. The enzyme migrates at around the same rate as the normal enzyme and is referred to as Gd(B –) as well as Gd(Mediterranean). Frequencies of 15–20 per cent occur in parts of Greece, Sardinia (see Fig. 10.7), the Middle East, and India. Among a Jewish isolate that formerly lived in Kurdistan a frequency of 50–60 per cent has been recorded! The deficiency in both this and the African form is due to the enzyme being unstable. Recently formed red blood cells are hardly deficient for Gd(A –), but the enzyme only has a half-life of 13 days as compared with 62 days for Gd(B +) and, of course, in red blood cells which lack a nucleus there is no ongoing synthesis. Gd(B –) is even more unstable and young red blood cells are affected as are leucocytes despite their possession of a nucleus.

Other enzyme variants are also associated with G6PD deficiency. Five per cent of southern Chinese show the phenomenon. It is due to at least three different variants the commonest of which is Gd Canton. Another variant, Gd Markham, seems to be fairly common in lowland areas of New Guinea. Although haemolysis due to G6PD deficiency is usually transient and elicited only by certain environmental agents, there are also several more variants that are associated with chronic haemolysis (non-spherocytic haemolytic anaemias). Two points emerge clearly from work on this enzyme: G6PD deficiency is a very heterogeneous condition, and

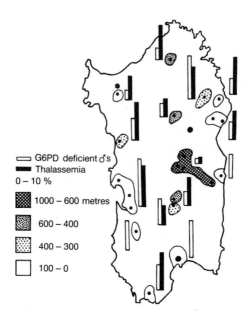

Fig. 10.7. The frequencies of thalassaemia and of G6PD-deficient males in various regions of Sardinia, showing the correlation between the incidences of these conditions and altitude. The lower regions are the most malarious. The incidence of thalassaemia is phenotype frequencies. (From Bernini, Carcassi, Latte, Motulsky, Romei, and Siniscalco, *Acad. Naz. Lincei*, 1960.)

some of the variants causing it are characteristic of certain regions of the world.

As we have seen some of the common forms of deficiency can be deleterious, especially in males. We may ask how some populations have attained such high gene frequencies and whether these frequencies are at equilibrium. The gene-frequency dynamics of sex-linked recessives are complicated because we have to consider the relative fitnesses of three female and two male genotypes. Whether a stable equilibrium at intermediate frequencies (i.e. between zero and 1.0) can be achieved depends critically on these fitness values and at present we can only estimate them very roughly for G6PD phenotypes. Balanced polymorphism is theoretically possible if there is heterozygous advantage in females or if the presence of the gene is advantageous in one sex, but disadvantageous in the other. Even so the differences of fitness must be within a certain range.

High frequencies of G6PD deficiency are usually found in areas that are, or were until recently, highly endemic for *P. falciparum* malaria. In East Africa and in Sardinia (Fig. 10.7), for example, the geographical correlation is impressive and suggests that the trait may have some protective value against this disease. In heterozygous females some red cells are G6PD-deficient and some normal because one X-chromosome is inactivated at random. It is therefore possible to study the susceptibility of the two cell types to infection by *P. falciparum* in the same individual and it has been found that the deficient cells are less often invaded. Presumably, both males and females are protected in this way although this advantage is offset to some extent in males and homozygous females by other deleterious effects of the mutant allele. It has been thought that malarial protection may arise from the lowered levels of reduced glutathione.

Isozymes and the magnitude of genetic variation

Study of isozyme variation within and between populations, particularly by H. Harris and his colleagues, has provided important insight into the overall extent of genetic diversity in the human genome. Not only is it possible to get some idea of the proportion of gene loci which are polymorphic, but also, from determination of the levels of heterozygosity at all these loci, of the average level of heterozygosity in the genome and the magnitude of individual uniqueness.

A rather limited example is shown in Table 10.3. Here, 20 arbitarily chosen enzymes were examined for polymorphism and if polymorphic the incidence of heterozygotes estimated in typical European and African populations. The 20 enzymes are controlled by 27 genetic loci. In both population groups six of the enzymes are polymorphic, though only five show the same polymorphism. By adding up the frequencies and dividing by 27 one has an estimate of the average heterozygosity over the loci concerned. The estimate for the two groups turns out to be very similar at between 5 and 6 per cent, but is made up rather differently. There is no reason to think that the loci examined are not representative of the genome, so the estimate may be expected to apply over the whole of it. However, all the enzymes variants were detected by electrophoresis and,

TABLE 10.3

Incidence of heterozygotes in 20 arbitary chosen enzymes representing 27 genetic loci

Enzyme	European	African
Red cell acid phosphate	0.51	0.28
PGM 1	0.35	0.33
PGM 3	0.38	0.47
Adenylate kinase	0.09	0
Peptidase A	0	0.16
Peptidase D	0.02	0.10
Adenosine aeaminase	0.11	0.06
Mean/27	0.054	0.052

since electrophoresis will only detect differences due to charge, and only about a third of amino-acid substitutions cause a change in charge, the likely level of average heterozygosity is 3 times that detected, i.e. around 15 per cent. An estimate based on 104 loci of which 24 were polymorphic gave an average heterozygosity level of 6.3 per cent $\times$ 3 = 18.9 per cent.

While the proportion of enzyme loci which are polymorphic is very similar at around 25–30 per cent in African and European populations it appears to be somewhat less in Indian and Far East groups, at about 10 per cent. This, however, may arise from bias through only looking in these populations for enzymes which are polymorphic in Africa and Europe!

It would seem that there is no difference in levels of polymorphism according to the type of reaction catalysed. Thus, oxidoreductase, transferase, hydrolase, and lyase enzymes all show similar levels. On the other hand, there is quite a precise relationship with enzyme structure with estimated percentage of polymorphisms in monomeric enzymes being 56 per cent, dimers 35 per cent, trimers 25 per cent, and tetramers only 21 per cent.

The extent of polymorphism as evidenced in these enzymes provides powerful insight into the level of genetic individuality. Considering just 14 enzymes which are polymorphic in European populations the most commonly occurring combinations of all of them will only be found in 0.06 per cent of a population and the probability of picking at random two people with the same combination is 1:32 000.

Rare variants (i.e. with gene frequencies less than 1 per cent) occur at enzyme loci which have polymorphic forms as well as at non-polymorphic loci. It is presumed that such variants are solely due to mutational processes and it may be queried whether there is any difference between the two kinds of loci in the number of rare variants which have been discovered. The answer to this question in general appears to be no, though one polymorphic enzyme — placental alkaline phosphatase — is characterized by having a large number of known rare variants. Excluding this system, the average frequency of heterozygotes for rare variants in 12 polymorphic systems was 1.10/thousand individuals. The equivalent frequency over 30 non-polymorphic enzyme loci was 1.16.

Serum proteins

The serum, which is the fluid remaining after removing the fibrin from clotted blood plasma, contains a very complex mixture of proteins. Electrophoresis on paper separates them into a few crude fractions, the albumin, and the $\alpha_1 \alpha_2 \beta$ and γ globins, examples of which are shown in Fig. 10.8. Electrophoresis of serum on starch or acrilamide gels gives better resolution of the various protein components because the fine pores of these media act as a sieve which separates them according to molecular

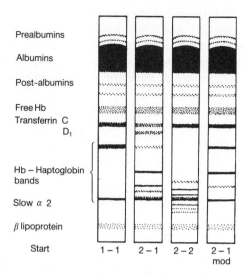

Prealbumins

Albumins

Post-albumins

Free Hb
Transferrin C
           D₁

Hb – Haptoglobin
bands

Slow α 2

β lipoprotein

Start   1 – 1   2 – 1   2 – 2   2 – 1
                                    mod

Fig. 10.8.   The pattern of haptoglobin, transferrin, and certain other protein bands as seen after electrophoresis of human serum in starch gel (tris-borate buffer system; horizontal method). Excess Hb has been added to the serum to saturate the haptoglobins. The three common haptoglobin types, 1-1, 2-1 and 2-2, and also a modified 2-1, are shown. In the type 2-1 specimen the transferrin variant Tf D_1 is shown, running behind Tf C. The globulin, much of which runs behind the starting-point, is not represented.
(From Barnicot 1961.)

size as well as electric charge. Either alone or in combination with other techniques starch gel electrophoresis has exposed a great variety in the serum proteins.

Haptoglobin (Hp)

One of the most genetically and anthropologically interesting of the serum proteins is haptoglobin. This is an α_2 globulin of molecular weight 85 000, which is concerned with transferring free haemoglobin to the liver where enzymes which will only react with the haemoglobin/haptoglobin complex begin breaking down the haem with the ultimate production of the bile-salt biliverdin. The Hb–Hp complex can be detected by staining electrophoretic gels for haemoglobin with reagents such as *o*-tolidine/ hydrogen peroxide.

Three simply inherited phenotypes are found in virtually all populations though in very varying frequencies. Type 1 is the homozygote for the allele Hp^1 and shows a single band. Type 2 the homozygote for the allele Hp^2 has a *series* of more cathodal bands and this seems odd if this gene codes for a single type of polypeptide chain in the orthodox manner. The heterozygote 2-1 has a weak band at the Hp 1 position together with a series of more cathodal bands that differ in position and relative strengths from those of type 2. It turns out that the *Hp²* allele does, in fact, code for a single kind of protein, but the molecules tend to stick together (polymerize) thus producing a series of polymers of increasing molecular weight. The haptoglobin molecule is composed of two light (α) and two heavy (β) chains. The essential structure of haptoglobin type 1-1 is shown in Fig 10.9.

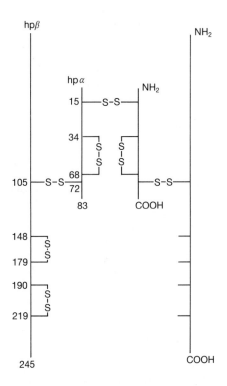

FIG. 10.9.   Structure of the Hp1 molecule.

Rare variants in the β-chain are known, but the polymorphism just described is due to differences in the α-chain. More refined electrophoresis using acid gels containing high concentrations of urea to break hydrogen bonds and mercaptoethanol to split sulphydryl bonds has shown that the hp1-1 phenotype is not homogeneous. The above procedure detects two kinds of α-chain: a fast moving one hpα1F and a slower moving one hpα1S. These are due to a pair of alleles *Hpα^{1F} Hpα^{1S}* so the hp1 as originally discovered may be any one of three genotypes *Hpα^{1F} Hpα^{1F}*; *Hpα^{1S} Hp^{1S}*; and the heterozygote *Hpα^{1F} Hp^{1S}*. This chain difference turns out to be due to a single amino-acid with lysine at position 53 in hpα1F and glutamic acid at the same position in hpα1S.

The hpα2-chain turns out to be of remarkable structure, made up of 142 amino-acids rather than 83 as in the hpα1-chain. The position of the key amino-acids is shown as being

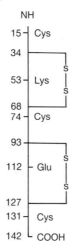

with lysine at position 53 and glutamic acid at position 112. Comparison with the hpα1-chain shows that the hpα2 is composed of most of an α1F-chain combined with an α1S-chain from which the first few amino-acids are missing. How can such a situation arise? The most likely explanation is that a chromosomal reciprocal translocation occurred in a heterozygote for *Hpα¹F Hpα¹S* as in Fig. 10.10.

An alternative explantion is that the *Hp²* gene arose through unequal crossing over in the *Hpα¹F Hpα¹S* heterozygote. This would afford a unique example of unequal crossing over within one gene, but it has been noted that the base-pair sequence for the first half of the *Hp¹* gene has an essential similarity to the sequence in the second half as if the gene itself may have arisen through a duplication process. Whatever the cause, *Hp²* is only found in man and has no equivalent in other primates. Since, however, it is found in practically all human populations it probably arose quite early in human ancestry.

In some tropical populations many individuals have no detectable Hp. This is usually a transient state due to haemolysis from malaria and other causes; the Hb–Hp complex is removed from the blood and Hp is only gradually replaced by synthesis in the liver. However, cases of a haptoglobinaemia due to an *Hp* null gene (or genes) in homozygous state *Hp⁰Hp⁰* are also known and tend to be relatively common in American blacks. The cause or causes of the 'defect' have not been determined.

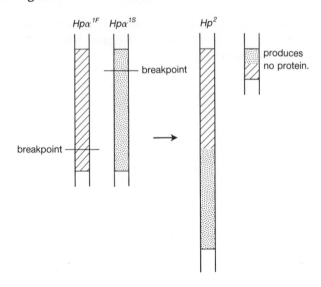

Fig. 10.10.   Formation of Hp^2 gene by reciprocal translocation.

Fig. 10.11 shows a world map of Hp^1 frequencies. High frequencies (60 per cent or more) are found in many tropical African peoples, in Central and South America, New Guinea, and Polynesia, while a zone of relatively low frequencies stretches from the Middle East to India and South-east Asia, and includes Australia. Far fewer populations have been tested for Hp^1 subtypes, but again there is evidence for considerable geographic variation. In western Europe the allele $Hp\alpha^{1S}$ is usually more frequent (20–25 per cent) than $HP\alpha^{1F}$ (10–15 per cent) and in Chinese, Japanese, Eskimos, and Australian Aborigines $Hp\alpha^{1F}$ is rare. Nigerians, on the other hand are found to have higher frequencies of $Hp\alpha^{1F}$ (47 per cent) than of $HP\alpha^{1S}$ (26 per cent).

It is still not clear whether the $Hp\alpha$ alleles are subject to selective pressures and, if so, how selection operates. Haemoglobin bound to Hp cannot pass the glomerular filter, and these gives some protection against damage to kidney tubules and loss of iron, but it is uncertain whether the Hp phenotypes differ in protective efficiency, and there is no clear evidence that the heterozygotes have an advantage over the homozygote in this respect. Haemolysis due to malaria and other diseases is common in Africa where Hp^1 frequencies tend to be high, but is equally prevalent in parts of Asia where the frequencies of this allele are conspicuously low!

A number of other serum proteins show polymorphisms and a few of those which are of special anthropological interest are now described.

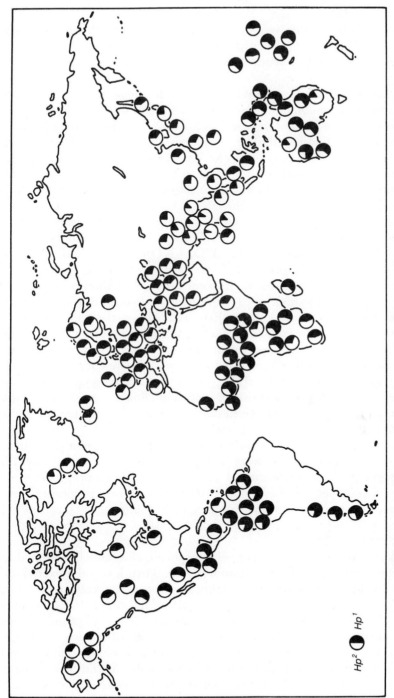

Fig. 10.11.   World distribution of the frequencies of the haptoglobin genes *Hp¹* and *Hp²* (From Walter and Steegmüller, *Human Heredity* **19**, 1969.)

Transferrin (Tf)

Transferrin (molecular weight 70 000) is a β globulin that transports iron from sites of red-cell destruction and from the intestine to the bone marrow, where Hb is synthesized. Each molecule can bind two atoms of iron.

About 20 inherited variants have been distinguished by differences of electrophoretic mobility. These variations probably depend on a series of alleles but, since most of the variants are rare, families segregating for two of them are seldom found. The commonest type, TfC, appears as a single band on gels while heterozygotes for one of the slow (D) variants or one of the fast (B) variants have an additional band. Transferrin bands can be identified specifically by adding radioactive iron (^{59}Fe) to the serum and, after electrophoresis, placing the cut gel surface on a photographic plate. When the plate is developed, after a suitable time of exposure, areas that were in contact with radio-labelled Tf show up as dark bands.

A slow (cathodal) variant, TfD_1, is quite common in African populations (gene frequency 1–5 per cent) and also in Australia, New Guinea, and adjacent islands. In some Australian aboriginal groups the gene frequency is as high as 20 per cent. Peptide analysis (finger-printing) shows that African and Australian TfD_1 have the same amino-acid substitution. Whether the same mutation occurred independently in these two regions and then increased in frequency due to selection or drift in each or whether both received it from some remote common ancestor is uncertain. Another variant, TfD_{chi}, with an electrophoretic mobility very much like that of TfD_1 was first described in Chinese, and has been found in various populations of South-east and East Asia, in the Veddahs of Ceylon, in some Indian tribes, and also in a number of American–Indian populations. Most of the B variants are infrequent, but Navajo Indians of the south-western USA were found to have 8 per cent of TfB_{0-1}. Experiments have so far failed to reveal any differences between the commoner types of Tf in their iron-transporting properties though different bacteriostatic properties have been claimed.

Serum albumin (Al)

To those who like to see a lot of polymorphism, albumin once seemed to be a rather boring protein that seldom varied. Then it was discovered that a fast (anodal) variant was remarkably frequent (gene frequency 9–14 per cent) in Naskapi and Montagnais Indians of Labrador. What appears to be the same allele (*Al^Naskapi*, now re-named *Al^Algonkin*) is found at lower frequencies in various other Algonkin-speaking tribes such as the Cree, Obijibwa, and Blackfoot, in Athabascan peoples, Sioux, Tlingit, and in Alaskan, but not Greenland Eskimos. Though the distribution of this gene is mainly in northern USA and Canada, it also occurs in the Navajo and Apache of the south-west who are linguistically related to the Athabascans.

The Apache and Navajo, and other south-western tribes, such as the Zuni and Pima, also have quite high frequencies of another albumin variant, which was discovered in Mexican Indians and is called albumin Mexico.

More than a dozen rare albumin variants have been reported from various parts of the world including Malaysia, India, Africa, Europe, and New Guinea.

α_1 anti-trypsin (Pi system)

This protein inhibits trypsin and certain other proteolytic enzymes. Its physiological role may be to block the proteolytic enzymes released from white cells at sites of inflammation. In acidic starch gels the protein separates as a series of bands anodal to the albumin. In a large sample of Norwegians four alleles were present at appreciable frequencies: Pi^m (95 per cent), Pi^s (2.3 per cent), Pi^z (2.0 per cent, and Pi^f (1.3 per cent) and two other rare alleles were also found. Little is known at present about the world distribution of these alleles, but in Lapps, Finns, and some Asiatic samples Pi^s was not found. Homozygotes for Pi^s show a marked deficiency of antitryptic activity, and it is also lowered to a lesser extent in the Pi^z/Pi^m heterozygotes and in subjects carrying the Pi^s allele. Pi^z homozygotes are unusually susceptible to chronic obstructive lung disease, tending to develop it in their late 30s on average, and it may be that susceptibility is also increased in heterozygotes. These deleterious effects presumably imply some selection against these genes, but the age of onset is fairly late in the reproductive span and this would diminish the selective pressure.

The C_3 component of complement

Complement was the name given long ago to a factor in fresh serum that is needed for the destruction of foreign cells that have been coated with antibody. It turns out to be a very complex system of 11 proteins that interact in sequence when the first in the series becomes attached to antibody molecules that have combined with the invading antigen. Polymorphism of one of the complement components, C_3, can be detected by electrophoresis of serum at high voltage in agarose gels. Two common alleles, C_3^1 and C_3^2 and about 10 rare alleles are known. C_3^1 has a frequency of 15–25 per cent in most European populations that have been tested. So far, these genetical variations have not been found to affect complement activity.

The Gc (group-specific) component

This protein can be seen as a series of two or more faint bands just cathodal to albumin on starch, or better acrilamide gels run at alkaline pH. However, identification depends ultimately on reaction with a specific anti-Gc antibody. Immunoelectrophoresis is commonly used to detect genetical variants. The serum sample is first crudely fractioned by electrophoresis in a thin layer of agar. A slit is then cut in the agar parallel to the direction of protein migration and filled with antibody. An arc of precipitated protein appears where antigen (Gc) and antibody meet by diffusion (Fig. 10.12). Three phenotypic patterns, due to two common alleles, Gc^1 and Gc^2, are found in all populations. Gc^2 frequencies vary around 25 per cent in Europe, but are lower (about 10 per cent) in Lapps, Africans, and Australian Aborigines. There is considerable variability in American Indians, the Navajo having only 2 per cent of Gc^2 and the Xavante of Brazil 69 per cent. Another allele $Gc^{Chippewa}$ was discovered in certain North American Indians and yet another, $Gc^{Aborigine}$ in Australia and New Guinea. The latter is not distinguishable by present methods from an allele that occurs in Africans. It is interesting that the Gc locus is closely linked to that for serum albumin, and it has been suggested that it may have been derived from the latter by gene duplication. It has been shown that the Gc protein is involved in the transport of vitamin D, and Gc^2 frequencies tend to be low in regions of low sunshine and high where levels of ultraviolet radiation are small.

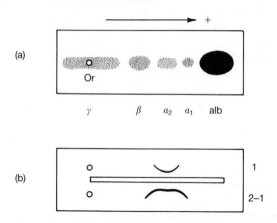

FIG. 10.12. Micro-immunoelectrophoresis. (a) A microscope slide has been coated with a thin layer of buffered agar. Serum is placed in a small hole cut in the agar near one end (Or). The slide is then connected to buffer tanks and a voltage is applied. Separation of the serum proteins into albumin and four globulin fractions is achieved. This slide has been stained to show the positions of the protein fractions. (b) Serum to be examined for Gc is placed in the two holes. After electrophoresis agar is cut from the central line to form a slot which is filled with anti-Gc serum. Precipitation in the form of arcs occurs where antigen and antibody meet by diffusion in suitable concentration. The slide shows a homozygous (type 1) with a single arc and a heterozygote (types 2–1) with two overlapping arcs.

Serum pseudocholinesterase

If alkaline starch-gel preparations of human serum are treated with a mixture of α-naphthylacetate and a suitable diazo dye, a series of four coloured bands develops. They are isozymes of pseudocholinesterase, an enzyme which hydrolyses certain organic esters and is present in various tissues, but not in red cells. It is distinct from acetylcholine-esterase which is active in splitting the neurotransmitter substance acetylcholine at neuromuscular junctions and which also occurs in red cells. It was noticed that some individuals develop prolonged paralysis and breathing difficulties when given the muscle-relaxant suxamethonium (succinyl dicholine) prior to surgery. Careful studies, using various substrates and enzyme inhibitors, showed that these individuals are usually homozygous for a rare pseudocholinesterase allele $E_1{}^a$ which reduces the activity of the enzyme. Heterozygotes for this and the normal allele $E_1{}^u$ show intermediate activity levels. The frequency of $E_1{}^a$ is about 1.5 per cent in various European populations, but considerably lower in Africans, East Asiatics, American Indians, and Australian Aborigines. A third allele, $E_1{}^f$, can be distinguished by its effect in lowering the sensitivity of the enzyme to inhibition by flouride, and there is a rare allele, $E_1{}^s$, which, in the homozygous state, is associated with virtual absence of enzyme activity.

Immunoglobins

The immunoglobins or γ-globulins form a highly complex and incredibly varied group of serum proteins. These are the humoral antibodies formed primarily by β-lymphocytes in response to the presence of foreign substances in the body and represent one of the main defensive mechanisms against infection by pathogens. The presence of a particular antibody in the sera depends upon the organism having been exposed to a particular antigen, but the capacity to synthesize that antibody is encoded in the genes. All that the antigen does is to stimulate a group of lymphocytes to synthesize the appropriate antibody which they are programmed to make and to release this into the sera. Any one lymphocyte can only make one antibody.

Electrophoresis distinguishes five major groups of immunoglobins with different physical and chemical properties: IgG, IgA, IgM, IgD, and IgE. The main features of these are summarized in Table 10.4. Most is known of the IgG group which has the highest concentration in serum and has provided examples of very large quantities of a single antibody to be found in sera in persons suffering from certain neoplastic diseases of the

TABLE 10.4
Some properties of Ig classes

Class	IgG	IgA	IgM	IgD	IgE
Chains	$(\varkappa \text{ or } \lambda)_2\gamma_2$	$(\varkappa \text{ or } \lambda)_2\alpha_2$	$\{(\varkappa \text{ or } \lambda)_2\mu_2\}_5$	$(\varkappa \text{ or } \lambda)_2\delta_2$	$(\varkappa \text{ or } \lambda)_2\epsilon_2$
Molecular weight	150 000	152 000 or 385 000	900 000	175 000	190 000
Carbohydrate content (per cent)	2.5	5–10	5–10		11.5
Mean adult serum concentration (mg/ml)	12.0	1.8	1.0	0.03	0.0003
Functional characteristics	Complement fixation; placental transfer	In external secretions	Complement fixation; early response; high agglutinating efficiency		Associated with certain types of allergy

lymphoid system (the myeloma proteins produced in myelomatosis). These have afforded material for detailed biochemical analysis.

An IgG molecule contains two sorts of polypeptide chain which differ in size. These are the light or L-chains containing about 220 amino-acids and the heavy or H-chains containing from about 400 to 750 amino-acids. Each molecule contains two identical L-chains and two identical H-chains and the chains are joined by disulphide bonds. There are also intra-chain disulphide bonds and they tend to separate groups of about 110 amino-acids. Thus, each L-chain contains two such groups. Amino-acid sequence analysis has shown that the first 110 or so amino-acids (from the amino-terminal end) in both L- and H-chains differ strikingly from molecule to molecule. This is known as the variable region and in it resides the high specificity of the antibody. Each antibody combines with just one antigen, and the antigen is bound in the variable regions of the heavy and light chains. Although variability characterizes all the region it particularly characterizes certain lengths of the sequence (the 'hypervariable' segments) and there are three such segments in L- and four in H-chains. The remaining lengths of both the L- and H-chains are much more similar from molecule to molecule and constitute what is known as the 'constant' region. These constant regions are concerned with the general effector functions of the molecule such as complement fixation and placental transfer. The typical structure of a molecule in very schematic form is shown in Fig. 10.13.

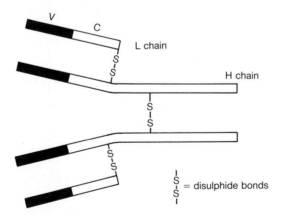

Fig. 10.13.   Diagram of an immunoglobulin molecule. Each half consists of a light (L) chain and a heavy (H) chain. The two halves are held together by disulphide bonds. Disulphide bonds also connect the L- and H-chains of each half. The black areas indicate regions of the L- and H-chains that are variable in amino-acid composition and the white areas indicate the constant regions.

Despite the relative invariance in the constant regions major sub-categories of both heavy and light chains are recognized, based upon differences in the regions. Two major types of L-chain occur, the kappa ($\varkappa$) and lambda (λ). The latter may be further differentiated in the constant region into three different kinds of λ light chain. Variation in the constant region of the heavy chains accounts for the major classes of the immunoglobin and there are four subtypes of IgG, two of IgM and two of IgA. What is more despite the variability of the V-regions different broad classes can be recognized and three of these occur in $\varkappa$ chains, five in λ chains, and four in heavy chains. Quite apart from the cause or the unique antigen specificity of immunoglobin molecules the above described situation requires some considerable genetic complexity. The minimum number of germ-line genes which would account for the situation as described is shown in Fig. 10.14.

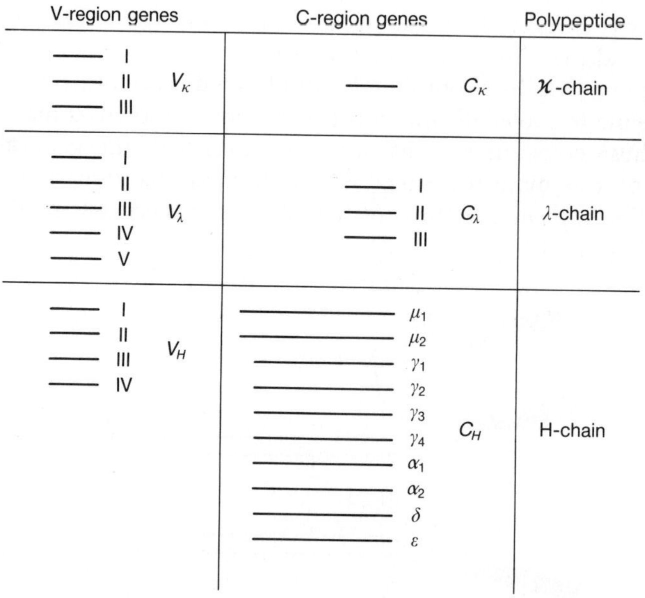

V-region genes		C-region genes		Polypeptide
— I				
— II $V_\varkappa$		— $C_\varkappa$		$\varkappa$-chain
— III				
— I				
— II		— I		
— III V_λ		— II C_λ		λ-chain
— IV		— III		
— V				
— I		——— μ_1		
— II		——— μ_2		
— III V_H		——— γ_1		
— IV		——— γ_2		
		——— γ_3		
		——— γ_4 C_H		H-chain
		——— α_1		
		——— α_2		
		——— δ		
		——— ε		

FIG. 10.14.   Genes determining immunoglobulin structure.

In fact, it now appears that the V-region is coded by at least two genes: one which codes for most of the amino-acid sequence and another which codes for a short sequence — the so-called J sequence which joins the main V section to the C section. It is thus evident that not only are a number of genes required in the production of a single immunoglobin molecule, but that even a single polypeptide in that molecule requires three gene loci — V-J-C — a clear exception to the one gene/one polypeptide hypothesis. There is, however, evidence that only a single strand of mRNA is involved and it is thought that in some way not yet understood the DNA strands responsible for the production of a particular polypeptide are brought into juxtaposition to form as it were a 'fused gene'. Such a structure is known as a 'translocon'. The three translocons for the two types of light chain and the heavy chain are not closely linked, and may well be on different chromosomes, but the genes for any one translocon are on the same chromosome.

The source of the variation in the V-regions is still a matter of controversy. Some consider that every variant is coded in the germ plasma. This demands an enormous number of genes though it may be noted that 1000 light chain variants and 1000 heavy chain variants permit 10^6 unique combinations. Within-gene recombination would generate equivalent diversity with fewer germ-line genes, but it has also been proposed that all the variety arises through somatic mutation.

Since any one lymphocyte is only capable of producing one antibody, clearly only the genetic code of one chromosome is being read. This phenomenon is known as allelic exclusion and is somewhat analogous to X-chromosome inactivation in females. Here, however, we are dealing with autosomal genes and the phenomenon is yet another example of the unique characteristics of immunoglobin determination.

So far, we have considered the genetic basis for the differences between antibodies from any one individual. However, there is also genetic variation between individuals, some of which is polymorphic. This is known for the λ and α heavy chains and for the $\varkappa$ and λ light chains.

Gm

The best investigated situation concerns the Gm specificities which are associated with the constant regions of the γ-chain particularly γG_1, γG_1, γG_2, and γG_3. Since antibodies are themselves complex proteins they represent antigens to individuals who do not have them and elicit antibody formation to them. Gm was first recognized through the varying capacity of sera from different people to react with a factor found in rheumatoid arthritis patients. This factor agglutinated red cells 'coated' with some incomplete rhesus anti-D antibody (p. 274), but was neutralized by some normal sera. In other words some people had an antibody which reacted with the rheumatoid arthritis antibody. Antibodies to antibodies can be obtained in various ways, specifically animal immunization and blood transfusion, but most illuminating is the fact that if a mother has a Gm factor not possessed by her baby, the baby will typically produce an antibody to it. This it does because maternal IgG antibodies are transferred to the baby and confer passive immunity on it, but as the child's own immune system develops it will synthesize antibody to the 'foreign protein'. Such antibodies tend, however, to disappear after the first few years of life.

The basis for some of Gm specificities is known. Thus, molecules of IgG_1 positive for Gm(1) — the first Gm factor to be discovered by Grubb in 1956 — have aspartic acid, glutamic acid, and leucine, at positions 356–358. Gm1 negative molecules have the sequence glutamic acid, glutamic acid, and methionine.

Gm(1) positive and negative, segregate as expected for traits determined by a pair of co-dominant alleles, as do other particular Gm specificities. However, if we test for several different factors we find that some are transmitted together in groups — a phenomenon well known in the Rh and MNS blood groups. In Europe for example, the patterns (Gm1,17,21), Gm(1,2,17,21), and Gm(3,5,13,14) are common and inherited as units. We know that factors 1, 2, 3, and 17 are G_1, whereas 5, 13, 14, and 21 are in G_3 and must therefore be produced by different loci. The loci are presumably closely linked since other combinations which could arise from crossing-over are seldom found in a given population. Cases of such cross-overs in families have been claimed and it is possible that some of the combinations present in different populations have been produced in this way. Since the combination of Gm factors are evidently transmitted by segments of DNA containing more than one gene it is convenient to speak of them as 'haplotypes'. The term indicates a set of factors regularly transmitted together by the haploid chromosome set of a gamete.

The polymorphism in the $\varkappa$ light chain is known as Inv and the variation is due to the substitution of Valine for Leucine at site 191. The Oz system occurs on λ chains and Am on the H chains of IgA.

Geographical distribution

Two systems of notation for these factors are still in common use, and this does not help to clarify a somewhat complicated subject; the relations between them are shown in Table 10.5.

The number of populations that have been tested for as many as 9 Gm factors is still not very large but the data assembled in Table 10.6 serves to show that there are some striking differences between the indigenous people of major geographical areas and in some cases between populations living in the same area. Certain haplotypes are frequent in some populations, but rare of virtually absent in others. Gm(3,5,13,14), for example, is common in Europeans (Caucasoids), but is infrequent or absent elsewhere unless there has been recent admixture with Europeans. There are four haplotypes that attain high frequencies in Africans or populations largely derived from them. It is interesting that about 10 per cent of Gm(1,5,13,14,17), an African haplotype, is found in Kurdish Jews of Iraq and about 2 per cent in Ashkenazi Jews. It will be noticed that the Bushmen have a deviant pattern, including one haplotype, Gm(1,5,17), that is peculiar to them and to peoples with whom they have mixed (Fig. 10.15). They also have another haplotype, *Gm(1,13,17)* that is found in eastern Asiatic peoples, but is not otherwise characteristic of Africans.

It is evident from Table 10.6 that testing for many factors increases the power of *Gm* studies for distinguishing between populations. For example, *Gm(1,5)* occurs in many parts of the world, but when additional factors are examined, it is seen to include four different haplotypes in Africa and a different combination again in eastern Asia and Oceania. It is an interesting fact that *Gm(1)* and *Gm(5)* segregate independently in Europeans.

Table 10.6 gives a broad view of regional contrasts in *Gm* haplotypes, but conspicuous frequency variations have also been reported in more restricted areas. A clinal increase in *Gm(1)* in passing northwards in Europe was noted some years ago. In work on Australian Aborigines it was found that certain tribes of northern Queensland have appreciable frequencies of *Gm(1,5,13,14)*, which is absent in tribes of the central and

TABLE 10.5

Relationship between original symbols and WHO symbols for Gm and Inv factors

Number	Original name
Gm 1	a
2	x
3	$b^w = b^2$
4	f
5	$b = b^1$
6	$c = c^5$
7	r
8	c
9	p
10	$\varkappa$
11	$\beta\, b^0$
12	γ
13	b^3
14	b^4
15	s
16	t
17	z
18	Rouen 2
19	Rouen 3
20	20
21	g
22	y
23	n
	b^3
	c^3
Inv 1	1
2	a
3	b

TABLE 10.6
Common Gm haplotypes in various populations tested for nine factors

	A	B	C	D	E	F	G	H	I	J.	K	Haplotypes
European (Caucasoids)	+	+	+									A (1,17,21)
Mongoloids	+	+						+	+			B (1,2,17,21)
Ainu	+	+		+				+				C (3,5,13,14)
New Guinea	+	+		+				+				D (1,5,13,14,17)
Australian	+	+								+		E (1,5,14,17)
Africans (Negroids)				+	+	+	+					F (1,5,6,17)
Bushmen	+			+				+			+	G (1,5,6,14,17)
												H (1,13,17)
												I (1,3,5,13,14)
												J (2,17,21)
												K (1,5,17)

(After Steinberg 1973)

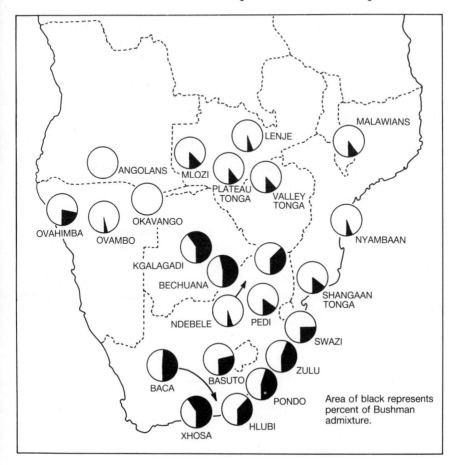

Fig. 10.15.   Map of southern Africa showing the estimated percentage of Bushmen admixture in various tribes as determined by the frequency of the *Gm1, 13* haplotype. (From Jenkins *et al., Am. J. phys. Anthrop.* **32**, 1970.)

western deserts. This, and certain other Gm data, are consistent with gene flow to Australia across the Torres Straits. In New Guinea itself there is a good deal of local variation in haplotype frequencies. In the Markham Valley area the frequencies of certain *Gm* haplotypes are correlated with the linguistic division between Melanesian- and non-Austronesian-speaking tribes and this evidence tends to support the view that the former came from South-east Asia. The frequency of *Gm(1,13)* has been used to chart Bushman–Hottentot admixture among the South African Bantu (Fig. 10.15).

Steinberg, who has been associated with much of the anthropological work on Gm factors, estimated the amount of Caucasoid and Khoisan (Bushman–Hottentot) admixture in Sidamo tribes of south-western

Ethiopia, and found 40 per cent and 12 per cent, respectively. He also studied the Ainu of Hokkaido, Japan, who have one haplotype *Gm(2,17,21)* peculiar to themselves, and he concluded that his sample of Ainu had about 30 per cent of Japanese admixture.

If the impression that *Gm* haplotype frequencies show greater contrasts between various populations than do other genetic markers is substantiated by further work, we must ask why these contrasts have been maintained in the face of gene flow and recombination tending to disrupt linkage associations. Strong regional selective forces would be an obvious, but at present speculative answer. The suggestion that certain haplotypes are associated with antibodies that are effective against particular diseases is attractive but unproved.

Three Inv factors, Inv(1), (2), and (3), are known. Inv(1) and Inv(2) are closely associated in inheritance and Inv(3) is almost antithetical to them. Most of the population work has been done using anti-Inv(1) or anti-Inv(2) sera. Inv(1,2) has been found at appreciable frequency in all populations examined. The frequency is relatively low (20 per cent) in Europe and the Middle East, but higher elsewhere, reaching a maximum of 94 per cent in certain Venezuelan American Indian tribes.

11   Population variation in qualitative traits (blood groups and others)

Blood groups

Blood-group variations depend upon differences in the chemical structure of substances which compose the membrane of the red-blood cell. In the case of the ABH and Lewis blood groups, about which most is known chemically, the differences reside in the carbohydrate components of complex molecules in which protein or lipid also occurs, i.e. these blood groups are glycoproteins or glycolipids.

Although blood groups represent just another kind of biochemical variety they are usually treated separately because of the techniques required for their identification. These are largely serological and involve immunological reactions between antigens and antibodies.

Blood-group substances are 'antigens', specifically 'haemaglutinogens', and elicit the formation of antibodies when introduced into an individual who does not himself possess the particular antigen in question. Many antigens are proteins, but other examples of carbohydrate antigens are known in bacteria. Specific antibodies that will detect blood-group variations are usually discovered by testing serum from pregnant women (who may have been immunized by the foetus) or from people who have received many blood transfusions. However, some occur 'spontaneously' in persons of particular blood-group constitution and a few (e.g. anti-M and anti-N) have been produced by immunizing rabbits with human red cells.

The principal technique used to detect blood-group variations of red cells, is 'agglutination'. When washed red cells are mixed with serum containing the specific antibody they become coated with antibody protein and clump together. However, some blood-group antibodies combine with the cells, but do not agglutinate them; they are said to be 'incomplete'. Even so agglutination can be produced by doing the reaction in a protein-rich medium, by pre-treating the cells with certain preteolytic enzymes, or by adding antibody that combines with human γ-globin. The last-named procedure (Coombs test) depends on the fact that the incomplete antibodies are themselves γ-globulins. The anti-γ globulin therefore combines with them and links the cells into clumps.

It is a curious fact that some useful blood-grouping reagents can also be obtained by extracting the seeds of certain leguminous plants. These 'lectins' are themselves globulins. A valuable anti-H is got from seeds of gorse (*Ulex europaeus*), an anti-A from *Dolichos biflorus*, and an anti-N from *Vicia graminea*.

Blood-group variations obey Mendelian rules of inheritance with gratifying precision. They are therefore very valuable in paternity testing, in distinguishing monozygotic from dizygotic twins, and in work on genetical linkage. Since many blood-group variations have high, but variable, frequencies in different populations, they are also valuable in anthropology. Moreover, most of them are detectable at birth and do not change thereafter except in a few rare diseases. It seemed at first that the rule 'one antigen, one gene' was a valid principle, but advances in molecular genetics and the discovery of certain odd blood-group phenomena have exposed its limitations. Structural genes are segments of DNA that code for the amino-acid sequences of protein polypeptide chains. The carbohydrate antigens detected in blood-grouping cannot therefore be direct gene products. The genes are believed to affect the enzymes (transferases) that build up these carbohydrate chains. Furthermore, genes contain numerous base sites each of which can undergo mutation. It is possible that a mutation affecting an enzyme could lead to a change in more than one antigenic specificity substance. We shall mention below several examples in which a group of antigenic specificities are inherited together. This situation can be formally explained by postulating a series of very closely linked genes; but if linkage is so close that cross-overs are virtually never observed this postulate is merely a convenience and throws no real light on gene structure. Serology is a sensitive and powerful tool, but less illuminating than chemistry from this point of view. It is unfortunate that very little is known about the physiological significance of the blood-group substances since knowledge of their functions might guide our search for selective agents acting on them. This ignorance may soon be dispelled by the intensive work now being done on cell membranes.

Race and Sanger in their classic book list 14 different blood-group systems, i.e. blood-group specificities or groups os specificities that appear to be controlled by independent gene loci. In addition there are a fair number of blood-group antigens that hardly every vary (public antigens) and some that have been found in only one or a few families (private antigens).

The A_1A_2BO system

This system was the first to be discovered; in 1900 Landsteiner showed that the serum of some individuals agglutinated the washed red cells of certain other individuals. This discovery laid the foundation of safe blood-transfusion and also gave geneticists and anthropologists a set of simply inherited variations that are quite frequent in most populations, are easy to detect, and are virtually unaffected in expression by the sex, postnatal age, and environment of the individual. It was not in fact until 1924 that Bernstein established the mode of inheritance of the ABO blood-groups. Table 11.1 shows the essence of the matter. Red cells can carry either A or B antigens or both or neither, so we have four phenotypes, groups A, B, AB, and O. Inheritance depends on three alleles A, B, and O which segregate to form six genotypes, AA, AO, BB, BO, AB, OO. The presence of the A antigen can be detected by mixing the cells with serum containing anti-A, whereupon they agglutinate; the B antigen is similarly detected with anti-B. No true anti-O is known, and we are therefore unable to distinguish homozygous A or B from heterozygous A or B (AO and BO) by serological tests. It also follows from this that we cannot determine the gene frequencies of a population simply by counting genes. We have to use indirect methods, making the assumption that the sample is in Hardy–Weinberg equilibrium for these alleles.

TABLE 11.1
Genetics and serology of ABO groups

Phenotypes	Genotypes	Cell antigens	Antibodies in serum
Group A	AA and AO	A	anti-B
Group B	BB and BO	B	anti-A
Group O	OO	—	anti-A and anti-B
Group AB	AB	A and B	neither

Anti-A occurs regularly in the serum of people who lack this antigen and anti-B in those who lack antigen B. It appears that these naturally occurring antibodies are formed in response to A-like and B-like antigens of bacteria and other organisms that invade the alimentary canal soon after birth.

Group A can be subdivided into Groups A_1 and A_2. A_2 cells react weakly with ordinary anti-A, but we have no specific anti-A_2 to identify this variation. However, some A_2 and A_2B people produce a specific anti-A_1 which can be used to subtype A cells. Nevertheless, we cannot distinguish serologically between the genotypes A_1A_2 and A_1A_1 or A_1O, or between A_2A_2 and A_2O. Various other quantitative variations of group A such as A-intermediate (A-int), which seem to be relatively frequent in Africans, have been described.

The secretor factor and the chemistry of ABH and Lewis antigens

In Western Europe about 75 per cent of group A, B, and AB subjects also have A and/or B antigens in their saliva and other mucous secretions. Secretion of these antigens in water-soluble form is inherited as a dominant trait depending on a gene *Se* which segregates independently from the ABO locus. In some populations, such as American Indians and Australian Aborigines, this gene is even more frequent and non-secretors (*se/se*) correspondingly rare.

Group-O secretors can be distinguished because they secrete an antigen known as H, which combines with the anti-H lectin obtained from gorse seeds. The H antigen is not specific to group O but is produced by other ABO phenotypes also, though in lesser amounts. It is a precursor substance which is converted to A or B antigen if the *A* or *B* genes are present and its formation depends on a gene *H* which is inherited independently from the *ABO* genes. There is thus an interaction between the *H* and *ABO* loci, the former producing a substance which can be acted upon by the products of the latter. Rare families are known in which some members have no ABH antigens either in their red cells or in their secretions (Bombay phenotype). Such individuals are apparently homozygous for *h*, a rare allele of *H*, and are therefore unable to form the H antigen.

The Lewis antigens Le^a and Le^b were first described as red-cell antigens determined by a locus *Le* inherited independently from the *ABO* locus. It was then noticed that people with Le(a – b +) cells are usually secretors of ABH and those with Le(a + b –) cells are usually ABH non-secretors. Nevertheless, Le(a – b +) individuals have both Le^a and Le^b in their secretions. It seems that the Lewis antigens are primarily water-soluble antigens that become attached to red cells only if their concentration in the plasma is high enough. Production of Le^a depends on a gene *Le* and if this gene is present Le^a is secreted whatever the ABH secretor type of the individual. Le^b specificity, however, depends on the presence of genes *H* and *Se* in addition to *Le*.

These rather complex relations between genotypes at the *H, Se,* and *Le* loci and the antigens found on red cells in secretions are summarized in Table 11.2.

They become easier to understand in the light of chemical work on the ABH and Lewis antigens. Most of this work has been done on the water-soluble forms since these are easier to handle than the lipid-associated antigens of the red-cell membrane. The water-soluble antigens are big molecules composed largely (80–90 per cent) of carbohydrate. The carbohydrate is in the form of chains built up of several different types of sugar united and attached at one end to a polypeptide core. Largely due to the work of Morgan and Watkins, it has been established that ABH and

TABLE 11.2

Relations between antigens on red cells and in mucous secretions and genotypes at the H, Se, and Le loci

Genotypes			Antigens on cells ABH Le^a Le^b	Antigens in mucous secretions ABH Le^a Le^b
H-locus	Secretor locus	Lewis locus		
HH or *Hh*	*SeSe* *Sese*	*LeLe* *Lele*	+ − +	+ + +
HH or *Hh*	*sese*	*LeLe* *Lele*	+ + −	− + −
hh	*SeSe* *SeSe*	*LeLe* *Lele* ⎱ Bombay types	− + −	− + −
hh	*sese*	*lele* ⎰	− − −	− − −
HH or *Hh*	*sese*	*lele*	+ − −	− − −
HH or *Hh*	*SeSe* *Sese*	*lele*	+ − −	+ − −

Lewis specificities depend on the pattern of carbohydrate residues at the free ends of the chains. The various genes involved are thought to code for enzymes (transferases) that catalyse the linking of particular sugar residues, but the structure of these enzymes is not yet known. There are two closely similar precursor substances, both of which can be converted to H antigen if the gene *H* is present. However, production of a fucosyl transferase that can form water-soluble H antigen depends in some way on the presence of the gene *Se*. The *A* and *B* genes code for glycosyl transferases that add, respectively, either N-acetylgalactosamine or galactose to the termination of the *H*-chain. If the gene *Le* is present fucose is added to the penultimate residue of precursor 1 to form Le^a antigen and Le^b is formed by a corresponding transformation of the H antigen.

The actual structural formula for the terminal sugars of A-specificity is

Galactose/N-acetylglucosamine is shown as a 1:3 linkage, but it also occurs as a 1:4 linkage. And, as can be seen from Fig 11.1, B substance has a terminal galactose in the place of N-acetylgalactosamine of A. H-substance lacks these terminal sugars. There are approximately 1 million ABH specificity sites on one red-blood cell.

It is now known that the loci for Lewis, Secretor, and H are closely linked on the short arm of chromosome 19. The *ABO* locus however is on chromosome 9.

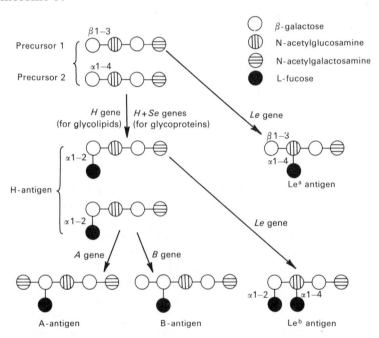

Fig. 11.1.   Structure and synthetic interrelations of the A, B, H, and Le blood-group antigens. Specificity depends on the structure of the terminal ends of the carbohydrate chains and only these are shown. The arrows indicate the actions of particular genes. (Based on data in Watkins, *Science, N.Y.* **152**, 1966.)

Geographical distribution of the A_1A_2BO blood groups

More is known about the distribution of this system than any other because it was the first to be discovered and because strong antisera for the relatively simple grouping techniques are easy to obtain. In medically advanced countries large numbers of people are ABO grouped by the blood-transfusion services so that relatively small regional variations in frequency may be detectable, for example in the UK (see Fig. 11.2).

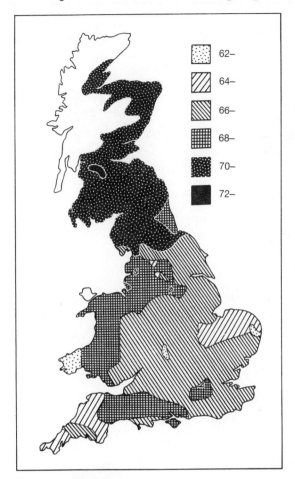

FIG. 11.2.   Frequency of the gene *O* of the ABO system in the UK. (Based on Kopec, *The distribution of the blood groups in the United Kingdom, 1970.*)

The broad pattern of the gene frequencies of the alleles *O*, *B* and *A* throughout the world are shown as contour maps in Figs 11.3, 11.4, and 11.5. Since the frequencies of the three alleles add up to 1.0 (or 100 per cent) there is some correlation between these frequencies; a marked rise in one must be accompanied by a fall in one or both of the other two. It is a striking fact that neither *A* nor *B* exceeds 50 per cent whereas *O* reaches 100 per cent in some South American tribes that have avoided intermixture. This suggests that there may be selective forces acting to limit the relative frequence of these alleles. We see in fact that *O* is above

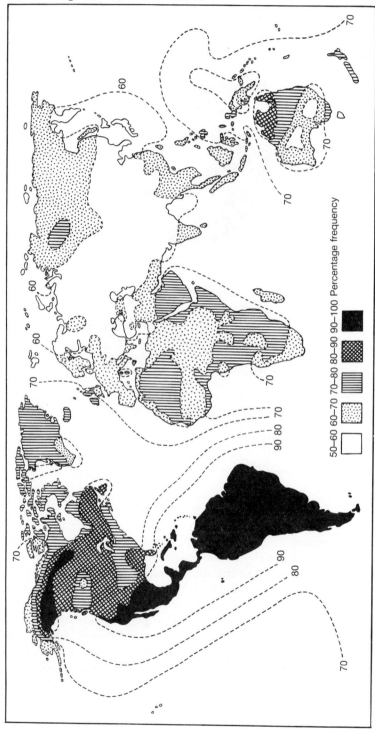

Fig. 11.3.   The frequency of the gene O of the ABO blood-group system. [Redrawn from Mourant, Kopec, and Sobczak, *The distribution of human blood groups* (2nd edn), 1976.]

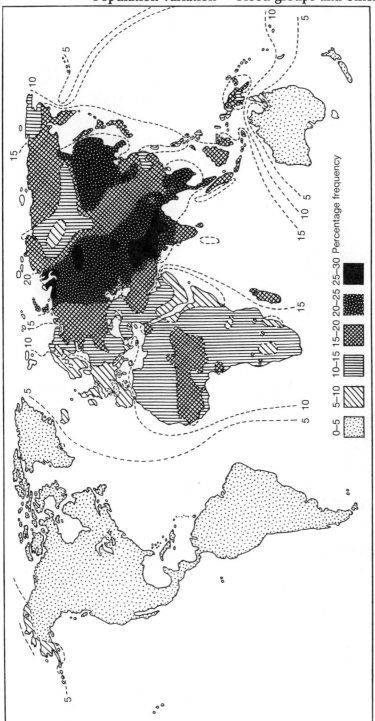

Fig. 11.4.   The frequency of the gene *B* of the ABO blood-group system. [Redrawn from Mourant, Kopec, and Sobczak, *The distribution of human blood groups* (2nd edn), 1976.]

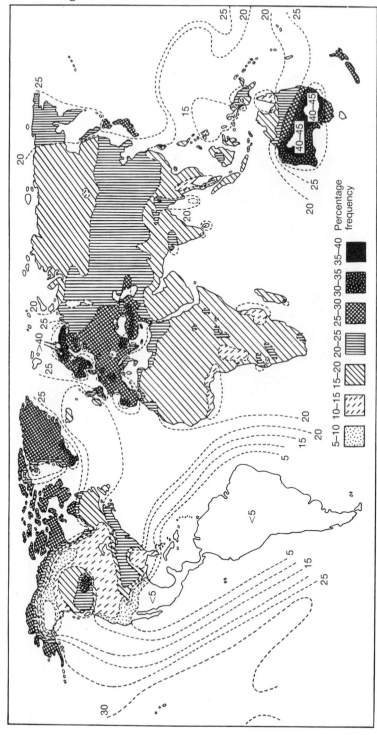

Fig. 11.5.   The frequency of the gene *A* of the ABO blood-group system. [Redrawn from Mourant, Kopec, and Sobczak, *The distribution of human blood groups* (2nd edn), 1976.]

80 per cent in most of the New World though the Alaskan and Canadian Eskimos in the far north have lower values. At the other extreme there is a broad zone of low O frequencies stretching from eastern Europe across much of central Asia and reaching part of the Pacific coast. In Europe we note that O frequencies in the British Isles are higher in the north-west and in Ireland. Comparably high frequencies (75 per cent) are found in Iceland, which we know from written records was colonized from Norway in the ninth century. However, no region of Norway has a frequency above 66 per cent. This discrepancy is probably due to the fact that the Norseman took Irish thralls with them, and these presumably contributed disproportionately to subsequent population growth. The frequency of the secretor gene is conspicuously low in Iceland (36 per cent), Northern Ireland (45 per cent), and Scotland (46 per cent) in comparison with Norway (53–59 per cent). Pockets of high O are also found in certain other peoples of Europe, such as the Basques, whose language is of obscure affinities, the Sardinians, Berber tribes of the Atlas Mountains, and people living in some parts of the Caucasus. We also see that O is relatively high in Arabia; however, Arabic-speaking tribes of Berber origin differ from the Arabians in other features of their blood-groups, apparently an example of a language spreading without much accompanying flow of genes. Before leaving the O map we should note the relatively high frequencies in northern Australia as compared with the rest of that country and with New Guinea. The island tribe, the Tiwi of the Gulf of Carpentaria has an O frequency of 95 per cent. This is no doubt an instance of genetic drift in a small community or of a founder effect; other examples of aberrant gene frequencies are known in various islands of this region.

The B gene map (Fig. 11.4) is to some extent the inverse of O; high frequencies are found in a large area of central east Asia with maximum of 25–30 per cent in the Himalayan region. In the New World and Australia, on the other hand, this allele is uncommon (5 per cent) or absent. The frequency tends to fall in passing from South-east Asia to Indonesia, and there is a marked drop between New Guinea and Australia, where B is found, in low frequency, only in the Cape York area. B is also rare or absent in Polynesia. Westwards from central Asia the gene frequency falls in an irregular cline and is notably low in the Basques. Presumably, the Asiatic peoples who first entered the New World had low B frequencies, in contrast to the modern inhabitants of eastern Asia; however, at least one east Siberian tribe, the Chuckchi, is known to have low B and A frequencies. In Africa B frequencies are low in Bushmen, but much higher in Hottentots.

Turning to the A map (Fig. 11.5) we see that frequencies are quite high in Europe, with some unusually high patches, notably in Lappland and Armenia. In the New World the A pattern is to some extent the inverse of

O. The gene is rare or absent in South and Central America, but much more frequent in North America, reaching 25 per cent or more in certain tribes such as the Blackfoot and Blood Indians. A is also frequent among Eskimos stretching from Alaska to Greenland. Again in Australia there is a resemblance of O and A patterns because the more southerly tribes, with low O have conspicuously high A frequencies, reaching 40 per cent or more in some cases.

The subgroup A_2 has a much more restricted distribution than A_1. The frequency of this gene is around 10 per cent in much of Europe and Africa, but falls to 5 per cent or less in India and South-east Asia. In other parts of the world it is rare or absent. The A_2 gene is conspicuously frequent (25–37 per cent) in the Lapps, though varying in different Lappish groups; it is also relatively frequent in the Finns. This is one of several examples of deviant gene frequencies in the Lapps, a peoples whose origins still remain obscure.

It is also interesting to consider not only the local frequencies of an allele, but where it occurs in greatest abundance. This means that population sizes have to be taken into account. Some Australian tribes have very high A frequencies, but these genes contribute a negligible fraction of A in the world because these tribes are small. McArthur and Penrose worked out the average frequencies for various regions and found that for the world as a whole they were $O = 62.5$ per cent, $A = 21.5$ per cent, and $B = 16.2$ per cent. They suggested that the average values would provide a useful standard against which the deviations of local populations could be assessed.

The ABO groups and natural selection

The fact that the great majority of the world's populations are polymorphic for two and usually three ABO alleles suggests that this variation has persisted during the many millenia of human dispersal when populations were small and liable to lose alleles by drift; however, we do not know how far gene flow between human groups may have hindered this process in these early times.

Antigens that closely resemble human ABH antigens occur in the great apes, which are also polymorphic for two or three alleles. It is therefore conceivable that the ABO polymorphism was already present much earlier, when the hominid and pongid lines diverged.

The recognition that the presence of a blood-group antigen in the foetus that was not present in the mother could lead to serious disease of the newborn in the case of the Rhesus system revived and stimulated a search for similar effects in the ABO blood-groups. In fact, comparable damage to the newborn due to ABO incompatibility does occur, but it is rare. The possibility remains, however, that immunological damage might take place at a much earlier stage of pregnancy and would pass unnoticed without special investigation. In general, apart from small chance fluctuations, we expect the *ABO* gene frequencies in a large sample of newborn children to be the same as in their parents. We can therefore look to see if there is a significant deficiency of certain phenotypes in the progeny of particular ABO matings. Some studies (usually on mother and child only) have shown remarkably large deficiencies in the births of children born to O mothers, but other investigations have not confirmed this. It must be remembered, of course, that very large samples, in the region of 15–20 thousand subjects, may be needed to give convincing evidence of selection in the 1–5 per cent range, a selective pressure which could, nevertheless, have substantial evolutionary effects. Selection by mother–foetus incompatibility necessarily involves the elimination of heterozygous progeny so that the gene-frequency equilibrium would not be stable unless there were some compensating mechanism such as heterozygous advantage in later life.

Another line of thought is that a person's ABO group may confer some protection against certain diseases. With this in mind the ABO frequencies of large samples of patients suffering from particular diseases have been compared with those of the control populations from which they were drawn: the best control being members of the same family segregating for the blood group. This has revealed a large number of statistically significant associations. The most striking is that between group A and carcinomas of almost every kind in peoples of European descent. Interestingly, carcinomatous tissue frequently contains an A-like antigen even in people who are blood group O and B. A individuals, of course, are unable to make an anti-A antibody. In contradiction people of blood group O tend to be prone to auto-immune disease, possibly because they possess both anti-A and anti-B antibodies. Various thrombotic conditions are more frequent in A group people. This is particularly evident in coronary thrombosis where the relative risk A/O is 1.29. Group O people, on the other hand, are more susceptible to haemorrhagic disorders. The difference is due to the lower level of clotting factor VIII in plasma from Group O than A. It was once thought that O people were more likely to develop duodenal ulcer than group A, but this difference is at least mainly due to the greater likelihood of ulcers haemorrhaging in the former and thus the sufferers being referred to hospital. However, non-secretors of ABH substances are more prone to such ulcers than secretors.

Cancers, ulcers, and thromboses tend to occur relatively late in life and after the years of reproduction. Their selective effect is likely to be small. Of greater evolutionary interest would be associations with infectious disease. Antigens that resemble human ABO antigens are found on the surfaces of pathogens, particularly bacteria, but the evidence for associations is slim. Claims for antigenic overlap between the plague bacterium *Pasteurella pestis* and H substance, and small-pox virus and A substance are probably not valid, but there are important associations between group A and haemolytic streptococcal infection, and its sequelae rheumatic fever and rheumatic heart disease. Virus diseases show a general tendency to associate with Group O. Mourant suggests that ABO frequencies were determined by compromises between maternal foetal imcompatibility and varying susceptibility to infections. He notes the high frequency to Group O in almost all genetic isolates; such isolates are also likely to be passed by epidemic disease and too small to maintain many endemic diseases.

The Rhesus (Rh) system

In 1940, Landsteiner and Wiener found that antisera obtained by immunizing rabbits with rhesus-monkey red cells agglutinated the cells of about 85 per cent of white Americans. Two phenotypes, Rh-positive and Rh-negative, could thus be distinguished, the former behaving as a dominant trait. Inheritance could be attributed to two alleles *Rh* and *rh*, which segregated to produce two dominant types, *Rh/Rh* and *Rh/rh*, and a recessive, rhesus-negative type, *rh/rh*. Subsequently, antisera were discovered in the blood of pregnant women or in multiple transfused subjects that gave essentially the same results. We now regard these latter antisera, derived from humans, as detecting an antigen D which is different from, but related to, the antigen in rhesus monkeys. The Rh-positive genotypes can thus be rewritten *D/D* and *D/d*, with *d/d* as the Rh-negative type.

The medical limelight was thrown on the Rhesus groups, when, as a result of work by Levine and Stetson in 1939, it was realized that they were connected with a severe disease of the newborn child. It emerged that this only occurred if the mother was Rh-negative (*d/d*) and the father Rh-positive (*D/D* or *D/d*). In a certain proportion of such matings the foetus is heterozygous (*D/d*) and thus carries the antigen D which the mother lacks. Disruption of the placenta at parturition allows some of these D-positive foetal cells to enter the maternal bloodstream and immunize her to the D antigen. In subsequent pregnancies anti-D antibody may cross the placental barrier and damage the red cells of the foetus, which is therefore born with a severe anaemia (haemolytic disease of the newborn, HDN). If the father is heterozygous (*D/d*) the risk is reduced because one-half of the progeny on average will be *d/d* and lack the D antigen.

Taking the frequency of Rh-negatives in this country to be approximately 16 per cent gives a *d* frequency of 40 per cent and about 9 per cent of random matings should be at risk for this disease. However, the incidence of HDN is only $\frac{1}{150}$–$\frac{1}{120}$, so that there must be other factors that reduce the risk. One of these is the ABO groups of the parents. If these are also incompatible, foetal cells entering the mother will be destroyed by her ABO antibodies and will have no chance to produce anti-D immunization.

For many years prompt blood-transfusion was the standard method of saving the lives of children with HDN. Then, in 1965 Clarke and his colleagues realized that foetal D-positive cells that had got into the mother's blood could be destroyed by giving suitable doses of anti-D. This method has been very successful in preventing maternal immunization.

The Rh system and natural selection

We have already mentioned immunological selection against heterozygous foetuses in relation to the ABO system. In the case of Rh it is D/d genotypes that are eliminated by this mechanism. Selection against heterozygotes only yields a balanced equilibrium if the frequencies of the two alleles are exactly 50 per cent; otherwise, the less frequent allele is gradually eliminated. In all populations, except the Basques, the frequency of d is below 50 per cent, and it may be that the polymorphism is transient and d is declining. If so we have to explain how it achieved such high frequencies locally in the first place. It has been suggested that the Basques are a remnant of a prehistoric western European population in which the d frequency was very high and which later mixed with eastern peoples with very high D frequencies. However, it is possible that the equilibrium is actually stabilized, perhaps by some unknown advantage of D/d heterozygotes in later life.

Further complexities of the Rh system

We have so far spoken of the Rh system as if it consisted solely of the antigens D and d, which are in fact the most important ones medically. However, intensive work on human sera soon showed that there are additional associated antigens. In 1943, noticing certain regularities in the serological results, Fisher put forward a scheme to explain the genetics of the system. He postulated three very closely linked gene loci, each of which could be occupied by one of a pair of alleles, D or d, C or c, and E or e. This would allow eight types of gene triplet, or haplotypes on a chromosome as shown below:

Fisher–Race notation	CDE	CDe	cDE	cDe	cde	Cde	cdE	CdE
Wiener notation	R_z	R_1	R_2	R_0	r	r'	r''	r_y

Fisher thought that each gene produced its particular antigen and the corresponding six antisera would ultimately be discovered. This prediction was fulfilled except that anti-d has never been found and may not exist. Fisher also noticed that $CDe(R_1)$, $cDE(R_2)$, $cDE(R_0)$, and $cde(r)$ are the commonest types, and suggested that the rarer ones could have been produced by very rare cross-overs between them. Indeed, unless cross-overs do occur we have no genetical way of distinguishing the hypothetical triplets from single genes. Wiener in fact did not accept Fisher's concept and preferred to envisage a set of eight alleles with complex antigen-determining properties. His notation, which is conveniently simple, is given above alongside the Fisher–Race symbols. No doubt the chemical structure of the antigens, which is still unknown, will resolve these controversies.

The reader will realize that since each gene-triplet is located on a single chromosome a person's genotype may, for example, be $CDe/cde(R_1/r)$ or $cDE/cDe(R_2/R_0)$ containing six genes. Tests with the five available antisera do not necessarily give unambiguous genotypes since they do not tell us how the six genes are arranged as a pair of haplotypes, except in certain cases such as $(c+)$, $(C-)$, $(D-)$, $(e+)$, $(E-)$, which must be the homozygote cde/cde. In individual cases the genotype can usually be resolved from family data, but in population material the presence of some of the rarer haplotypes may be in doubt.

It has been noted that three of the common haplotypes cDE, CDe, and cde are each connected to the fourth cDe by mutation with regard to a single antigenic specificity. This may indicate that cDe was the ancestral condition.

Research on the Rh factor has revealed many curious serological phenomena not all of which are fully understood. We can mention only a few, especially those that affect population studies. Certain D antigens react weakly with some anti-D sera and are designated D^u; moreover, the reaction strength varies suggesting a series of different D^u antigens, D^u is common in Africa and may be mistaken for D-negative by the unwary.

An antiserum anti-V was discovered which gave positive reactions with the cells of less than 1 per cent of white Americans, Orientals, and American Indians, but reacted with those of 40 per cent of West Africans. The interpretation of anti-V seems to be that it reacts with cells that have both c and an allele of e, e^s, on the same chromosome. Other antisera of this kind such as anti-ce, and anti-Ce have been discovered.

Rare individuals have been found who appear to have only the D antigen (-D-) and some which give no reactions for any of the Rh antigens (Rh null). In at least one of the latter cases the parents were positive for Rh antigens and these must have been suppressed in the child.

There is another blood-group system — LW — which is very closely associated with Rhesus: indeed the antibody discovered by Landsteiner and Wiener in 1940 (p. 274) was an anti-LW hence the name of the system. Its first recognition as different from anti-D was when Murray and Clark found that heat extracts from rhesus-negative red cells could stimulate the production in guinea pigs of an apparent anti-D. This was first called 'D-like'. Most people are positive for the LW antigen, but occasionally one finds the rare homozygote *lwlw*. The latter have normal rhesus antigens, but Rh^null is also lacking in LW. It would appear that Rhesus is a precursor to LW.

Geography of the Rh groups

The Rh system with its eight haplotypes and variety of alleles at each locus gives great scope for variation as we see from Table 11.3 in which the frequencies in a selected range of populations are given. $CDe(R_1)$, for example, ranges from less than 5 per cent in some African populations to over 90 per cent in many tribes of New Guinea. There is a tendency for this haplotype to be more frequent in the Mediterranean zone than elsewhere in Europe and for $cde(r)$ to be correspondingly low. This trend, which is marked in the Sardinians, continues into northern India and in the Far East $cde(r)$ is rare. An allele of the gene C, C^w, is unusually frequent in Lapps, Finns, and Latvians. $cDE(R_2)$ is conspicuously frequent in American Indians and it is fairly high in Polynesia and South-east Asia.

TABLE 11.3
Frequencies of variants at the Rh locus in various populations

	No. tested	CDE (R_z)	CDe (R_1)	cDE (R_2)	cDe (R_0)	Cde (r')	CdE (r_y)	cdE (r'')	cde (r)
English	1798	0.27	41.43	14.50	2.62	1.19	0.00	0.97	39.03
Italians (Milan)	772	0.35	47.56	10.77	1.63	0.67	0.33	0.69	38.00
Basques	383	0.00	37.56	7.07	0.50	1.47	0.00	0.25	53.15
Sardinians	107	—	66.84	8.84	2.12	0.00	—	0.00	22.19
Norwegian Lapps	183	0.00	52.46	18.37	10.34	0.00	0.00	0.00	18.83
N.W. Pakistanis	253	0.63	66.69	7.78	0.00	0.50	0.00	0.00	24.40
S. Chinese	250	0.49	75.91	19.51	4.09	0.00	0.00	0.00	0.00
Australians	234	2.08	56.42	20.09	8.54	12.87	0.00	0.00	0.00
New Caledonians	147	0.00	80.27	12.71	7.02	0.00	0.00	0.00	0.00
Blood Indians	241	4.06	47.81	38.84	0.00	0.00	0.00	3.42	9.87
Chippewa Indians	161	2.04	33.67	53.03	0.00	0.00	0.00	2.22	8.03
S.W. Nigerians	145	0.00	9.51	8.28	61.91†	1.87	0.00	0.00	18.43
S. African Bantu	644	0.00	2.81	4.27	73.95†	7.07	0.00	0.00	11.84
Bushman	232	—	9.04	1.96	89.00†	0.00	—	0.00	0.00

† Including $cD^u e$.

$cDe(R_0)$ is remarkable because the frequencies in Africa greatly exceed those in any other region. Frequencies of over 90 per cent have been recorded in peoples as physically contrasted as certain Nilotic tribes and the Bushmen. As we approach the Middle East, R_0 frequencies fall, but are still 50–60 per cent in Ethiopia, 15–20 per cent in some tribes of the Yemen, and 10–16 per cent in the area south of the Caspian Sea. Moderately high R_0 frequencies have also been found in a few other scattered peoples such as Negritos of Malaya, and some tribes of northern Australia.

Among the rarer haplotypes we find that $CDE(R_z)$ is relatively high among some of the so-called Veddoid tribes of India, and in some Australian tribes, though not among the Veddah themselves. The rare type $cdE(r'')$ is surprisingly frequent among the Ainu of northern Japan.

We have mentioned the 'compound' antigen V(ce^s). Like cDe high frequencies of this antigen are characteristic of Africa, where gene frequencies range from 20–30 per cent, falling to 5–10 per cent in Arabia.

The MNS system

The MN blood-groups were discovered in 1927 by Landsteiner and Levine as a result of immunizing rabbits with human red cells. Rabbit antisera are still in general use for work on this system, but various purification procedures are necessary, and it is not easy to prepare antisera which are both strong and specific for the required antigen. The inheritance is very simple; three phenotypes M, MN, and N can be recognized corresponding to the three genotypes MM, MN, and NN (or $Ag^M Ag^M$, $Ag^M Ag^N$, and $Ag^N Ag^N$, cf. p.156).

The M map of the world (Fig. 11.6) shows some major regional contrasts, ranging from less than 20 per cent in New Guinea and parts of Australia to 90 per cent in some areas of the New World. Areas of relatively high M are also seen in Arabia, north-east Siberia, and parts of South-east Asia. In much of western Europe frequencies range around 50 per cent, but are lower in some Lapp groups and higher in nearby Finland. The Sardinians, though they show the high O of the Basques, differ from them greatly in their high M frequency (75 per cent).

In 1947 the simplicity of the MN system was shattered by the discovery of an antiserum, anti-S. This detects an antigen which is clearly different to M or N, but is in some way associated with them. If S were independent of MN groups inheritance we should expect to find the same proportion of S-positives in each of the phenotypes M, MN, and N within the limits of error; but as shown below in data from a British sample this is not the case and the proportion of S-positive M is much higher than would be expected by chance.

		M + MN	N	
anti-S	obs	93	15	108
	exp	82	26	
	obs	52	30	82
	exp	63	19	
		145	45	190

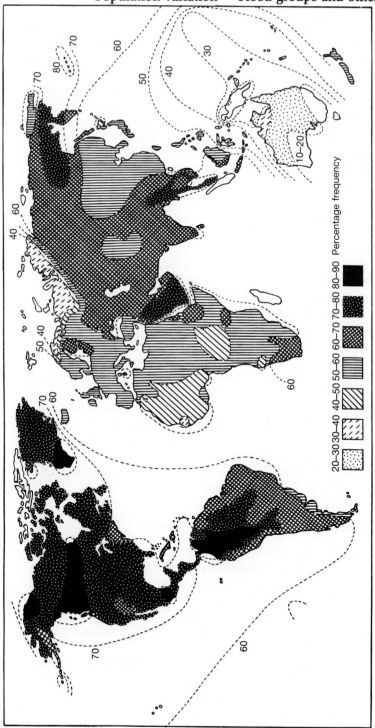

FIG. 11.6.   The frequency of the gene *M* of the MN blood-group system. [Redrawn from Mourant, Kopec, and Sobczak, *The distribution of human blood groups* (2nd edn), 1976.]

Studies of families show that M and S segregate as a pair. We therefore have a situation like that in the Rh groups. The antiserum anti-s, which detects the allelic antigen s, was discovered later. Using four antisera we can therefore distinguish four gene pairs or haplotypes, MS, Ms, NS, Ns. Anti-s is a fairly rare antiserum and is not often used in survey work. In this case we cannot distinguish heterozygotes, such as $MSMs$, but we can identify $MsMs$ because it does not react with anti-S. The frequency of S in Europe is usually about 30–35 per cent, and it is mostly associated with M. S frequencies are even higher in the Middle East (40 per cent) and here, also, as in India, MS is the commonest association. S frequencies are much lower in Eastern Asia (5–20 per cent), and here M is usually with s. S is virtually absent in Australia despite the fact that this gene reaches high frequencies in parts of New Guinea.

A further development was the discovery of anti-U. This antiserum was found not to react with the cells of some American Blacks and these individuals were usually, but not invariably, negative for both S and s. It is now usual to think of an allele S^u at the S locus which suppresses the expression of the other allele when it is present. S^u seems to be very common in parts of Africa; 36 per cent of a sample of Congo Pygmies were found not to react with anti-U. The gene S^u has not been detected in Australian Aborigines.

The MNS system is certainly even more complicated, but we need not go into details. On the other hand, there are various rare genes that appear to be alleles of M, N, or S and on the other there are additional specifications associated with the system as MN is with Ss. One of these, Henshaw, was detected by rabbit antisera in a Nigerian subject. It appears to be widespread in Africa, but not frequent (5 per cent or less), and, at least in West Africa, is associated with NS.

A summary of some of the features of other important blood-group systems is given in Table 11.4.

The histocompatibility (HL-A) system

It is well known that skin and other tissues grafted from an individual to another usually evoke an immunological reaction leading to destruction of the graft. This graft-rejection is in general less severe if host and donor are closely related. Inherited differences in the HL-A antigens of the tissues are an important factor in these incompatibilities and the system has been intensively studied with a view to improving graft-tolerance by matching the antigens of donor and recipient.

Antisera to HL-A antigens are usually obtained from pregnant women who have become immunized by the foetus; however, unlike some of the blood-group antibodies they do not damage the unborn child. The antigens are detected by a cytotoxicity test. Lymphocytes from the person

TABLE 11.4

Characteristics of some anthropologically important blood-group systems

System	Genetic characteristics	Geographical distribution
P	Common antigen Tja can be differentiated into P_1 and P_2 by certain antisera. This differentiation determined by a pair of autosomal genes.	Technical difficulties made early survey work unreliable, but P_2 varies from less than 5% in West Africa, through about 50% in Europe to 75–80% in the Far East.
Lutheran Lu	Two autosomal alleles Lu^a and Lu^b with three genotypes phenotypically distinguishable. Rare Lu negatives due to gene at a different locus.	Lu^a is found in low frequency in Europe and Africa: at still lower frequencies in India and virtually absent from Australia.
Kell K	Three pairs of autosomal closely linked alleles K/k; Kp^a/Kp^b; and Js^a/Js^b (Sutter antigen). Incompatibility for the K/k antigens may cause haemolytic disease of the newborn.	Only half of the expected eight haplotypes found because K and Kp^a are almost exclusively found in peoples of European and Near East decent whilst Js^a is confined to Africa and peoples of African descent. There is marked linkage disequilibrium in hybrid groups.
Duffy Fy	Three autosomal alleles Fy^a, Fy^b, and Fy^4. Fy^4 appears to provide protection against vivax malaria.	Fy^4 is extremely common in Central Africa reaching 100% in a sample of Pygmies. Also high around the Red Sea and in the Kurds of N Iraq and Iran. Elsewhere it is rare being no more than 3% in Europe. Fy^a reaches 100% in Australians, and is between 35% and 45% in western Europe.
Diego Di	Autosomal with only one antibody known	Highest frequency of Di in South American Indians is 30%; somewhat lower in North America and lower still in Japan and China. But absent elsewhere.
Xg	Sex linked on the short arm of the X-chromosome. Xg^a.	Highest frequency of Xg^a in New Guinea and Australians 85–79%, but common in most populations studied. Lowest known frequency 38% among Taiwan aborigines.

to be tested are mixed with a specific anti-HL-A antiserum in the presence of complement. If the cells carry the corresponding antigen they are damaged and can be distinguished from normal cells by staining with certain dyes.

The HL-A antigens are determined by five closely linked loci *A, B, C, D*, and *Dr* on the short arm of chromosome 6. Their relative positions and recombination fractions can be represented as follows and it may be seen that they are closely linked with the region of chromosome 6 concerned with the second and fourth component of complement, and Factor B in the complement system.

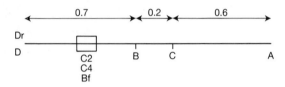

Other loci which are in the close vicinity carry the genes for PGM3, glyoxylase, and pepsinogen 5. Loci controlling the immune response (the *Ir* genes) are also believed to map in the HLA region — these, interestingly, determine the level of antibody response.

HLA A, B, and C are found on the surfaces of almost all cells (but not red blood cells, sperm, and placental trophoblast). Dr has a more restricted distribution and is found, for example, on β-lymphocytes, but not T-lymphocytes. The molecular organization of A, B, and C antigens consists of two polypeptide chains, one with a molecular weight of 43 000 and the other with mol wt 12 000. The HLA antigenicity lies in the larger chain and the gene coding for the smaller one — the so-called β_2 microglobulin — is on chromosome 15 and appears invariant. HLA Dr has a different molecular architecture with two polypeptide chains (33 000 and 28 000 mol wt) both of which span the cell membrane unlike β microglobulin.

At least 19 alleles (antigens) have been described for the *A*-locus, 27 for *B*, 7 for *C*, 11 for *D* (which determines the mixed leucocyte response), and 7 for *DR* yielding an enormous number of haplotypes. Furthermore, not all the variety has yet been recognized for antibodies to some antigens have not detected. In the case of the *C* locus there are around 50 per cent 'blank' alleles. Furthermore, none of the alleles are extremely common though a number reach frequencies of 10 per cent or over in some populations. Notwithstanding the close linkage most of the combinations considered in pairs are not in linkage disequilibrium. Among 300 pair-wise combinations only between 10 and 20 pairs show such disequilibrium in Europeans, e.g. A, and B8.

A number of diseases have been found to be very strongly associated with HLA antigens. Thus, ankylosing spondylitis is much more frequent among possessors of the B27 antigen than among other types. Similarly, coeliac disease is associated with DW3 and multiple sclerosis with DW2. A3 are prone to haemochromatosis. Less marked associations have been reported for Hodgkin's disease, acute lymphatic leukaemia, juvenile insulin-dependent diabetes, thyrotoxicosis, and even schizophrenia. Typically, there is at least a suggestion of autoimmunity in the aetiology of the diseases which show HLA association. A particularly interesting possibility is an association between HLAB8 and intolerance to gluten: a protein contained in wheat and barley. The frequency of the allele appears to be inversely geographically related to length of time since adoption of agriculture by populations in Europe.

Geographical distribution

The amount of variation which is being uncovered in the HLA system is enormous, more than in all the known red-cell blood-group systems put together. Indication of the extent of the geographic variety is given in Table 11.5 in terms of allele frequencies at three of the loci in three major geographical groups of people. Here it will be noted that the frequencies of 'blanks' due to appropriate antisera not having yet been found, are particularly high in African and Japanese populations.

The alleles broadly fall into three groups: those that are present in all populations at quite high frequency such as A_2; those which tend to be present, but are occasionally totally absent from some population such as A_1 in Africans, and those which are confined to a geographically restricted group such as *Bw42* in Africans. The HLA system is turning out to be of immense value in searching for genetic relationships between populations.

Variations in sensory perception, genetic disease, and total genome

Taste-deficiency for phenylthiourea

A chance observation by Fox in 1931 showed that some people are unable to taste the synthetic compound phenylthiourea (phenylthiocarbamide or PTC) which others describe as very bitter like quinine. It was later shown that the ability to taste this substance is inherited as a simple Mendelian dominant. The simplest form of test is to give people crystals of PTC, or paper impregnated with it. A better method is to get the subject to taste a

TABLE 11.5
HLA-A -B, *and -C gene frequencies (%)*

Allele		European (228)*	African (102)	Japanese (195)
A1		15.8	3.9	1.2
A2		27.0	9.4	25.3
A3		12.6	6.4	0.7
Aw23	A9	2.4	10.8	—
Aw24		8.8	2.4	36.7
A25	A10	2.0	3.5	—
A26		3.9	4.5	12.7
A11		5.1		6.7
A28		4.4	8.9	—
A29		5.8	6.4	0.2
Aw30		3.9	22.1	0.5
Aw31		2.3	4.2	8.7
Aw32		2.9	1.5	0.5
Aw33		0.7	1.0	2.0
Aw43		—	4.0	—
Blank		2.2	11.0	4.2
B5		5.9	3.0	20.9
B7		10.4	7.3	7.1
B8		9.2	7.1	0.2
B12		16.6	12.7	6.5
B13		3.2	1.5	0.8
B14		2.4	3.6	0.5
B18		6.2	2.0	—
B27		4.6	—	0.3
B15		4.8	3.0	9.3
Bw38	Bw16	2.0	—	1.8
Bw39		3.5	1.5	4.7
B17		5.7	16.1	0.6
Bw21		2.2	1.5	1.5
Bw22		3.6	—	6.5
Bw35		9.9	7.2	9.4
B37		1.1	—	0.8
B40		8.1	2.0	21.8
Bw41		1.2	1.5	—
Bw42		—	12.3	—
Blank		2.4	17.9	7.6
Cw1		4.8	—	11.1
Cw2		5.4	11.4	1.4
Cw3		9.4	5.5	26.3
Cw4		12.6	14.2	4.3
Cw5		8.4	1.0	1.2
Cw6		12.6	17.7	2.1
Blank		46.7	50.2	53.5

The data in this Table is taken from the *Seventh HLA Workshop* (Bodmer *et al.* 1978) series of dilutions of the substance in water and this test shows that tasters vary considerably in sensitivity. The diagnosis of a taster or non-taster is, however, essentially subjective, depending on the subject's description of his sensations. The test can be further improved by Harris and Kalmus's procedure in which the subject is asked to pick out the PTC solutions from a number of samples, half of which are water. By repeating the test with increasing concentrations of PTC the threshold concentration at which the substance is first detected can be determined, and the distribution of thresholds in a population can then be plotted as a histogram (Fig. 11.7). A bimodal distribution is obtained, one mode for the tasters and one for the non-tasters, but there is usually some overlap so that a few subjects are

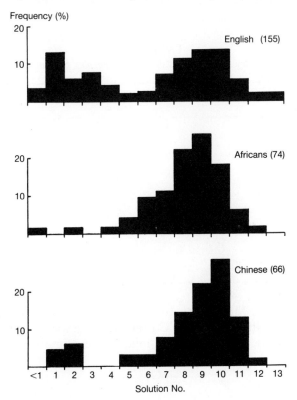

FIG. 11.7.   The distributions of taste-thresholds for PTC in three populations. The solution numbers represent serial dilutions by one-half of a 0.13 per cent solution of PTC. The tests were made by the sorting technique. The form of the non-taster distribution, occupying the region from solution 5 to less than 1 is clearly shown in the case of the English sample. (From Barnicot, *Ann. Eugen.*, 1950.)

difficult to classify. The test can be further refined by correcting for the slightly greater sensitivity of females, the decline of taste acuity with age, and for general taste sensitivity to bitter materials by control tests with brucine. Inability to taste PTC is inherited as a recessive trait, but there is some evidence that the threshold is higher in the heterozygous tasters (Tt) than in the homozygotes (TT).

Distribution   A limited number of populations have been tested by the sorting technique and some results are shown below (Table 11.6). The frequency of the non-taster phenotype in north-western Europe is 35–40 per cent, but appreciably lower in Mediterranean samples. In Africans, Chinese, Japanese, American Indians, and also in the Lapps it is very much less frequent.

TABLE 11.6
Frequency of non-tasters of PTC in various populations

	Number tested	Non-tasters (per cent)
Hindus	489	33.7
Danish	251	32.7
English	441	31.5
Spanish	203	25.6
Portuguese	454	24.0
Negritos (Malaya)	50	18.0
Malays	237	16.0
Japanese	295	7.1
Lapps	140	6.4
West Africans	74	2.7
Chinese	50	2.0
South American Indians (Brazil)	163	1.2

Selective influences   Many other compounds chemically related to PTC show taste bimodality, though PTC itself gives the widest separation of the two phenotypes. It appears that the following configuration is essential:

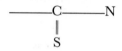

The corresponding urea derivatives in which oxygen (O) is substituted for sulphur (S) are inactive in this respect.

The fact that certain thioureas, notably thiouracil, are used clinically for the suppression of thyroid activity led to the idea that the polymorphism for the taster gene might be in some way connected with thyroid function.

Taste-testing of patients with various thyroid diseases has shown that the taster frequency in samples with certain types of goitre is divergent.

Observations on the reactions of chimpanzees to PTC solutions suggest that taste-deficiency may also occur in these apes. If this is so, it has been argued, the polylorphism may be one which has been maintained, presumably by heterosis, since a remote stage of human evolution.

Colour blindness

Defective colour vision of the kind known as 'red-green colour blindness' is a familiar example of human X-linked inheritance. Affected people tend to confuse red and green hues and, in the most severe types, they seem to have no perceptions of these colours. Critical testing shows that there are two physiologically different forms of red-green defect, 'protan' and 'deutan', and each of them varies in severity. The most severe types are called protanopia (red blindness) and deuteranopia (green blindness). The corresponding milder defects are known as protanomaly and deuteranomaly.

It is almost certain that protan and deutan defects are due to mutations at two different, but quite closely linked, loci on the X-chromosome. If this is the case we would expect to find rare males with both types of defect. We would also expect to find occasional families in which a woman with normal colour vision has protan, deutan, and normal sons; this situation could arise if the mother were heterozygous at both loci and produced a normal chromosome by crossing-over. Both these situations have been reported, though the diagnosis of mixed defects is not easy. Differences in the severity of red-green defects are thought to be due to two or more allelic variants at each locus.

The mode of action of the colour-defect mutations is obscure; it is difficult to get material for detailed biochemical work on the human retina. However, some elegant optical experiments on living subjects suggest that protanopes lack the red-sensitive pigment in the cone receptors; their sensitivity to light in the red range is in fact very low. Deuteranopes, on the other hand, have normal light sensitivity in both the red and green ranges, and it may be that the defect is at some higher level of neural organization. We must also mention that there are autosomally inherited defects of colour perception in the blue-yellow region ('tritan' defects), but they are rare. Cone monochromatism, in which perception of colour is entirely lacking, is even rarer. The approximate frequencies of these various deficiencies in Western European populations are shown in Table 11.7. It will be noted that deuteranomaly is the commonest type. The frequencies are considerably higher in males than in females as expected for X-linked recessive traits. The frequency in males gives a direct estimate of the gene frequency q, and, at equilibrium, the frequency of affected females (homozygotes) is expected to be q^2.

TABLE 11.7

Frequencies of various types of colour-vision defects in Europeans

Type of defect	Frequency in population (per cent)	
	Male	Female
Monochromatism	Very rare	Very rare
Dichromatism	2.105	0.06
Protanopia	1.0	0.02
Deuteranopia	1.1	0.01
Tritanopia	0.005	0.003
Anomalous trichromatism	5.0	0.40
Protanomaly	1.0	0.02
Deuteranomaly	4.9	0.38
Tritanomaly	Fairly rare	Fairly rare
Colour-vision defects	(approx.) 8.0	(approx.) 0.46

Modified from Wright (1953).

A thorough investigation of colour-vision defects involves measuring sensitivity to light and ability to make colour matches and discriminate between hues throughout the spectrum. This requires plenty of time and quite elaborate apparatus. Simpler and quicker methods are therefore needed for screening large samples of populations. Colour-confusion charts, such as the well-known Ishihara set, are generally used, though they tend to miss some of the milder defects. The subject is asked to name a series of coloured numerals on a coloured background or to trace a wavy line if he is illiterate. Both numerals and background are composed of colour dots, mostly in shades of red or green, and these are chosen so that colour-blind subjects either fail to see the figure or see only parts of it and so mistake it for another numeral. The tests are designed for use in diffuse daylight and can give false results in other illumination.

The Nagel anomaloscope is a useful instrument for detecting the less severe defects and it has been adapted for work in the field. Normal subjects can match a spectral yellow with a mixture of monochromatic red and green. In fact, they can match any coloured light by mixing not more than three coloured lights, usually suitably chosen red, green, and blue. They are therefore known as 'trichromats'. Protanopes and deuteranopes match all colours with only two colours and are called 'dichromats'. In using the anomaloscope the subject looks through an eyepiece at a circular field one half of which is illuminated with a spectral yellow. By turning calibrated knobs he can add red or green to the other half-field until a match is obtained. The proportion of red and green chosen by normal people varies little but dichromats are much more erratic. Protanomalous subjects require more red and deuteranomalous subjects more green than normal. Red and green perception is evidently weak rather than absent and they are therefore referred to as 'anomalous trichromats'. The interpretation of anomalscope results may need special care when studying unsophisticated populations who are unfamiliar with such gadgets and unaccustomed to making precise colour matches.

Table 11.8 shows the frequencies of red-green defects as a whole in male samples from selected populations tested by rapid methods. Values for Europe are around 7–8 per cent; they are somewhat lower in various peoples of eastern Asia. In India, as far as we can tell from relatively small samples, there is a good deal of variation between the caste groups of some regions, and between certain tribal and Hindu populations. Values for sub-Saharan Africa, the New World, and Australasia are relatively low (1–3) per cent. It has been suggested that colour blindness in males is selectively disadvantageous in hunting–gathering peoples and that with the coming of agriculture and civilized life this negative selection was relaxed. However, the data on hunter–gatherers are meagre, and it is questionable whether mutation pressure could have produced such a large increase in gene frequency in a period of less than 10 000 years.

TABLE 11.8
Frequency of red-green colour blindness in males of various populations

Population	Sample size	Colour-blind males (per cent)	Author
Arabs (Druse)	337	10.0	Kalmus (1961)
Norwegians	9047	8.0	Waaler (1927)
Swiss	2000	8.0	Von Planta (1928)
Germans	6863	7.7	Schmidt (1936)
Belgians	9540	7.4	De Lact (1935)
British	16 180	6.6	Grieve (1946)
Iranians	947	4.5	Plattner (1959)
Andra Pradesh (India)	292	7.5	Dronamraju (1961)
Chinese (Peiping)	1164	6.9	Chang (1932)
Chinese (all)	36 301	5.0	Chun (1958)
Tibetans	241	5.0	Tiwari (1969)
Japanese	259 000	4.0	Sato (1935)
Mexicans	571	2.3	Garth (1933)
Navajo Indians	535	1.1	Garth (1933)
Eskimos	297	2.5	Skeller (1954)
Tswana	407	3.0	Squires (1942)
Hutu	1000	2.9	Hiernaux (1953)
Tutsi	1000	2.5	Hiernaux (1953)
Zairians	929	1.7	Appelman (1953)
Australian Aborigines	4455	1.9	Mann (1956)
Fiji Islanders	608	0.8	Geddes (1946)

Few populations have been studied with the anomaloscope and there is therefore little information about the relative frequencies of different types of defect throughout the world. Deuteranomaly seems to be the most variable type, ranging from about 5 per cent in Europeans to 1–2 per cent in Uganda, New Guinea, and Thailand; but if selection has been relaxed one might have expected the frequencies of the more severe defects to be most affected.

Distribution of rare inherited conditions

Many of the serological and biochemical variants which we have discussed are common, at least in certain populations, and this makes it easy to compare regional frequencies by examining quite small samples. It is more difficult to get reliable frequency estimates for rare conditions, but it is, nevertheless, evident that some of them are by no means uniformly distributed and they present some interesting problems in population genetics.

Inherited diseases causing serious disability are likely to come to medical attention despite their rarity, and even rare benign variations may be detectable in mass screening of blood or urine samples for other purposes. Although many populations are too remote from highly developed medical services for reliable data to be collected, geographical variations in the incidence of inherited disease may sometimes be inferred from experience with immigrants to better-equipped countries. In systematic attempts to determine the frequencies of rare conditions in the general population, methods of ascertainment must be carefully considered so as to avoid false estimates due to biased sampling.

Cases of rare recessive disorders are often found to be concentrated in restricted localities in which the small size and isolation of the community inevitably raises the incidence of consanguineous unions. If an isolate is descended from a small number of ancestors amongst whom a given gene happened to be present, the local gene frequency may be much higher than in the population as a whole. Homozygosis due to inbreeding exposes a recessive gene with deleterious effects to more stringent selection and a small advantage of the heterozygote which might otherwise suffice to maintain equilibrium at a low frequency might be inadequate to prevent its local extinction under these conditions. Changes due to chance in the transmission of genes also becomes significant if the population is very small.

The infantile type of amaurotic idiocy is a severe neurological disorder inherited recessively, which leads to premature death. Its incidence in certain European countries is estimated at 1 in 10 000, but a high proportion of the affected cases are found to be Jews. The very rare and apparently harmless anomaly known as 1-xyloketosuria, in which an unusual sugar is excreted in the urine, is also found mainly in Jewish subjects, most of them derived from a certain region of Russia. Another form of inherited mental defect, phenylketonuria, on the other hand, is certainly very uncommon in Jews. It has been found in many parts of Europe, with some suggestion of regional frequency differences, but is much rarer in American blacks and very probably in African populations and also in Japan. A rare inherited deficiency in the enzyme catalase in the red cells is so far known only in Japanese. The depigmentation of albinos, who are homozygous for a recessive gene, makes them especially easy to detect in pigmented populations and they have been recorded in most parts of the world. Defects in eyesight generally bring them to medical attention in places where facilities are sufficiently advanced. The frequency of cases of albinism has been estimated at 1 in 10 000 to 1 in 20 000 in some European countries, but in West Africa it is probable that it is three to five times as high and even higher incidences (1 in 200) have been reported in certain Indian tribes of Panama who live in relatively small village communities on islands.

Recent concern regarding the genetical effects of atomic radiations has emphasized the need for more exact knowledge about the incidence of the less common inherited abnormalities including those lumped together under the heading of congenital malformations. This group includes, among others, deformities such as hare-lip, club-foot, congenital hip dislocation, polydactylies, and central nervous defects such as anencephaly and spina bifida. The rates of occurrence of congenital malformations as a whole seem to be similar, namely 1–1.5 per cent in European populations, American blacks, and Japanese, but there is evidence that these populations differ in the incidence of particular defects. The reasons for such differences must undoubtedly be complex; though there is a tendency for certain malformations to be concentrated in families, no clear genetic basis has been established for most of them. Some types which are grouped together because they resemble one another phenotypically are no doubt genetically heterogeneous. Differences in population incidence may reflect not only differing gene frequencies, but also the effects of environment on the manifestation of genes and on the liability to produce phenocopies.

The distribution in Europe of anencephaly, a severe congenital defect in which the brain is grossly underdeveloped, has been summarized by Penrose, who showed that the frequency varies from 1.0 to 0.32 per cent in Ireland and certain western regions of the UK, falls to about a tenth of this in much of western continental Europe, and is even lower in certain regions of south-eastern France. Searle examined records for various ethnic groups in Malaya and found frequencies varying from zero in the Hakka group of Chinese, to 0.65 per cent in Sikhs. The causes of anencephaly are still largely obscure, but a consideration of population distributions along with other evidence suggests that both environmental and genetical factors are involved.

Alleles which are very rare also occur at loci responsible for polymorphisms. We have mentioned, for example, rare variants at the β-haemoglobin locus where Hb^A, Hb^S, Hb^C, and Hb^D, also occur (p. 223). Rarity is often associated with selective disadvantage as in the cases we have just been considering but in many cases no obvious disadvantage is apparent. Rare enzyme variants are usually apparently neutral as are 'family blood groups'. In these cases the presence of the variant is probably due to a relatively recent mutation. It has been suggested that study of the distribution of such variants may be particularly informative of population movement. If rarity is essentially a function of the rarity of the mutation then all the examples of a variant may derive from a unique event, and all persons carrying it must be genealogically related. This is clearly not the case with common variants where the 'same' gene is likely to have arisen separately many times. A condition of 'woolly hair' is known to be due to a single dominant gene. The gene arose in a Norwegian family and all cases of it, including some in North America, can trace descent to the first known case.

Human chromosome polymorphism

With the ever greater resolution of chromosome morphology which has occurred during the past 30 years, evidence of variety in structure has been increasing. The most striking of these variations is in the length of the Y-chromosome which can be as large as a member of the G(13–15) group or as small as a short acrocentric. This difference has been shown to be due to the amount of Q-staining heterochromatin (cf. p. 151) in the distal region of the long arm of the Y which may be massive or virtually absent. Not much comparative work on populations has been done but it appears that a long Y is relatively common among Japanese and a very short one in Australian Aborigines. Q-staining also shows strongly fluorescent bands in the centromere region of chromosomes 3, and in the short arms and satellites of the D and G groups. Polymorphic variations in the size of these bands has also been observed. Methods have also been devised that stain mainly heterochromatin around the centromere (C-banding); particularly large C-bands are seen in chromosomes 1, 9, and 16, and these also show polymorphic variation. What these differences mean in functional terms is very far from clear.

Variation in the DNA

The value of the discovery of endonuclease restriction enzymes in revealing variation in the DNA has already been mentioned in considering the globin genes (p. 227). Other enzymes of this kind in conjunction with various techniques of recombinant DNA technology are beginning to reveal a considerable diversity of further variation.

There is now a rapidly growing literature on restriction fragment length polymorphisms of nuclear DNA (RFLPs). Restriction enzymes can detect genetic variety which is not evident in proteins, such as single base changes in redundant systems, changes in the 5' or 3' regions of the genes which although translated into mRNA are not represented in proteins and changes in the intervening sequences or introns (p. 154). Various estimates have been made of the number of RFLPs which range from 1 restriction enzyme site in 85 nucleotides to 1 in 500. However, it has also become evident that they do not occur randomly over the genome, but are more common in the non-coding sequences. Thus, of 12 RFLPs found in the β-globin gene cluster seven are in flanking DNA, three in introns, one in the pseudogene, and only one in a coding sequence. Another analysis of a 16.5 kb region of β-globin showed only two polymorphic restriction sites in 885 nucleotides of six exons, but 36 sites in 6216 nucleotides of flanking sequences and introns.

Overall, human DNA does not vary greatly from one member of the species to another: about 999 bases out of over 1000 are the same — they make us *Homo sapiens*. There appear, however, to be certain hotspots of hypervariability which are particularly to be found in minisatellite DNA — regions of repeat sequence DNA which, unlike satellite DNA, consist only of short lengths of common base pairs. Such minisatellites occur in the region of the α-globin gene and the myoglobin gene. Jeffries and his colleagues have used probes obtained by restriction enzyme analysis of the myoglobin minisatellites, to locate other similar hypervariate regions. One of the clones obtained consisted of 29 repeats of an almost perfect 16 base core sequence. On purification this was used to screen the DNA of 54 individuals from a single family pedigree. Every one of these people produced a different hybridization pattern — a unique DNA fingerprint. It is concluded that in hypervariable regions there is essentially no homozygosity. Clearly, work of this kind is going to reveal a wealth of new anthropological variability.

Mitochondrial DNA

A particularly rewarding approach appears to be in the analysis of the DNA of mitochondria. Mitochondria are organelles within living cells which contain enzymes for the provision of cellular energy. These enzymes are produced by the mitochondria's own DNA and this appears to come solely from the egg. It therefore shows maternal inheritance and there is no recombination. It occurs as closed circles of about 16.5 kb. Heterogeneity from individual to individual is detected by restriction endonucleases, the radioactive labelling of restriction fragments, and separation by electrophoresis. One informative endonuclease is Taq which recognizes the nucleotide sequence TCGA and provides information on

112 nucleotides per individual. Fig. 11.8 shows, in linear form, the structure of human mtDNA and polymorphism within it as located by 12 different restriction endonucleases. In 112 people 163 different forms have been detected indicating the high variability and high mutability of mtDNA. Most of the mutations are single base changes and do not affect amino-acid sequences in the enzymes. Some do, however, and additions and deletions also occur. Evolutionary rates appear to be about 10 times faster than for the nuclear gene. Early assessment indicates notable geographical heterogeneity, with some variants found only in one continental region. However, other variants are widely shared and there is suggestive evidence for considerable parallel evolution.

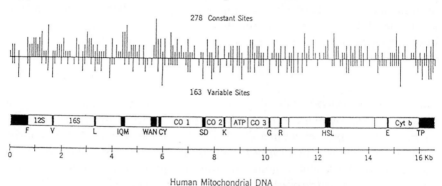

Human Mitochondrial DNA

FIG. 11.8. Location of cleavage sites and functional regions of human mtDNA. The 16 569 bp circular genome is drawn in linear form. The major bar shows the regions of known function: 22 tRNA genes, each represented by a single letter and black shading, two rRNA genes (12s and 16s), and 13 genes coding for proteins, eight of which are unassigned and five of which are known (three cytochrome oxidase subunits, one ATPase subunit, and cytochrome B). Diagonal lines represent the large non-coding region, extending from 16 024 to 576 bp. The upper panel shows the locations of cleavage sites found in mtDNAs from 112 humans plus the Cambridge reference sequence with the aid of 12 restriction enzymes. Vertical lines below the horizontal line show the variable sites, or those present in some but not necessarily all of these mtDNAs. The vertical lines above the horizontal show those sites present in all the human mtDNAs examined. Height of the vertical lines is proportional to the number of sites found within an 80 bp segment. (This figure is from Cann, Brown, and Wilson, Polymorphic sites and the mechanism of evolution in human mitochondrial DNA, *Genetics*, 1984. Reprinted with permission.)

Surnames

A 'character' whose relevance to human biology has been strongly developed recently by G. W. Lasker is a person's inherited name. In many 'Western' societies where most of the research has been conducted this amounts to a surname which is transmitted from a man to all his children, but is lost by his daughters on their marriage. Its appearance then, at least in married couples is like that of a Y-chromosome linked trait. Since it is indicative of male ancestral lineages and can often be established, unlike genes, in past populations, it provides a powerful tool of analysing population genetic structure. One, of course, needs to be aware of various assumptions that one may be called to make in any particular analysis concerning such matters as the multiple origin of many surnames, distant common ancestry, and name changes.

One of the most common purposes of using surnames is in establishing population inbreeding levels. For this one needs to know the frequency of cases where bride and groom have the same surname. This is known as the marital isonomy rate I. Crow and Mange have shown that the population inbreeding coefficient $F = I/4$. The constant 4 comes from the relationship between surname inheritance and gene inheritance, e.g. brothers and sisters share the same surname, but the inbreeding coefficient for the offspring of sibs is 1/4. First cousins have a 1 in 4 chance of possessing the same surname and F for their offspring is 1/16. F can be divided into that part which will randomly occur in populations of limited size 'F_r' and the non-random element F_n. $F_r = \Sigma p_i q_i)/4$ where p_i is the frequency of the ith surname in fathers and q_i is the frequency of the same surname as the maiden name of mothers. $F_n = (I - \Sigma p_1 q_1)/4(1 - \Sigma p_i q_1)$.

Surname distribution has been particularly thoroughly studied in England and Wales. No true isolates have been detected but there is much isolation by distance. Certain names concentrate in south Wales and others in north Wales whilst a north–south cline of English names appears to exist. Rare surnames are more localized and show steeper clines, but also more sampling variance, than common names.

12 Population variation in quantitative traits

While ever more attention is being devoted to identifying and analysing biochemical, molecular, and other simply inherited variation, much of human biological value can still be obtained from study of the quantitative complexly inherited variation which one finds within and between human populations. Included in this variation are the characters first examined by biological anthropologists because they are visually obvious, such as body shape and size, skin and hair pigmentation, and hair form. The genetic basis to this kind of variation has been considered in Chapter 9, but as emphasized there quantitative traits frequently have an environmental component to their variation. This can be of at least as much human biological interest as the genetic component, especially in ecological and epidemiological terms (see Part IV). On the other hand it can clearly be a source of confusion when one is considering population affinities (i.e. the genetic similarity between human populations).

The form of quantitative within-population variation has been fully described. It is essentially characterizable by the distribution curve which typically is of 'normal' or Gaussian form. Differences between populations are then expressed in terms of differences in mean values and differences in scatter, as reflected in either the variance or standard deviation. Whether or not an observed difference in the means for two populations is statistically significant depends upon the magnitude of this difference and the standard error of each of the two means. (The standard error is itself a function of the scatter, i.e. the standard deviation and the number of observations on which the estimate of true population means is based. The larger the number of observations and the smaller the standard deviations, the smaller will the difference between two means need to be to reach statistical significance at a particular probability level.)

J. C. Trevor demonstrated a number of years ago that in hybrid populations formed by the intermixture of populations which differed in a number of skeletal quantitative traits (some of them quite markedly) the character distribution was also Gaussian. In no case was there any tendency for a bimodal (two-peaked) distribution, although the hybrid mean value lay intermediate between the two parental means. We may therefore conclude that quantitative variation always shows essentially the same form of variability whatever the nature of the population. The key requirement is random mating for the character concerned.

Anthropometry

The need for exact measurements to supplement verbal descriptions of bodily size and form was already felt in the early nineteenth century when physical anthropology was becoming a distinct discipline, and until quite recently anthropometry has been the mainstay of the subject. It is true that the proportions of the body had interested artists long before this, but their aim was the definition of ideal types by numerical rules rather than accurate and systematic description of natural variation. It is not surprising that anthropologists, preoccupied with problems of evolutionary descent, have laid special emphasis on the measurement of the skeleton, for skeletal remains are the most direct evidence of the physique of earlier populations. This bias is reflected in the standard measurements used in comparing living peoples, many of which are essentially measures of skeletal dimensions. It was also felt that differences in individual environment were less likely to affect the bones than the soft tissues, and that skeletal characteristics being stable in this sense were therefore a better guide to genetic affinities. This, as we shall see, is a very doubtful assumption. From a wider biological standpoint soft-tissue measurements from which the amount of fat and muscle can be estimated are as important as skeletal ones, while in applied anthropometry, which deals with problems of fitting into clothes, chairs, aeroplanes, etc., other measurements outside the usual anthropological repertoire may be required.

Both the living body and the skeleton are very complex forms. Given enough ingenuity and time, many of the finer points of shape can be expressed metrically, but for most purposes we use relatively few measurements chosen so as to represent only the major features.

In principle nothing would seem easier than to take linear or circumferential measurements with rulers, calipers, or tapes, but in practice meticulous attention to details of technique is needed to ensure reliable results. The terminal points on the body or bone, between which the measurement is to be taken, must be clearly described and attention must be paid to the posture of the subject since this may have a large effect on some dimensions. The accuracy of the instruments must also be checked periodically. In spite of many efforts to secure international agreement on points of anthropometric technique, it is all too often impossible to compare the data of different observers reliably because of uncertainties about their methods.

Stature

It is convenient to begin with stature because no measurement of the body is taken more often and because it serves to illustrate various principles which apply to metrical characters in general. Military service records provide us with a mass of data on the heights of young adult males in many countries and over a considerable period of time.

Method of measurement

A fixed vertical scale with a cross-bar which can be brought into contact with the top of the head is the normal equipment. The subject should stand erect, without shoes and with heels together, the line of vision directed horizontally. Any tendency to stoop can be corrected by the observer holding the subject's head behind the ears and exerting firm upward pressure.

Although stature is fairly simple to measure, it is anatomically complex since it includes the dimensions of the legs, pelvis, vertebral column, and skull, and the contribution of each of these to the total varies in different individuals and also, on average, in different populations. A man's height gives us some idea of his general size, but weight is a better measure of this since it depends also on transverse measurements. Obviously, some people are tall and thin while others are both tall and broad. The relationship between stature and weight, or between stature and one or more transverse diameters, can be used to assess such variation in body-build or physique (Part III). The empirical fact that body-weight in excess of the population average for a given height is associated with a lower expectation of life has made this relationship an important one in assessing life-insurance premiums.

The interrelationships between stature and other body measurements can be conveniently summarized in the form of a correlation table (Table 12.1). The correlation coefficient r, which can have values from $+1.0$ to -1.0, is a convenient measure of association. It is not surprising that leg length and trunk are highly correlated with stature because they are themselves the major components of height; the correlation in this case is said to be spurious. However, the correlation with arm length is also fairly high ($+0.68$). The correlations with transverse measurements are distinctly lower. Weight, as might be expected, shows fairly high positive correlations with all the other measurements.

In all populations women are on average shorter than men, the magnitude of the difference lying between about 5 and 8 per cent.

Geographical variations in stature

There is a great deal of information about stature variation throughout the world, but not infrequently it is based on samples which are either very small or were selected in a way which may render them unrepresentative of the general population. In many countries there are not only stature differences between localities, but between socio-economic classes, and between the inhabitants of rural and urban environments.

TABLE 12.1

Correlations between body measurements 2400 adult male RAF volunteers, ages 17–38 years. Averaged correlations for all ages

	Standing height	Sitting height	Arm length	Leg length	Thigh length	Abdomen girth	Hip girth	Shoulder girth	Weight
Standing height	—	0.732	0.677	0.864	0.608	0.321	0.490	0.386	0.627
Sitting height	0.732	—	0.421	0.498	0.201	0.173	0.417	0.384	0.548
Arm length	0.677	0.421	—	0.683	0.447	0.266	0.483	0.363	0.466
Leg length	0.864	0.498	0.683	—	0.817	0.312	0.424	0.304	0.556
Thigh length	0.608	0.201	0.447	0.817	—	0.515	0.440	0.234	0.525
Abdomen girth	0.321	0.173	0.266	0.312	0.515	—	0.667	0.526	0.709
Hip girth	0.490	0.417	0.483	0.424	0.440	0.667	—	0.562	0.785
Shoulder girth	0.386	0.384	0.363	0.304	0.234	0.526	0.562	—	0.681
Weight	0.627	0.548	0.466	0.556	0.525	0.709	0.785	0.681	—

From Burt and Banks, *Ann. Eugen.*, 1947.

On the whole the pattern of stature variation throughout the world shows no very striking regularities. Both tall and short peoples are to be found in most of the major regions. Relatively high averages (1.70–1.75 m) are centred in north-western regions of Europe such as Scandinavia, and are also found in Albania, while in some parts of Spain and southern Italy the figures fall below 1.60 m. The Lapps, in common with various indigenous peoples stretching eastward across Siberia, are even shorter. In Africa we find extreme contrasts of stature, from the very tall Tutsi and Nilotic tribes such as the Nuer and Dinka to the diminutive Pygmies and Bushmen. In both the latter cases, however, there is considerable variation from one group to another. The Bushmen of the southern Kalahari appear to have been the shortest (1.45–1.50 m), but farther north the average rises to 1.55 m or more. Among the shortest peoples of the world we may also mention the Andamanese, and certain aboriginal groups of central Malaya and of the Philippines, who are sometimes classified together as Negritos on account of their short stature and a number of other morphological similaries. In the New World, especially North America, there has been much relatively recent movement of tribes from their original homelands, and many have become extinct or much intermixed. It seems, however, that the tallest peoples (1.70 m or more) were in the Great Plains, the East Coast, and Patagonia, while the shortest were in tropical central America and the northern parts of South America. This distribution of stature has been interpreted as an example of a cline with adaptive significance in relation to climate (Bergmann's rule). A substantial negative correlation between body weight and mean annual temperature has been demonstrated for various regions of the world.

Estimation of stature from long bones

It is useful in forensic work and in examining archaeological remains to be able to estimate the stature of an individual from the length of the arm and leg bones. In order to do this we need data on the relationship between stature and, for example, femur length for a large random sample. If we then plot the two variables we find that the points do not lie exactly on a straight line (or some other regular curve) because of individual variability in the relationship. The statistical method of 'regression analysis', however, enables us to calculate the equation for a straight line, or if need be some other curve, which fits the set of points most closely. The criterion of fit is that the variability of the values about the line shall be minimal. The regression equation can then be used to calculate the most probable stature of an individual given the length of a particular bone. Trotter and Gleser, whose work was based on a large collection of American material, give the following linear equation for estimation of male stature (age 18–30 years) from the maximum length of the dried femur:

stature (m) = 2.38 × femur length (m) + 0.6141, SE ± 0.0327 m

The estimate has a certain margin of error, indicated by the standard error ± 0.0327 m. In fact the stature would be estimated correctly within 0.06 m in 68 per cent of cases. The accuracy can be improved a little by including tibia length together with femur length, but the inclusion of the lengths of the arm bones makes little difference.

It should be emphasized that this and other estimates based on regression methods give the best predictions when applied to material derived from the population for which the equations were originally calculated. Different equations are needed for the two sexes, and the equations derived from European data may not provide an accurate prediction if they are applied to other populations. American blacks, for example, have on average longer legs in relation to stature than do Europeans and also relatively longer distal limb segments. The use of regression equations based on a modern population to estimate the stature of fossil hominids of widely different structure is clearly a dubious procedure.

The length of trunk and limbs

The most generally used measurement of trunk-length is the 'sitting height'. The subject is seated on a horizontal surface (e.g. a table top) and the measurement is taken from the vertex of the head to the sitting surface using a graduated rod with a movable cross-piece (anthropometer) or a fixed scale. Particular care must be taken to see that the back is stretched as straight as possible by applying upward pressure to the head or jaws; the thighs should be horizontal to avoid any tendency to lean backwards or forwards.

The difference between sitting height and stature is often taken as a convenient, but arbitrary measure of lower-limb length, which is difficult to obtain directly because the greater trochanter of the femur is an ill-defined subcutaneous point. Obviously leg length derived in this way is less than the anatomical length because the level of the acetabulum is some centimetres above the seat level.

There is some variation between populations in the average contribution of trunk length to stature. In Australian Aborigines and in many African peoples the trunk is relatively short; the 'cormic index' or 'relative sitting height' [(SH/St) × 100] lies between 45 and 50 per cent, whereas in some Chinese, Eskimo, and American Indian samples the value may be as high as 53 or 54 per cent.

Measurement of the lower limb segments, especially the thigh, in the living subject is not entirely satisfactory, but the length of the dried bones can easily be measured. The usual apparatus is an osteometric board with a fixed upright face at one end against which the head of the bone can be pressed and a sliding upright which is applied to the other end of the bone.

The length of tibia in relation to femur length (tibio-femoral index) tends to be high (85 per cent or more) in those populations with relatively long lower limbs.

The total length of the upper limb can be measured from the acromial point, the most lateral point on the margin of the acromion process of the scapula, to the tip of the longest (3rd) digit. The arm is held by the side of the body with hand pointing down, and with elbow, wrist, and finger joints fully extended. Arm length measured in this way is longer than the anatomical length since the acromion lies above the head of the humerus. The length of the upper arm can be measured from the acromial point to the proximal margin of the head of the radius, which can be palpated at the elbow, and the length of the forearm from this radial point to the tip of the styloid process of the ulna at the wrist. In general the forearm tends to be long in relation to the humerus in those populations in which the distal (tibial) segment of the lower limb is also relatively long.

Measurements of breadth

Three transverse diameters may be mentioned because they are fairly easy to measure. The width of the shoulders is taken as the distance between the two acromial points (biacromial breadth). This diameter changes if the shoulders are braced back and one should aim to control posture so as to get a maximal measurement; this is done by relaxing them downwards and somewhat forwards.

Two measurements are useful for the hip region. The bi-iliac or bicristal diameter is the distance between the tuberosities of the iliac bones, which lie close to the most lateral points on the iliac crests. The bispinal breadth is measured between the anterior superior spines of the ilia, which can be felt as fairly sharp prominences at the anterior ends of the iliac crests. Both measurements must be made with considerable pressure so as to compress covering fat as much as possible; even so they are somewhat inaccurate as measures of skeletal dimensions alone.

Circumferential measurements are sometimes taken with a tape at specified levels on the thorax, abdomen, or hips.

There are quite pronounced average differences between populations in the relations between transverse measurements and stature. Figure 12.1 shows the average proportions of males and females of a central Australian tribe compared with Maya Indians of Yucatan. The Australians, with their relatively narrow shoulders and hips, are said to be linear in build; the sex difference in hip proportions is less marked than in Europeans.

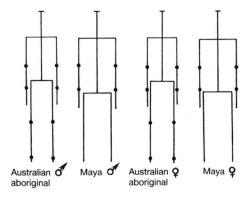

Fɪɢ. 12.1.   Diagrams to show contrasting average body proportions of two populations. All measurements are as proportions of stature. The axial line represents sitting height. Bi-acromial width is drawn in at an arbitrary level, the same for each. The relative lengths of the upper limb segments are shown. The total length of the lower limb is taken to be stature minus sitting height. (Australian data from Abbie 1957; Maya data from Steggerda 1932.)

Anthropometry of the head

The braincase

No measurements in anthropology are more sanctified by usage than those of the length and breadth of the head. The 'cephalic index' [(B/L) × 100], which is derived from these, was introduced by the Swedish anatomist Retzius over a century ago as a measure of skull shape and rapidly gained prestige as a means of classifying populations.

The maximum length, usually taken with spreading calipers, is measured from the glabella, the bony prominence between the eyebrows, and above the nasal depression, to the most posterior point on the occiput (opisthocranion). The maximum breadth, taken at right angles to the sagittal plane, is generally in the vicinity of the parietal eminences.

The statistical distribution of the cephalic index is continuous and approximately Gaussian in form. On average, it is two units higher in the living subject than for the skull. It is a common practice to divide the range into three arbitrary groups, dolichocephalic (below 75), mesocephalic (75–80), and brachycephalic (above 80), and to compare the frequencies of these classes in different populations, but it is misleading to chop a continuous distribution into arbitrary categories in this way, and there is little to recommend it.

The index is only a crude measure of skull shape in the horizontal plane and skulls with the same index value may differ considerably in their overall form. It is hardly surprising that studies on the inheritance of this arbitrary character have yielded no clear results.

In Europe average values between 75 and 80 are found in the British Isles, parts of Scandinavia, and some regions of the Mediterranean zone. In west-central and central Europe the means are usually above 80 or even above 85 (Table 12.2). Most of the figures for Africa are below 75, but some central African peoples have values of 80 or more. Averages around 80 are found in many parts of China, Japan, Indonesia, and the East Indian islands, but in many peoples of New Guinea and other parts of Melanesia the indices are rather lower, while in Australian Aborigines particularly low values are found. The Greenland and Canadian Eskimos have mean indices around 75 in contrast to the more brachycephalic Alaskan ones. Archaeological finds from many different regions have shown a tendency for modern populations to be more brachycephalic than their predecessors. It should be emphasized that, as in the case of other body measurements, all populations are very variable, the standard deviation of the index usually being about points on either side of the mean.

TABLE 12.2

Mean cephalic index (C.I.) taken on living males in various populations

	Number measured	Mean C.I.		Number measured	Mean C.I.
Montenegrins	100	88.6	Kikuyu (Kenya)	384	76.0
Albanians	112	86.4	Somalis	244	74.3
Norwegian Lapps	254	85.0	Moroccans	5 210	74.3
Armenians	234	83.5	Koreans	522	83.4
Germans	925	82.5	Chukchi (Siberia)	148	82.0
Dutch	4 600	80.3	E. Chinese	359	81.7
Norwegians (Troms)	548	81.0	Japanese	6 000	80.8
Norwegians (Hedmark)	988	77.7	Ainu (Japan)	95	77.3
Scottish (N.E.)	320	78.1	Maya (Yucatan)	133	85.8
Portuguese	11 658	76.4	Eskimos (S.W. Alaska)	61	80.7
Rwala Bedouin	270	75.0	Sioux (central USA)	537	79.6
Negritos (Philippines)	147	82.7	Otomis (S. Mexico)	178	77.9
Semang (Malaya)	103	79.0	Eskimos (E. Greenland)	225	76.7
Bengal Brahmins	100	78.7	Hawaiians	203	84.0
Nayar (Malabar)	175	73.2	Fijians	133	81.5
Baluba (Congo)	367	81.6	Maori	421	77.7
Ituri Pygmies	386	76.5	New Caledonians	185	76.5
Ibo (Nigeria)	2 603	76.4	Australians (Arnhem Land)	236	71.8

Cranial capacity

In view of the low average cranial capacities of early Pleistocene hominids, this measurement is important in palaeontology, but it is doubtful whether the labour of measuring it in modern populations has been very rewarding. There has been much disagreement about the technique to be used in estimating the capacity of skulls. A common method is to fill the cranial cavity with mustard seed and then measure the volume of seed. Estimates of the cranial capacities of living persons can also be obtained by formulae based on measurements of length, breadth, and height of the braincase.

The figure of 1450 cm^3 is often taken as an average for European males, and the figure for females is about 10 per cent lower. Somewhat higher means have been reported in some non-European peoples, e.g. Eskimos, while mean capacities somewhat below 1300 cm^3 for Australian Aborigines are among the lowest values.

Although increase of brain-size is a notable feature in earlier phases of hominid evolution, the significance of brain-volume in relation to mental functioning is by no means clear. Within a European population the range of cranial capacity is very wide and values of 400 cm^3 or more above or below the mean are compatible with normal or even outstanding intellectual ability. Individual differences in this measurement are partly a function of general body-size. It seems most unlikely that the average differences between modern populations have any relevance to the problem of variation in mentality.

The braincase height, which is needed in estimating volume from external dimensions can be measured on the skull from the most anterior point of junction of sagittal and coronal sutures. Alternatively, the skull can be orientated in the Frankfort plane (bounded by lines drawn between the lowest points on the orbital margins and the uppermost points on the tympanic ring surrounding the auditory meatus) and a measurement of auricular height can be taken from the meatal point (porion) to the vertex. With special instruments a roughly comparable height measurement can be taken on the living subject.

The face

The length of the face is generally measured from the nasion at the root of the nose to the lowest point of the chin in the midline (total facial height). The nasion is really the junction between the upper margins of the nasal bones and the frontal in the midline, and this suture is difficult to locate in the living person. A horizontal line tangential to the fold defining the top margin of the upper eyelid passes close to the nasion and provides a useful substitute. The upper facial height can be measured from nasion to the alveolar point, the lowest point on the bony wall between the upper incisor teeth, or the lowest point on the gum in the living.

'Nasal height' is taken from nasion to the point of junction of the nasal septum with the upper lip, or on the skull to the midpoint of a line tangential to the lowest points on the pyriform aperture on either side of the nasal spine.

The most useful facial breadth measurements are the bizygomatic, between the most lateral points on the zygomatic arches in a plane at right angles to the sagittal, and the bigonial between the bony angles of the mandible. 'Nasal breadth' is taken as the distance between the most lateral points on the alae of the nostrils in the living and the maximum breadth of the pyriform aperture in the skull. The average nasal breadth index $[(NB) \times 100/(NL)]$ is as high as 104 in Ituri Pygmies and values above 90 are found in many African Negro peoples, in Melanesia, and in Australian Aborigines. The noses of Eskimos, on the contrary, particularly as measured on the skull, are unusually narrow. It has been shown that when American blacks are grouped according to the proportions of European ancestry there is a corresponding proportional difference in the average nasal indices. In India the index shows a general, but irregular correspondence with status in the caste hierarchy; it is lowest in Brahmins, higher in lower castes, and highest of all in some of the tribal peoples. It must be remembered, however, that measurements of the nose are liable to considerable errors.

Population variation and the nature/nurture problem

Attempts to understand the extent to which differences in environment are directly responsible for population differences in anthropometric traits have been made by comparing migrants to new environments with parental sedentes (cf. p. 179). There have been several studies of this kind since the pioneer work of Boas in 1911 in which he showed that the progeny of Sicilians, Russian Jews, and others reared in the USA differed anthropometrically from the parent populations in Europe. The most ambitious study is that of Shapiro and Hulse on Japanese immigrants to Hawaii. The men migrated to become labourers in the plantations and later imported brides from their villages in Japan. After making due allowance for age difference it was found that the offspring reared in Hawaii were, on average, considerably different and generally larger in several body measurements than their parents (Table 12.3). The males, for example, were 4.1 cm taller, but a few measurements such as head length were actually lower. It was also shown that there were systematic differences in size and proportions between the first-generation immigrants and samples from the villages in Japan from which they came. Whatever the interpretation of this may be, it shows that comparisons of groups born and reared in the new country with those remaining at home can be misleading if they are taken to show exclusively the effects of maturation in a different environment. Lasker's work on Mexican immigrants to the USA indicated that, as might be expected, anthropometric changes were most pronounced in those who migrated while still immature, so that effects on growth are certainly involved. It is obvious that migration to a different country may involve changes in

TABLE 12.3

Body measurements of Japanese immigrants to Hawaii (B) compared with those of Hawaiian-born Japanese (C), and of Japanese in Japan (A)

	(A) Male sedentes			(B) Male immigrants				(C) Male Hawaiian-born			
	Number	Mean	±S.E.	Number	Mean	±S.E.	B − A	Number	Mean	±S.E.	C − B
Age (years)	172	35.55	0.64	178	40.60	0.55	+5.05	188	26.15	0.26	−14.45
Weight (lb)*	143	119.80	0.77	174	124.00	0.84	+4.20	185	127.40	0.93	+3.40
Stature (cm)	171	158.39	0.28	178	158.72	0.26	+0.33	188	162.83	0.26	+4.11
Sitting height (cm)	171	84.50	0.20	178	83.10	0.19	−1.40	187	85.48	0.18	+2.38
Upper leg length (cm)	170	35.78	0.17	177	37.52	0.15	+1.74	186	37.72	0.16	+0.20
Lower leg length (cm)	171	31.71	0.10	178	32.65	0.10	+0.94	188	33.93	0.10	+1.28
Biacromial (cm)	171	39.53	0.10	178	40.28	0.09	+0.75	187	41.35	0.09	+1.07
Head length (mm)	172	189.70	0.35	178	189.38	0.32	−0.32	188	186.54	0.32	−2.84
Head breadth (mm)	172	151.90	0.26	178	152.72	0.27	+0.82	188	155.08	0.28	+2.36
Upper facial length (mm)	172	72.66	0.23	178	81.40	0.25	+8.74	187	81.62	0.25	+0.22
Nasal length (mm)	172	47.88	0.19	176	50.24	0.20	+2.36	188	50.22	0.18	−0.02
Nasal breadth (mm)	172	36.50	0.14	176	35.80	0.15	−0.70	188	34.56	0.14	−1.24

* 1 lb = 0.45 kg.

From Shapiro and Hulse (1939).

many environmental variables; it is not clear which are the most important in producing changes in bodily dimensions, but better nutrition and perhaps less illness due to infections and parasites may be the most significant.

Pigmentation

No normal variations are more immediately striking to the eye than those of pigmentation. Although the geographical distribution of skin, hair, and eye colour has been known in a general way for a long time, technical difficulties have hindered the collection of exact data. Genetical analysis has lagged behind for the same reason and also because of the complexity of the mode of inheritance. The pigments themselves are difficult to analyse chemically, and the biochemical approach has therefore made limited headway.

Skin colour

The colour of the skin depends on two main factors; the blood in the smaller vessels of the dermis and the amount of dark pigment, melanin, in the epidermis. In skin with little melanin the colour depends chiefly on the blood and varies according to the amount in the vessels and the state of oxygenation of the haemoglobin. In dark skin the contribution of the blood-supply to colour is masked to a greater or lesser extent by melanin.

The melanin is formed by specialized dendritic cells, the melanocytes, situated in the basal layer of the epidermis. These cells make granules of pigment about 0.5 μm (1 μm = 1/1000 mm) in diameter and pass them into epidermal cells by way of their processes. The number of melanocytes per unit area of skin varies somewhat in different regions of the body, but counts on dark- and light-skinned people give essentially the same means for a given body region so that major population differences in skin colour do not appear to depend on variation in melanocyte numbers. In black Africans the granules are more numerous and larger, whereas in a light-skinned European only a few brownish granules can be seen in the basal layer of the epidermis. In an African this layer is densely black, and granules are also conspicuous in the Malphigian layer and in the stratum corneum.

Melanins are a widely distributed group of pigments, found in both plants and animals, which are formed by oxidation of phenolic compounds followed by polymerization of the products into highly insoluble dark complexes. In man and many other organisms melanin is formed from the amino acid tyrosine in a sequence of stages, some of which are catalysed by an enzyme tyrosinase. The indole-5, 6-quinone produced in this way polymerizes to form the pigment. The resistance of melanins to degradation makes it difficult to study them analytically, but there is evidence that melanins made in the laboratory vary in structure according to the conditions under which they are formed. The melanin granules of the tissues contain protein and this makes the chemistry of the natural pigments even harder to work out.

Methods of measurement. Techniques for measuring pigmentation quantitatively are to be preferred both in surveys for comparing populations and also in genetical work where we need to compare members of a family. Direct chemical estimation of the pigment, in the way in which we determine the concentration of a substance in blood or urine, is not feasible, but a measurement of the contribution of melanin to skin colour can be obtained by reflectometry (reflectance spectrophotometry). The principle is to shine a beam of light on the skin and to compare the amount diffusely reflected from it with that reflected from a pure white standard under the same conditions. From a series of readings throughout the visible spectrum (400–750 nm), or if need be, the ultraviolet and infra-red, a reflectance curve can be drawn representing the percentage of light reflected over the chosen range. Examples of reflectance curves from light and dark skin are shown in Fig. 12.2). The former rises fairly steeply from the blue to the red end of the spectrum and shows a pronounced trough in the green region owing to absorption of light by haemoglobin. In the latter this trough is hardly detectable, and the curve is smooth and much less vertical, rising less steeply towards the red end. The large amount of melanin dominates the reflectance and the curve

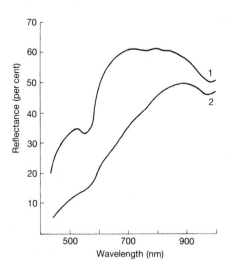

Fɪɢ. 12.2.   Reflectance curves of fair European (1) and Negro (2) skin in the visible and near infrared range. Note the marked trough between 500 nm and 600 nm due to haemoglobin in (1) which is less pronounced in (2). The curves approach one another closely in the infrared. (From Barnicot, *Man*, 1957.)

is similar to that obtained from a suspension of this pigment. Changes in blood-supply affect the reflectance at the red end relatively little and readings in this region give a convenient measure of the contribution of melanin.

Incidentally, it is amusing to note that, at longer wavelengths in the infra-red, the amount reflected from European and African skin is very similar. At 1200 nm there is considerable reflectance, i.e. both would be judged 'white' were these radiations visible. Beyond 8000 nm all incident radiation is absorbed and the human body behaves like a perfect black body radiator, irrespective of the amount of melanin in the skin.

Exposure to sunlight increases the amount of melanin pigment in skin, and both whites and blacks tan. The most effective wavelengths for tanning are around 295 nm. This is because within the ultraviolet they have the greatest capacity to penetrate the outer layers of the skin and stimulate the melanocyte. Different parts of the body tan to different degrees and, of course, some are more protected from exposure to sunlight than others. In genetic studies it is customary to make measurements on some part of the inner arm to reduce the contribution of variable tanning. Preliminary cleansing of the skin is a precaution not to be ignored!

Inheritance of skin colour. Apart from the contribution of a few major and rare genes, such as those for albinism, skin colour differences have remained largely intractable to genetic analysis. As described at some length previously (cf. p.175) a biometric analysis of European white and African black differences suggests that only a small member of gene loci, three or four, may be involved. Similar conclusions were reached from studies which examined the observed distribution of skin colour in mixed populations with the theoretical distributions to be expected from different numbers of gene pairs. These theoretical distributions had, of course, to take into account the extent of the intermixture (cf. p.201). Such studies both in the USA and in Brazil indicated that a small number of genes were involved in black/white differences. Research on hybrids between other parental groups differing in skin colour, such as European/Indian and European/Australian aborigines, have also suggested the involvement of small numbers of genes. However, none of these have been localized and nothing is known of the basis for variation within unmixed populations.

In some societies mating with respect to skin colour is clearly not random. In South Africa unions between blacks and whites have been prohibited by law. In Hindu society the higher strata in the complex system of endogamous castes are on the whole lighter-skinned and marital restrictions work to maintain these differences. There would appear to be a preference for light-skinned wives in a number of societies.

Skin pigmentation and natural selection. The geographical distribution of skin pigmentation broadly follows that of ground-level ultraviolet radiation. There are exceptions: many peoples in equatorial ecosystems are quite heavily pigmented yet these systems often have low ultraviolet because of atmospheric humidity and volcanic ash, and a heavy leaf canopy. The dark Tasmanians lived in a temperate climate. Some anomalies are explicable in terms of recent migrations, and it is broadly true that increasing pigmentation is concordant with increasing ultraviolet. Moderate pigmentation in Arctic peoples is associated with quite high ultraviolet from snow and sky. The fact that all people, except albinos, tan on exposure to ultraviolet also implicates this environmental factor as a cause for skin colour difference.

Ultraviolet radiation is damaging to tissues. It does not have a strong penetrative power, but can reach the lower levels of the epidermis and outer layers of the dermis. It is known in white groups to be a cause of skin cancer, especially on the parts of the body exposed to sunlight. It is strongly absorbed by melanin and would not seem to be a cause for skin cancer (epithelioma) in genetically dark-skinned peoples. Skin cancers tend to develop rather late in life and are rarely fatal — at least in modern circumstances. Perhaps just as important as the carcinogenic effect is the immediate damage caused by ultraviolet radiation in poorly pigmented people to skin cell integrity including that of the blood capillaries. Sunburn with its associated erythema would be a severe disadvantage in daily life.

Vitamin D is formed from certain steroids near the skin surface under the influence of ultraviolet radiation and it has been argued that depigmentation in northern temperate regions favours this beneficial effect of sunlight. It has been observed that black migrants into northern cities are particularly prone to rickets, which is due to inadequate Vitamin D. However, it is difficult to disentangle the direct effects of nutritional conditions which are often bad in such migrants. Vitamin D is toxic in excessive amounts and it has been suggested that heavy pigmentation in areas of high ultraviolet radiation may prevent the over-production of the vitamin.

Black surfaces, by definition, absorb more visible light than white ones. It therefore follows that heavily pigmented peoples will absorb extra energy on exposure to sunlight. The noticeable effect of this has been measured by monitoring the thermal tolerance of naked black and white subjects exercising in the sun. Evidently heavy pigmentation in the tropics is not an unmixed blessing. However, it would appear that skin colour probably does not affect heat loss from the body by radiation. The human body radiates in the wavelengths 5000–20 000 nm and in this region is very nearly a perfect 'black body' radiator irrespective of colour.

Hair colour

The colour of hair is due to granular pigment contained in the keratinized cortical cells of the hair shaft. The granules are formed by melanocytes situated in the epithelium of the hair bulb adjacent to the dermal papilla.

Objective measurements of hair colour are easily obtained by examining small samples of hair by reflectometry. As the reflectance curve in Fig. 12.3 shows, even hair from highly depigmented albino subjects is far from white. With the exception of red shades, hair colours can be arranged in a continuous series from dark to light, representing in effect increasing dilutions of the same pigment, namely melanin. While we cannot exclude the possibility of variations in the chemical structure of the pigment, differences appear to be mainly quantitative; in the lighter shades of hair, melanin granules are less numerous, smaller, and less completely impregnated with the melanin polymer than in darker ones.

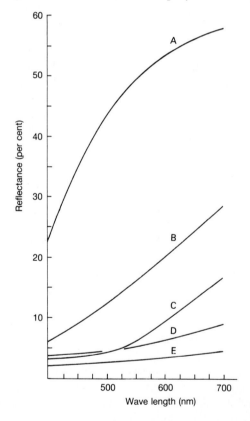

FIG. 12.3.   Reflectance curves of human hair of various colours. (A) A European albino, (B) blonde European, (C) a European with strongly red hair; note upward deflection at about 520 nm, (D) brown-haired European, (E) African Negro. (From Barnicot, *Man*, 1957.)

The reflectance curve for strongly red hair is distinctive in having a sharp upward deflection in the green region. The pigment is granular, but very probably different in composition from melanin.

Distribution. In most parts of the world hair colours are dark, and it is mainly in north-western Europe that the incidence of blonde and red shades is high. We should note, however, that even in dark-haired populations there may in fact be considerable variation in the actual amount of pigment in the hair, but it is not easy to detect differences by eye or even by reflectometry when the shade approaches black. Interesting regional variations in hair colour can be found within particular countries. The gradient of increasing blondeness from south to north Italy is a well-known example and parallels the distribution of various other physical characters. The data were collected long ago by subjective methods, but probably give an acceptable general picture of regional differences. In the UK red hair (measured by reflectometry) is more frequent in the north and west (Fig. 12.4), and this distribution resembles that of blood-group O to some extent (Fig. 11.2), though there is a discrepancy in East Anglia.

Hair colour often changes during childhood. In parts of southern Europe, where the adult population is predominantly dark, lighter shades are quite frequent in young children, but in Chinese and Africans, on the other hand, the hair is darkly pigmented even in infancy. Blonde hair is by no means uncommon in young Australian aboriginal children.

Inheritance. With the exception of red hair, the genetics of hair colour have been inadequately studied. Change of colour in the earlier and later phases of life complicates analysis. Variation in lightness and darkness due to melanin content probably depends on many genes which control the rate and extent of darkening.

The more extreme examples of red hair are certainly striking, but more detailed work on large population samples shows that this colour is not a sharply distinguishable category, but merges into reddish-brown and reddish-blonde shades. Nevertheless, red hair colours can probably be regarded as a separate class, although they cannot be unambiguously distinguished by visual inspection or by reflectometry. The characteristic shape of the reflectance curve can be used to drive numerical indices of redness. Statistical methods can then be used to obtain the optical sorting into red and non-red shades and this procedure is an improvement on subjective classification by eye. It is possible that chemical criteria may ultimately provide better means of discrimination. The difficulty of deciding whether certain specimens should be regarded as red naturally causes trouble in genetical analysis. Red hair behaves approximately as a recessive character depending on a single gene, but there seems to be manifestation in some heterozygotes. It is not improbable that similar phenotypes may be produced by a number of different genes which we

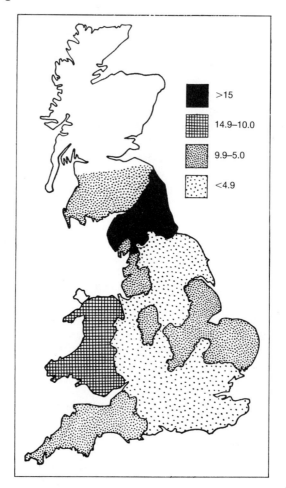

FIG. 12.4. Frequency of red hair in the UK expressed as the index of redness R. The statistics calculated from reflectance curves of hair samples from young soldiers. Note high frequencies in Wales and border countries. (From Sunderland, *Ann. Human Genet.*, 1956.)

cannot at present distinguish and that modifiers may also influence the expression of the character.

Eye colour

The colour of the iris depends in part on granular melanin pigment and in part on optical effects. In dark irises there are abundant pigmented cells in the anterior layer of tissue (Fig. 12.5). In lighter types the pigment in this situation is reduced in amount; some of the incident light is reflected from the residual pigment and also from the colourless, but turbid anterior layer

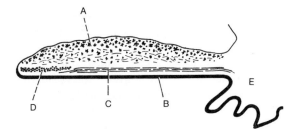

Fɪɢ. 12.5.   Diagram of transverse (i.e. radial) section of one half of the human iris. A, the anterior layer containing pigment cells, especially near the surface, and loose connective tissue. B, the posterior pigmented epithelium. C, the radial, dilator muscle. D, the circular constrictor muscle. E, the ciliary body.

tissue. However, some passes through this layer and is absorbed by the dark pigment of the posterior epithelium. If light passes into a turbid medium, the particles of which are comparable in size to the wavelengths of the incident beam, the shorter wavelengths are selectively reflected (Rayleigh scattering). This optical effect accounts for the predominance of blue in the light reflected from a depigmented anterior layer of the iris. A progressive decrease in the amount of pigment yields a graded series of colours from brown to green, and finally to blue and grey. The iris is not, however, a uniformly coloured area; in the lighter types the tissue of the anterior layer is thinner in some regions than others, producing characteristic radial patterns and the melanin pigment is not evenly distributed, sometimes forming well-marked spots. It is not surprising, in view of these complexities, that eye colour is difficult to record accurately and difficult to analyse genetically.

Sets of glass eyes have been used as standards for matching eye colours, but the fact that the iris is a very small area, varying in extent with the degree of contraction of the pupil, and exhibiting a wide range of variation in detail, presents problems in recording that do not arise in the case of skin or hair colour.

It is often stated that dark eye colour is dominant to lighter shades, and while this may be true in a general way, clear segregation is only obtained if minor degrees of variation are ignored and colours are grouped in definite categories such as brown, mixed, and blue. The apparent simplicity is partly artificial, depending on the initial classification of phenotypes. It is probable that there are many genes involved, some exerting relatively large and some smaller effects either on the amount of pigment or on the pattern of tissue atrophy. It has sometimes, but not invariably, been found that in western European populations dark eyes are more frequent in females than in males, and it is claimed that sex-linked genes are involved. It should be noted that in some regions light eyes frequently occur in dark-haired individuals. Some genes appear to affect mainly the colour of the eyes, while others may have a more general effect on pigmentation.

The iris pigment restricts the entry of light into the eye to the pupil and help to protect the retina against damaging ultraviolet radiation. It is by no means clear whether depigmentation within the range normally occurring in the population has any perceptible influence on visual functions, though the more extreme pigment deficiency of albinism is associated with avoidance of strong light (photophobia) and various visual defects.

Hair form

Little attention has been devoted in recent years to variation in the form of human hair. This is partly because of difficulties in objectively measuring the variation and partly because of the widespread habit, particularly in females of artificially altering head hair for social reasons. Nevertheless, natural head hair form shows striking geographical differences and distinguishes human groups on a continental basis better than almost any other character. It was indeed sometimes used in the past as the primary basis for the classification of humankind.

Hair form is usually first categorized according to its degree of curvature from 'straight' through 'wavy', and 'helical' to 'spiral'. This curvature, which depends on cross-sectional shape can be roughly quantified by laying single hairs on a sheet of paper and measuring the diameter of the circle formed or partially formed. However, the curvature of hair is not always continuous or uniform; in spiral hair, in particular, the curvature increases along the length of the hair from its root.

Spiral head hair is found in its extreme form in Bushmen of south-west Africa where the coils are so tight and intermeshed that they leave quite large patches of naked scalp. Such hair has been termed 'peppercorn'. Helical hair is found throughout most of the rest of sub-Saharan Africa and, though of different microstructure, in Melanesia. Straight hair characterizes the peoples of eastern Asia and Amerindians, whilst wavy hair is most frequently found in Europe and the Middle East, India, Polynesia, and Australia.

Despite the qualitative distinctions that are made, the genetics of the hair form differences are complicated, and in hybrid groups there is no clear segregation into countable categories. The cause for the variation is also unknown. It has been surmised that hair form may play some part in the thermoregulation of the head but there is little hard evidence for this. Clearly hair form today plays an important part in sexual attraction.

Head hair varies in the natural length to which it will grow, but for obvious reasons this is a difficult trait to study. The amount of hair on the body also shows marked differences both between individuals and between populations. The peoples of eastern Asia tend to have little body hair.

Finger- and palm-prints

On the palmar surfaces of the hands and plantar surfaces of the feet there are numerous fine epidermal ridges which form regular but complex patterns. Similar systems of 'dermatoglyphs' are found in other Primates besides man and also in some arboreal marsupials. Sweat ducts open at regular intervals along the ridges and when the skin contacts a smooth surface residual secretions leave an imprint of the pattern, but to get clear and permanent records suitable for detailed study the skin must be 'inked' with special compounds and pressed on a prepared paper. Francis Galton in 1892 was one of the first to classify finger-prints and to investigate their inheritance. Although most prints of the volar surface of the finger can be classified as one of three types — arches, loops, or whorls — there is great individual variation in detail, so great indeed that no two people have identical patterns, not even monozygotic twins.

The three main volar pattern types are illustrated below (Fig. 12.6). The distinctions between them are not absolutely sharp for transitional

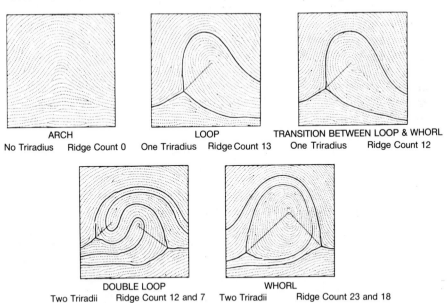

ARCH	LOOP	TRANSITION BETWEEN LOOP & WHORL
No Triradius Ridge Count 0	One Triradius Ridge Count 13	One Triradius Ridge Count 12

DOUBLE LOOP
Two Triradii   Ridge Count 12 and 7

WHORL
Two Triradii   Ridge Count 23 and 18

FIG. 12.6.   The three main finger dermatoglyph types, arch, loop, and whorl, a transitional type between loop and whorl, and a more complex type, the double loop. The main lines which meet at a triradius are shown as continuous lines. A lighter line is drawn from triradius to core; the number of ridges crossed by this line is the ridge-count. (Drawn from original photographs by courtesy of Dr. Sarah Holt, Galton Laboratory.)

forms can be found and complex types which cannot readily be fitted into the classification also occur. The arch is the simplest form with a parallel series of curved ridges passing transversely across the finger pad. In the case of a loop these adjacent systems of lines meet to form a Y-shaped point, known as a triradius or delta on either the ulnar or radial side; in a whorl there are two such triradii and the main lines which participate in them surround a central core.

In the British population (according to Scotland Yard records) loops are the commonest pattern (70 per cent), whorls next (25 per cent), and arches the rarest (5 per cent). The scoring of finger-print patterns is complicated by the fact that they are often not the same on different fingers. In Europeans whorls are most frequent on the 1st and 4th digits (35 per cent) and least frequent on the 5th, while arches occur most often on the 2nd (11 per cent) and are rare on the 5th (1 per cent). Loops may be of the radial type with the opening directed to the radial side or they may have the reverse, ulnar, arrangement; radial loops are commoner on the right hand than on the left.

A count of the number of digital triradii on both hands is sometimes used as a measure of 'pattern intensity'. An arch typically lacks a triradius and therefore scores 0 (though in some in which the arch is much compressed laterally a triradius is formed at the centre); loops, with a single triradius, score 1; and whorls 2. The range of pattern intensity can therefore vary from 0, where there are arches on all digits, to 20 where there are only whorls.

A measure of so-called 'pattern size' is obtained by ruling a line from each triradius to the core of the pattern and counting the number of ridges it crosses (Fig. 12.6). In this method arches again have a count of zero while whorls, with two triradii, have higher counts on average than loops.

Variation in the general shape of the pattern can be assessed by measuring the proportions of length and breadth by standard methods.

The major individual variations in finger-print patterns are largely under genetic control, but they have so far remained inexplicable on any simple scheme of inheritance. Holt's studies on the inheritance of the ridge-count, the distribution of which for a random population is continuous, but non-Gaussian, showed a correlation close to 0.5 between sibs and between parents and children, with no evidence of dominance. The correlation between monozygotic twins was found to be 0.95 ± 0.02 as compared with 0.46 ± 0.10 for dizygotic twins. The non-Gaussian form of the ridge-count distribution may perhaps mean that much of the variation can be attributed to a small number of genes with major effects, but if so the metrical differences between genotypes overlap and are not distinguishable as separate modes in the distribution.

When we compare the dermatoglyphs of populations we can safely assume that the phenotypic differences reflect variations in the distributions of genes, but as with metrical traits in general, we cannot analyse the results in terms of particular genes or specify their frequencies. The intricate nature of dematoglyphs and the wealth of variation which they may show makes it difficult to summarize the facts in a concise, quantitative form. The usual method in population studies has been simply to compare the percentage of the three main pattern types and to give the index of pattern intensity in the way already described above. Some of the more extreme variations are shown in the diagram (Fig. 12.7). In some Bushman populations, for example, arches are unusually common and pattern intensity correspondingly low, whereas in Chinese and North American Indians whorls are very frequent and pattern intensity high.

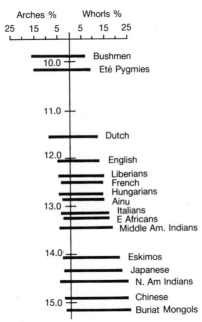

Fig. 12.7. The percentage of arches and whorls, horizontal scale, and the pattern intensity index for various human populations. (Modified from Cummins and Midlo 1943.)

The ridge patterns of the palm also show both individual and population differences. At least five triradii are usually discernible, four of them at the bases of digits 2–5 and the fifth near the centre at the junction of palm and wrist. The usual position of these triradii and of the lines which originate from them are shown in a diagram (Fig. 12.8). Arches, loops, whorls, or more complicated patterns are found with varying frequency in the areas of the interdigital pads, on the thenar eminence overlying the

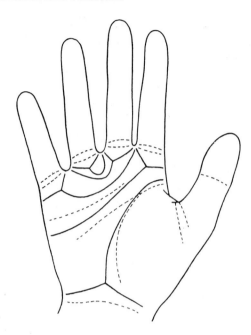

FIG. 12.8.   The main lines and triradii of the palm. The palmar creases are shown as dotted lines.

first metacarpal, and on the hypothenar area over the fifth. In Chinese and various other Far Eastern peoples the main lines are said to run more longitudinally than in Europeans, and the incidence of patterns in the first and third interdigital areas is relatively low.

Mental attributes

A group of quantitative characters which are of particular interest in anthropology, but which are especially difficult to analyse are mental attributes. Included here are the various components of intellectual ability, such as cognitive and verbal skills, more specific talents such as those required to succeed in mathematics or music, the various elements of personality, and even aspects of mood, though these may change quite markedly in the same individual from time to time.

A metric of intelligence which has been much used by psychologists and educationalists is the 'intelligence quotient' or IQ. As initially developed for children this is the ratio of the age level at which a child performs intellectually to its actual age. A child of 4 who performs at the level of the average 5 year old has an IQ of $\frac{5}{4} \times 100 = 125$. This logic is not applicable to adults and IQ's are determined nowadays by converting actual scores in tests, through tables to score distributions which have a mean of 100 and a standard deviation of 15 (or 16) at any given age. The tables, sometimes called tables of norms, are based on the actual performances of a sample (standardization sample) which is representative of the population from which it comes.

There has been much debate of what constitutes intelligence and how far it is measured by IQ, but there can be no doubt that IQ tests applied to populations similar to those on which they were developed provide interesting and useful information. They predict quite well the educational and career success of groups of children in European and North American countries and can even do this for many individuals.

The extent to which IQ (and the various components to it which tests measure) is heritable has been an issue of much controversy. Far too often judgement seems to have been contaminated by political considerations. However, even the scientific problems are more formidable than for most physical traits. These arise primarily from the considerable magnitude to be expected of gene–environment interaction, and gene–environment correlation (cf. p. 173). Cultural inheritance as well as biological inheritance are important for mental traits. High-IQ parents are likely to transmit whatever high-IQ genes there may be, but they will also create for their children environments which stimulate IQ development such as plentiful books. High-IQ individuals will seek out educational experience and be most affected by it.

We cannot measure the magnitudes of these interactions and rigorous scientific analysis is therefore not possible. Notwithstanding, few without prejudice would be surprised to discover (through future genetic neurobiochemistry) that there was a considerable genetic basis to IQ and other mental variation. Within populations, family, twin, and adopted children studies point that way. This, of course, is not to say that future work will establish some fixed heritability. As has already been emphasized, heritabilities are expected to vary from population to population and time to time.

These predictions also only relate to the situation within populations; that between populations is quite another. It is a general tenet of human biology that if there is genetic variation within groups, there will be genetic variation between groups, but again intelligence is a rather special kind of character; it is likely to be selected for in much the same kind of way in all human groups whatever their environment may be.

IQ has certainly been shown to vary between human populations, but the meaning of the differences is far from clear. Many of the studies have used tests developed in Western white societies and these are completely unsuitable for other cultures. Attempts have been made to design so-called 'culture free' tests, but it is doubtful whether these have been successful or indeed can be successful if they are to measure anything important. Even the motivation of intelligence is likely to be very different in different cultures. The outstanding feature of human behaviour in all its forms is its flexibility.

13 Anthropological genetics and peoples of the world

In the last few chapters we have systematically considered the nature and magnitude of the variation found in the human species. Space, of course, has limited the number of systems that could be considered, but a representative proportion of those of complete or high heritability have been described. Genetic variation in man turns out to be of similar form and extent to that found in other species.

We have also considered the functional significance of the variation, where it is known. It would appear that this tends to be more evident in the quantitative morphological and physiological characteristics than in the qualitative biochemical and serological ones. In part this may be because they have been longer known, and are more fully and easily studied. Even within the qualitative variation it is in the widely studied systems like ABO blood groups and haemoglobins that the firmest associations with disease have been found. However, this can hardly be the complete explanation. Two other possibilities exist: either the genes responsible for biochemical variation are more likely to be neutral than those for morphological variation, or adaptiveness is more difficult to detect in them. The first of these possibilities seems inherently unsound in general terms, since it must be biochemical variation in development which ultimately leads to morphological variation! Perhaps herein lies the clue. If many biochemical systems are contributing to the variation in a single morphological system the effect of a single gene substitution would indeed be hard to detect. It has been suggested that one should be considering arrays of genes over many loci, rather than individual substitutions when examining biochemical polymorphic systems for fitness variation (cf. p.208). Of course, if these are the units of natural selection, then the biochemical systems also represent multifactorial quantitative inheritance when viewed in such terms!

Finally, we have considered, in a rather broad way, the patterns of geographical variety that are to be found in the various systems and it is with these patterns that we will be mainly concerned in this chapter.

A number of general features emerge from examining between-population differences. Firstly, as evidenced by the qualitative simply inherited systems where populations can be characterized in terms of gene frequency, populations tend to differ from one another mainly in terms of the frequency with which the same genes are present. One commonly finds situations where one population is polymorphic and another monomorphic for a biochemical trait. However, there are practically no situations where one population is monomorphic for one trait and another population monomorphic for the alternative trait. Generally, populations, particularly of course neighbouring ones, differ in but a few percentage points in which genes are present. Put in another way this means that variances between populations are rather small by comparison with variances within populations, which we have already seen are enormous (cf. p.241). It has been estimated that the differences between the great continental groups of mankind represent only around 5 per cent of human variability.

Another striking feature of geographical variation is that it tends to occur gradually and evenly over extended distances. In other words gene frequency variation is clinal. There are places where quite sharp changes occur, particularly where there are barriers to human movement such as oceans, deserts and high mountains. At least in qualitative traits these are the exception rather than the rule. Some quantitative traits, such as skin colour, hair form, facial form, and other features of physique, show more marked and abrupt geographical changes. This is especially so at continental boundaries. Even they, however, may also show gradual changes over considerable distances. These gradients are termed phenoclines (in comparison with genoclines) because one cannot identify the responsible genes and has to refer to the character variation itself.

Another very important feature of geographical variation is the discordance in the distribution of different traits. No two characters have identical distributions and many characters have little or nothing in common in their geography. This is well exemplified by comparing the distributions of the ABO, MN, Hp and Hb^S. Nor do morphological characters show much closer correspondence with one another or with any simply inherited trait. Clearly, whatever may be the cause of geographical variety, this cannot be explained solely in terms of some distinctive ancestral groups mixing in various combinations with one another. On the other hand, the point does need to be emphasized that all the systems are not randomly distributed with respect to one another and constellations of characters are recognizable, particular on a continental basis, as we shall later see.

The form of genetic geographical variation has long been of great interest to anthropologists. Often it is referred to as racial variation though that prejudges its form and often presumes its origin. Its essential interest is in what it tells us about the relationship or affinity of populations to one another. Unfortunately, the concept of affinity is a complex one, often with various distinct elements intertwined. When we speak of the relationship between two people we usually do so in the genealogical sense of the steps in descent to a common ancestor. Such a view is applicable to the relationship between two populations, i.e. the number of generations or length of time since they shared the same ancestral population. This of course sees population descent as a process of division (or fission) and subsequent isolation so common ancestry becomes progressively remote. However, it needs to be remembered for populations that they can gain common ancestry by exchanging with one another, as well as lose it. Affinity based on descent is sometimes referred to as cladistic affinity.

Another kind of affinity, particularly common in evolutionary discussion, arises from the genetic similarity that populations possess as a result of being descended from the same population. In other words, two populations which are genetically identical to one another and derive this identity from common ancestry have the higher level of affinity, no matter how long ago that ancestry was. Populations who are different from one another as a result of evolutionary diversification have low affinity even if they have had a recent common ancestry. This kind of affinity is referred to as patristic affinity.

Genetic similarity is the basis of a third kind of affinity. Here genetically similar populations have high affinity *however the similarity arises*. Thus, populations which have attained their similarity by evolutionary convergence have just as high affinity as ones of equal similarity due to ancestry. This is known as phenetic affinity and of course can be established without any recourse to evolutionary history or evolutionary assumptions. If all evolution were diversification and diversification at a constant evolutionary rate, then the comparative cladistic, patristic, and phenetic affinities of a group of populations would be the same, but clearly these assumptions are not valid.

An enormous amount of work has been done in anthropology measuring population similarities on the expectation that it would reveal 'evolutionary relationships'. It is constantly worth remembering that this work only means anything if ancestry is by far the most important determinant of similarity. Such a proposition is more likely to be true if drift rather than selection is the main cause for diversification, since drift divergence is a function only of time and effective population size.

Because different genetic systems vary between populations both in magnitude and geographical pattern, it is evident that population affinities need to be assessed over a number of systems. One is after all concerned with the overall similarity between populations. Just how many systems should be examined is a moot point, but clearly only systems which vary between at least some of the populations being considered will be informative. Practical considerations usually impose a rather modest number, since it is rare for a group of populations all to have been tested for more than a few of the same systems.

There are various mathematical and statistical approaches to combining data from different genetic systems. When considering polymorphic and other simply inherited traits the calculations are made using gene frequencies and affinities expressed in terms of the genetic distances between the populations being considered. The comparative positions of a number of populations lie in multi-dimensional space, but there are ways to represent them in two or three dimensions without excessive distortion. Pairs of populations (or already grouped populations) can be put together in cluster analysis procedures and these may be represented diagrammatically as dendrograms. An example of such a dendrogram is shown in Fig. 13.1 in which the affinities of 15 'populations' are shown as established from five blood-group loci (*ABO, Rh, MNS, Fy, Di*). This particular dendrogram is presented as a phylogenetic tree representing the descent patterns of the populations on the basis that they have been diverging from each other and a common ancestor at a constant rate under genetic drift. The vertical axis represents the number of gene substitutions which have occurred, but this is determined by effective population size and time.

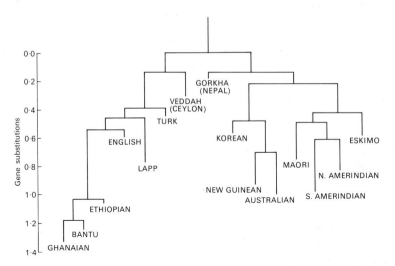

FIG. 13.1.   Most probable tree of relationships between 15 populations based on the gene frequencies for five blood-group systems. The scale is in terms of number of genes substituted over the given span of time. (From Cavalli-Sforza *et al., Cold Spring Harb. Symp. quant. biol.* 1964.)

Different approaches are required for the analysis of quantitative traits. Here, since one cannot work in terms of gene frequencies, it is necessary to deal with character measurements themselves and their distributions. A particularly important factor that needs to be taken into account is the intercorrelation that occurs between anthropometric traits. However, dendrograms of relationship can be constructed in comparable fashion to those based on gene frequencies, and on the whole, the results are surprisingly similar, both when used for expressing microscale and global variation.

The 'race' concept

So far we have deliberately avoided the issue of race (and racial names) since it is on the one hand a contentious issue, yet, on the other, of very little biological consequence. Race as such *explains* practically nothing. The collective unit of evolution is the population and it is in populations that all the forces we have considered operate. Whether these forces end up producing races is only of significance in that races are a step in the speciation process.

Because of the nature of geographical variation in man it has been stated that there are 'no races only clines'. However, a well regarded definition of races is 'populations differing in gene frequency' and clearly such populations exist. Indeed, on such a definition practically every human breeding population is a race! The real question is whether some larger grouping in the population hierarchy can be recognized, i.e. whether taxonomic grouping within the species is meaningful and helpful or whether groupings of any kind above the breeding population are arbitrary if not artificial? Some authorities see 'race' as being of more or less equivalent status to subspecies in animal taxonomy. Mayr considers that a subspecific category is recognizable if 75 per cent of individuals in it can be assigned to it unequivocally. By this criterion and if, as they should, quantitative traits are considered as well as qualitative traits, human races almost certainly exist, and do so on a broad continental basis. This is not to say that some populations, like some individuals, are not hard to categorize; hybrid groups in particular are unassignable, but for many purposes a racial name can be very convenient and convenience after all is the principle (and some might say the only) purpose of classification. Some authors, while damning the race concept in man, are more than happy to use terms like mongoloid, negroid, and caucasoid! If such terms carry useful and accurate information there is nothing evil in them and certainly they accord well with lay judgements about human variety.

In considering, as we now shall, the characteristics shared by peoples of major regions, a factor not previously mentioned but contributing importantly to geographical patterns needs to be raised. We have discussed the roles of migration and intermixture but have done so in isolation from other demographic features. Migration is often associated with, and indeed caused by, population growth. When a population grows the genes it contains also obviously grow in number and increase in frequency in the world as a whole by comparison with those in populations which are not growing. The differential growth and expansion of populations may well then have played as crucial a part in determining gene distribution patterns in pre-Columbian days as it clearly has since. This important consideration was first developed by F. S. Hulse who also noted that the expansion of populations had nothing to do with most if any of the genes they contain. Technology and economy would dictate which prehistoric and historic groups expanded and which declined, so a principal cause for the pattern of biological variety might well be non-biological. Mere comparative changes in population size profoundly alter the genetic composition of the human species.

As already indicated geographical variety is most manifest on a continental basis and we shall here consider some of the distinctions that can be drawn between populations grouped on this basis. Limitations of space prevent any detailed consideration of variation within continents but such variation can be striking and reference will be made to some of the major events which appear to have contributed to it.

Africa

From a human biological viewpoint the main continental unit here is not the land mass of Africa, but that part of the land mass south of the Sahara desert. This desert appears to have acted as a major isolating barrier at least in the west and centre. In the East, the Nile valley and proximity to Arabia have allowed relatively easy exchange of peoples and the categorization of peoples in the Horn of Africa is difficult.

Sub-Saharan populations tend to be characterized by heavy pigmentation of skin, hair, and eyes, helical and spiral hair, thick lips, wide noses, and some projection of the face (prognathism) especially in the alveolar region. Stature is very variable, but Africans tend to have high surface area/volume relationships either by being small, like the pygmies and bushmen, or by being linear in physique. Such linearity is extremely marked in the Nilotic peoples of southern Sudan, e.g. Dinka and Nuer. Steatopygia — large deposits of subcutaneous fat on the buttocks — is widely found in the South of the continent especially in Khoisan groups.

ABO blood-group frequencies are not very different from those in Europe and as in that continent M and N are usually of about equal frequency. However, S is much rarer in Africa and there is less linkage disequilibrium in the MNS system as a whole. In the Rhesus system the haplotype cDe is extremely common, almost everywhere exceeding 50 per cent and often reaching 90 per cent. The V-antigen is also typically African, the Rhesus negative gene d is usually around 20 per cent, less than in Europe, but more common than in the other continents. Other common blood group genes are Js^a in the Kell system, Fy^4 in the Duffy system which can reach over 90 per cent and the P_1 and Jk^a genes. Haemoglobin S genes reach high frequencies in some African populations, and Hb^c is only polymorphic in West Africa. Some of the DNA restriction fragments in the region of the β haemoglobin genes also seem to be uniquely African. Whilst β-thalassaemias and G6PD are widespread in other malarious regions, some variants are characteristically African as is hereditary persistence of foetal haemoglobin. A variant of red-cell acid phosphatase, P^r, is peculiar to the peoples of southern Africa.

An issue that has long intrigued anthropologists is the relationship between the Khoisan peoples (Bushman and Hottentots or San and Khoikhoi as they are usually called today) and other sub-Saharan African populations. They were once regarded as a distinct major race on the basis of their lighter coloration, peppercorn hair, facial and cranial form, and steatopygia. Serologically, however, they are far less distinctive, though they do have low frequencies of Rhesus cde and haptoglobin Hp^1, and high frequencies of acid phosphatase P^r. It is possible that they are a relic group which once had a wider distribution over southern Africa, but except in the Kalahari have lost their distinctiveness through intermixture. Some of the fossil evidence favours such an interpretation. On the other hand, it may be that they represent but one form of the various diversifications which have clearly occurred within Africa.

A feature which dominates the anthropology of Africa in the last few millennia was discovery of iron smelting and manufacture of iron tools. This enabled tropical forest to be cut down and the spread of swidden cultivation. One of the great migrations following from this was that of Bantu-speaking peoples from a quite localized area in the Sudanese savannah right through central and east Africa into the south. These migrations were still underway at the time of the European discovery of the Cape and clearly had a most profound influence upon the genetic composition of Africa.

The spread of swidden agriculture, within West Africa, has been associated by F. Livingstone with the spread of falciparum malaria, and the various genes which confer protection from it. Slash-and-burn provides many small pools of water to form close to human habitation. These are ideal breeding grounds for *Anopheles gambiae*, the main vector of *Plasmodium falciparum*.

Recent analyses, particularly of mitochondrial DNA, have suggested that all non-African populations of present-day man are descended from groups which migrated out of some part of Africa in quite recent times (i.e. around 50 000 years BP).

Europe

Geographically, Europe is but a part of the Asiatic land mass, but its complex history and prehistory give it an anthropological distinctiveness. Included within Europe, however, from this perspective must be much of

North Africa, the Near and Middle East, and the western parts of the Soviet Union.

Europeans are characterized by relatively low skin pigmentation and lighter shades of hair and eyes. This is particularly true of northern Europe where melanin in unexposed skin is very low, and there are high frequencies of blue eyes and blond hair, and in some groups red hair. Hair form is commonly wavy, lips tend to be narrow, and noses are usually narrow and high bridged. Physique is very variable, but weight and stature tend to be greater in the north than in the south.

Most European populations have high frequencies of *A* genes of the ABO system which typically exceed 25 per cent. They are also characterized by possessing A_2 genes which apart from Africa are rare elsewhere. *M* and *N* genes tend to be very nearly equal throughout the region, but *S* is mainly linked to *M*, and *s* to *N*. In Rhesus, the commonest positive haplotypes are *CDe* and *cDE*, but *d* also tends to be very common, especially in the west. Around the Mediterranean a number of genes associated with resistance to falciparum malaria are to be found, particularly those causing β-thalassaemias and G6PDD and some variants of these are unique.

There are a number of small groups in Europe, often rather isolated, which have distinctive and anomalous gene frequencies. The Basques, for example, of the Western Pyrenees have extremely high frequency of the *O* gene (a feature of most European isolates) and of Rhesus *d* which can, uniquely, exceed *D*. The Basques also speak a non-Indo-European language. The Lapps of northern Scandinavia, have exceptionally high frequencies of A_2 (up to 40 per cent), the *N* gene at over 50 per cent and mainly linkage with *S*, the lowest frequencies of *d* in Europe and high *Fy*. They speak a form of Finno-Ugric which is also the language group of the Finns and Hungarian Magyars.

Within Europe gene frequency variation is extremely clinal with most of the long gradients running east–west. This is spectacularly seen in the *B* gene which is at high frequency throughout all eastern European populations and rare in the West (cf. p.269). It also occurs in other genetic systems and in some morphological traits like cephalic index. There is, however, evidence for north/south clines as well as in skin colour, and Rhesus *CDe* versus *cDE* with the latter highest in the north.

The east–west patterns are explicable through the series of invasions from Asia to which Europe appears to have been exposed throughout time. One of the first was associated with the Neolithic revolution and the population expansions that arose from greater food supply. These brought people of the Middle East around both shores of the Mediterranean and up the Atlantic seaboard. Western Europe would then have been occupied by Palaeolithic and Mesolithic hunters, and it has been suggested that the Basques may today best represent the genetic constitution of these peoples.

There were, of course, many later movements up to the Slav and Turkish invasions of the Balkans, and most were of an east–west direction. Some, however, like the Norse movements were essentially north to south.

Asia

Vast areas of Asia are poorly known genetically, including much of the Soviet Union and China! However, broadly speaking most of the people of the continent can be considered as falling into two rather distinct groups: those in the west and India, and those in the centre and east. The former share many of the features of Europeans, including head, face, and hair form, but tend to be more heavily pigmented especially in India. Arab characteristics, most clearly found in peninsula Arabia, include high frequencies of O, M, S and K blood-group genes. India shows its affinities with Europe in the presence of the A_2 gene and high MS in disequilibrium frequency, but has comparatively low frequencies of A and high B. Anti-malarial genes also occur in the subcontinent.

The peoples of eastern Asia are very different and, though variable amongst themselves, are conveniently distinguished as Mongoloid. They are characterized by a yellowish hue to their skin — though this is still due to melanin — black hair and eyes. Head hair is straight, but body and facial hair is sparse. They tend to be brachycephalic, with low noses, and prominent cheekbones and have a distinctive fold of skin covering the upper eyelid — the so-called epicanthic fold. These features are most evident in Mongolia and China, and become less striking in South-east Asia and Indonesia. In blood groups Mongoloids tend to be characterized by high frequencies of the B gene, and a near total absence of Rhesus d with the haplotypes being essentially either CDe or cDE. M and N are in about equal frequency in China but in the south M is higher. Haemoglobin E has high frequency centred on Thailand.

As in the other continents there are small groups of special interest. One of these is the tribal populations of southern India, sometimes mistakenly referred to as 'Veddoid'. They are distinctive from other Indians in facial features, high frequencies of the A gene, and unusually high Rhesus CDE. In these they resemble Australian aborigines.

In northern Japanese islands of Hokkaido and Sakhalin (now USSR) are to be found the Ainu who were once throughout the islands. They differ considerably from the Japanese in not having any mongoloid facies, develop a lot of facial and body hair and have a uniquely high frequency of Rh cdE, with a small frequency of RhV (not with ce as in Africa, but with CDE) and with an unusually high Asian frequency of the N gene. The serogenetics of modern Japanese indicates an important Ainu component in their ancestry.

Finally, one must mention the Negritos of central Malaya, the

Andaman islands, and parts of Indonesia. In stature, pigmentation, and body build they resemble African pygmies, but their serogenetics is firmly not African. Their closest affinities are probably with Melanesians, with high frequencies of *CDe*.

The geographical patterns in Asia are dominated by the effects of the barriers of the Himalayas and other central mountain chains. To the north and east of these are the Mongoloids, while to the south and west are European affinities. India has been colonized through a series of migrations mainly from the north-west, and culminating with a series of invaders speaking Indo-European languages. It is these, presumably, which give the similarity to Europe, and the tribal groups and Dravidian speakers of the south who provide the distinctiveness. Mongoloids may very well have had a quite local northern origin during the last phases of the Pleistocene and then rapidly expanded and migrated southward throughout eastern Asia, intermixing with indigenous groups as they went. Ainu and Negrito, though very different from one another, could well represent such groups.

Australia and the Pacific

Whereas hominids have been in Africa and Eurasia throughout all or most of their evolutionary history, they only entered Australasia comparatively recently and as anatomically modern man. However, evidence for ever greater antiquity for the first colonizations is accruing with dates of 50 000 years BP seeming likely.

Present-day Australian Aborigines are heavily pigmented in skin, hair, and eyes, have robust skulls, typically characterized by marked supra-orbital bony thickenings, and heavy jaws. Noses are broad and low, hair form is wavy, and body hair well developed. Physique is characteristically linear and arms and legs slightly muscled.

ABO gene frequencies are variable, but *O* tends to be high in the north and *A* in the south while *B* is absent except round the Gulf of Carpentaria. The *N* gene has its highest frequency anywhere reaching 95 per cent in parts of Western Australia and the *S* gene is totally absent. Rhesus-negative genes are also absent, but the Rh system is in less linkage disequilibrium than elsewhere and, in addition to the haplotypes *CDe* and *cDE, CDE,* and *cDe* are quite common. Fy^a is universal whilst there are a number of distinct rare genes such as Gc^{Abo} in the Gc plasma proteins.

The people of New Guinea tend to share many of these characteristics, but there is also immense diversity within that island. Physique also is shorter and more robust, the *B* gene is present as is *S* in the MNS system whilst *CDe* may exceed 90 per cent. A blood-group system, Gerbich, which is essentially invariant elsewhere, has an amorph allele in New Guinea which reaches frequencies of 80 per cent. Haemoglobin Tongariki, β-

thalassaemias, and G6PD occur in coastal regions where malaria is rife. Island Melanesia shares many New Guinean characters, especially those of the northern coast where Austronesian languages are also spoken.

The far islands of the Pacific including New Zealand are mainly inhabited by Polynesians who are distinguishable in many traits from Melanesians and Australians. They are much lighter skinned, taller, and more muscular. They tend to have high frequencies of A and the S gene is mainly in haplotypic combination with N which is less common than in Melanesia. The most striking blood-group feature is the high frequency of *cDE* which exceeds *CDe*. Only in the Americas are comparable frequencies found, but in Polynesia there is little *CDE*. The secretor gene is also of distinctively low frequency, i.e. around 40 per cent in Polynesia. Polynesians are particularly prone to obesity and diabetes when provided with affluent conditions. It has been surmised that the genes responsible may have been advantageous in the nutritional rigours of the long sea journeys which Polynesians had to make. These genes have been termed 'thrifty genes'.

Clearly, Australasia has experienced a whole set of migrations out of Asia. The ancestors of the Aborigines must have been among the first movements, and similarity in some aspects of the serogenetics of present-day groups with the tribal populations of south India has been pointed out. The now extinct Tasmanians also showed morphological affinity with Negritos. It is presumed that the next waves, from similar homelands, colonized New Guinea and their descendants are to be found in the Highlands much diversified through isolation and small population size. Then there were various waves of Austranesian speakers who moved into island Melanesia and coastal New Guinea. Evidence from linguistics and archaeology, which in many ways are more revealing than genetics in tracing movements, indicates a series of complex migrations with some 'back tracking'. Finally, came the great ocean migrations of the Polynesians. The original homeland for these is debatable but the final stages seem to have taken off from Tonga and Samoa around 1000 BC. A question that has been much discussed is the extent of the connection between Polynesia and South America. That there was contact is indisputable and the fact that genetic linkage disequilibrium is so marked in Polynesian serogenetics suggests that an Amerindian contribution to ancestry might be considerable.

The Americas

The Americas were also colonized from Asia and probably more recently than Australia. The major movements came across the Bering Sea at times during the Pleistocene when sea level lowered with the accumulation of water in the arctic ice sheets.

American Indians covering the immense latitudinal distances from Alaska to Tierra del Fuego, vary considerably in morphology, particularly in physique. Many plains Indians are robust and tall, while those of the Amazon jungles are slight and gracile. Then there are the highland tribes of the Andes showing physique characteristics of high-altitude adaptation. Typically though, American Indians show morphological similarity with northern mongoloids, especially in their face form (though many have high bridged noses), hair form, and pigmentation levels. In blood groups they all probably lacked B prior to European entry, except for the Eskimo. A is common in North America, but is also absent from Central and South American populations. M is at high frequencies throughout (thus producing an MN cline around the Pacific) as is S. The main rhesus haplotypes are CDe and cDE, but CDE and cDe also occur in moderate frequencies. Rhesus negative d is absent. In Nicaragua, a V antigen rather like that in Ainu has been found. ABH secretor and PTC taster are at very high levels. The Diego blood group has a frequency of a few per cent in North America, rises to around 10 per cent in Central America, and reaches frequencies of up to 40 per cent in some South American populations (it has not been found in Polynesians). No abnormal haemoglobins or G6PD deficiency occur among the indigenous inhabitants of the New World.

The Eskimo (or Inuit as they prefer to be known) who with the Aleuts were the last to enter the Americas before the Europeans are more like extant northern Asiatics than Amerindians. They probably arrived at least 10 000 years ago, and spread across Northern Canada and into Greenland. Harper estimates the separation of 'Bering Strait Mongoloids' from Siberian mongoloids as occurring 19 000 years ago, and the division into Aleuts and Eskimo about 9000 years ago. Just how many migrations there were before that and their dates is a matter of some controversy. Archaeological and skeletal evidence indicates human antiquity of 30 000 years or so in many sites in the continent. It would be particularly helpful to know the history of events since this would allow ideal opportunities for analyses of population diversification in the absence of intermixture with other major groups. In recent years extensive work has been done by Salzano and Neel and their colleagues on microdifferentiation within Brazilian Indians. Here the evolutionary effects of endless separation and coming together (fission and fusion) have been demonstrated to be extremely important.

Suggestions for further reading (Part II)

Molecular and Mendelian genetics

BODMER, W. F. and CAVALLI-SFORZA, L. L. (1976). *Genetics, evolution and man*. W. H. Freeman, San Francisco.

Boyce, A. J. (ed.) (1976). *Chromosome variations in human evolution.* Symposium of the Society for the Study of Human Biology. Taylor and Francis, London.

Cavalli-Sforza, L. L. and Bodmer, W. (1971). *The genetics of human populations.* Freeman, New York.

Emery, A. H. (1983). *Elements of medical genetics* (4th edn). Churchill Livingstone.

McKusick, V. A. (1969). *Human genetics* (2nd edn). Prentice-Hall, New Jersey.

(1985). The molecules of life. *Scient. Am.* **253**, 4.

Strickberger, M. W. (1985). *Genetics* (3rd edn). Macmillan, New York.

Weatherall, D. J. (1985) *The new genetics and clinical practice* (2nd edn). Oxford University Press.

Quantitative variation

Armitage, P. (1971). *Statistical methods in medical research.* Blackwell Scientific Publications, Oxford.

Falconer, D. S. (1981). *Introduction to quantitative genetics* (2nd edn). Oliver and Boyd, Edinburgh.

Halsey, A. H. (ed.) (1977). *Heredity and environment.* Methuen, London.

Mather, K. and Jinks, J. L. (1982). *Biometrical genetics* (3rd edn). Chapman and Hall, London.

Osborne, R. and de George, F. V. (1959). *Genetic basis of morphological variation.* Harvard University Press.

Sneath, P. H. and Sokal, R. R. (1973). *Numerical taxonomy* (2nd edn). W. H. Freeman, New York.

Snedecor, G. W. and Cockran, W. G. (1980). *Statistical methods* (7th edn). Iowa State University Press, Ames.

Population genetics — mating systems

Cavalli-Sforza, L. L. and Feldman, M. W. (1981). *Cultural transmission and evolution: a quantitative approach.* Princeton University Press.

Harrison, G. A. (ed.) (1977). *Population structure and human variation.* Cambridge University Press.

—— and Boyce, A. J. (1972). *The structure of human populations.* Oxford University Press.

Mascie-Taylor, G. C. N. and Boyce, A. J. (1987). *Human mating systems.* Symposia of the Society for the Study of Human Biology. Cambridge University Press.

Schull, W. J. and Neel, J. V. (1965). *The effects of inbreeding on Japanese children.* Harper and Row, New York.

Wallace, B. (1981). *Basic population genetics.* Columbia University Press, New York.

Population genetics — gene frequency changes

BOYCE, A. J. (ed.) (1984). *Migration and mobility.* Symposia of the Society for the Study of Human Biology. Taylor and Francis, London.

DOBZHANSKY, T. (1962). *Mankind evolving.* Yale University Press, New Haven.

JACQUARD, A. (1978). *Genetics of human populations.* Freeman Cooper & Co., San Francisco.

KIMURA, M. (1983). *The neutral theory of molecular evolution.* Cambridge University Press.

LEWONTIN, K. (1982). *Human diversity.* Scientific American Books, New York.

Population variation (proteins)

DICKERSON, R. E. and GEIS, I. (1983). *Haemoglobin: structure, function, evolution and pathology.* Benjamin/Cummings. Menlo Park, California.

GIBLETT, E. R. (1969), *Genetic markers in human blood.* Blackwell Scientific Publications, (Oxford).

HARRIS, H. (1980). *The principles of human biochemical genetics* (3rd edn). Elsevier/North Holland.

KAN, Y. W. (1982). *Haemoglobin abnormalities: molecular and evolutionary studies.* Harvey Lecture Series, 76. Academic Press Inc.

LIVINGSTONE, F. B. (1967). *Abnormal haemoglobin in human populations.* Aldine, Chicago.

Population variation (blood groups, etc.)

CANN, R. L., BROWN, W. M., and WILSON, A. C. (1984). Polymorphic sites in the mechanism of evolution inhuman mitochondrial DNA. *Genetics* **106**, 479–99.

KALMUS, H. (1965). *Diagnosis and genetics of defective colour vision.* Pergamon, Oxford.

LASKER, G. W. (1985). *Surnames and genetic structure.* Cambridge University Press.

MOURANT, A. E., KOPEC, A. C., and DOMANIEWSKA-SOBEZAK, K. (1976). *The distribution of human blood groups and other polymorphisms* (2nd edn). Oxford University Press.

RACE, R. R. and SANGER, R. (1975). *Blood groups in man* (6th edn). Blackwell Scientific Publications, Oxford.

THOMSON, G. (1981). A review of theoretical aspects of HLA and disease associations. *Theoret. Pop. Biol.* **20**, 168–208.

Population variation (quantitative traits)

CUMMINS, H. and MILDO, C. (1962). *Fingerprints, palms and soles.* Dover, New York.

EVELETH, P. B. and TANNER, J. M. (1976). *World wide variation in human growth.* Cambridge University Press.

HARRISON, G. A. (ed.) (1961). *Genetical variation in human populations.* Pergamon, Oxford.

LOEHLIN, J. C., LINDZEY, G., and SPUHLER, J. N. (1975). *Race differences in intelligence.* W. H. Freeman, San Francisco.

WEINER, J. S. and LOURIE, J. A. (1981). *Practical human biology.* Academic Press, London.

Anthropological genetics and peoples of the world

HARRISON, G. A. and PEEL, J. (eds) (1969). Biosocial aspects of race. *J. Biosoc. Sci.,* Suppl. 1.

HIERNAUX, J. (1974). *The people of Africa.* Weidenfeld and Nicolson, London.

HULSE, F. S. (1971). *The human species* (2nd edn). Random House, New York.

KIRK, R. L. (1981). *Aboriginal man adapting. The human biology of Australian Aborigines.* Oxford University Press.

—— and SZATHMARY, E. (eds) (1985). *Out of Asia: peopling of the Americas and the Pacific.* The Journal of Pacific History Inc., Canberra.

MOURANT, A. E. (1983). *Blood relations: blood groups in anthropology.* Oxford University Press.

NURSE, G. T., WEINER, J. S., and JENKINS, F. (1985). *The peoples of Southern Africa and their affinities.* Oxford University Press.

Part III

Human growth and constitution

J.M. TANNER

14   The human growth curve

The study of growth is important in elucidating the mechanisms of evolution, for the evolution of morphological characters necessarily comes about through alterations in the inherited pattern of growth and development. Growth also occupies an important place in the study of individual differences in form and function in man, for many of these also arise through differential rates of growth of particular parts of the body relative to others.

In Fig. 14.1 is shown the growth curve in height of a single boy, measured every 6 months from birth to 18 years. Above is plotted the height attained at successive ages; below, the increments in height from one age to the next. If we think of growth as a form of motion, then the upper curve is one of distance travelled, the lower curve one of velocity. The velocity, or rate of growth, naturally reflects the child's state at any particular time better than does the height attained, which depends largely on how much the child has grown in all the preceding years. The blood and tissue concentrations of those biochemical substances whose amounts change with age are thus more likely to run parallel to the velocity than to the distance curve. In some circumstances acceleration rather than velocity may best reflect physiological events; it is probable, for example, that the great increase in secretion from the endocrine glands at adolescence is manifested most clearly in acceleration of growth (see Fig. 15.3, p.366).

The record of Fig. 14.1 is the oldest published study of the growth of a child; it was made during the years 1759–77 by Count Philibert Guéneau de Montbeillard upon his son, and was published by his friend Buffon in a supplement to the *Histoire Naturelle*. It shows clearly that in general the velocity of growth in height decreases from birth onwards, but that this decrease is interrupted shortly before the end of the growth period. At this time, from 13 to 15 years of age in this particular boy, there is a marked acceleration of growth, called the adolescent or pubertal growth spurt, which will be discussed in the next chapter. A slight increase in velocity also occurs between about 6 and 8 years, providing a second wave on the general velocity curve, known as the juvenile or mid-growth spurt. Unlike the adolescent growth spurt, the mid-growth spurt is not present in all children, and its biological significance is at present unclear.

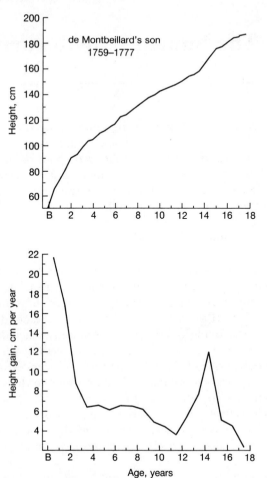

Fig. 14.1.   Growth in height of de Montbeillard's son from birth to 18 years, 1759–77. *Above*, distance curve, height attained at each age; *below*, velocity curve, increments in height from year to year (from Tanner 1962).

Prenatal growth

The velocity curve of growth in height begins a considerable time before birth. Fig. 14.2 shows the distance and velocity curves for body length in the prenatal period and first postnatal year. The peak velocity of length is reached at about 18 weeks post-menstrual age. (Age in the foetal period is usually reckoned from the first day of the last menstrual period, an average of 2 weeks prior to actual fertilization, but as a rule the only locatable landmark.)

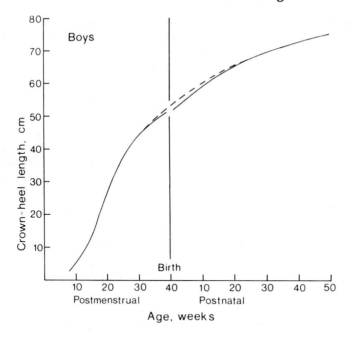

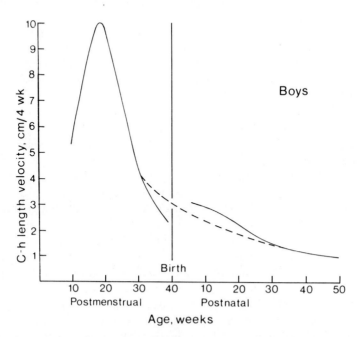

FIG. 14.2. Distance (*above*) and velocity (*below*) curves for growth in body length in prenatal and early postnatal period. Diagram based on several sets of data. The solid lines represent actual length and length velocity; the interrupted line represents the theoretical curve if no uterine restriction took place (from Tanner 1978).

Growth in weight in the foetus follows the same general pattern, except that the peak velocity is reached later, usually at the 34th post-menstrual week. From about 36 weeks to birth (at 40 weeks), the rate of growth of the foetus slows down particularly, due to the influence of the maternal uterus, whose available space is by then becoming fully occupied. Twins' growth slows down earlier, when their combined weight is approximately the 36-week weight of the singleton foetus. Birth weight, and birth size in general, reflect the maternal environment more than the genotype of the child. The slowing-down mechanism enables a genetically large child developing in the uterus of a small mother to be delivered successfully. Directly after birth the growth rate increases again, particularly in genetically large children, and in weight reaches its peak approximately 2 months after birth. The slow-down and catch-up in length can be seen in Fig. 14.2 where the solid lines represent what actually happens, and the dashed line what would perhaps happen if no restriction was exercised by the uterus.

The velocity of growth in length is not very great during the first 2 months of foetal life. This is the period of the embryo. During this period, differentiation of the originally homogeneous whole into regions, such as head, arm, and so forth, occurs ('regionalization'), and also histogenesis, the differentiation of cells into specialized tissues such as muscle or nerve. At the same time each region is moulded by differential growth of cells or by cell migration into a definite shape. This process, known as morphogenesis, continues right up to adulthood, and indeed, in some parts of the body, into old age. But the major part of it is completed by the eighth post-menstrual week and by then the embryo has assumed a recognizably human or child-like appearance.

The high rate of growth of the foetus compared with that of the child is largely due to the fact that cells are still multiplying. The proportion of cells undergoing mitosis in any tissue becomes progressively less as the foetus gets older, and it is generally thought that few, if any, new muscle- or nerve-cells (apart from neuroglia) appear after 6 foetal months, the time when the velocity in linear dimensions is sharply dropping. The muscle- and nerve-cells of the foetus are considerably different in appearance from those of the child or adult. Both have very little cytoplasm around the nucleus. In the muscle there is a great amount of intercellular substance and a much higher proportion of water than in mature muscle. The later foetal and postnatal growth of muscle consists of building up the cytoplasm of the muscle-cells; salts are incorporated and proteins formed. The cells become bigger, the intercellular substance largely disappears, and the concentration of water decreases. This process continues quite actively up to about 3 years of age and slowly thereafter; at adolescence it briefly speeds up again, particularly in boys, more substances being incorporated into the fibres under the influence of androgenic hormones. During the same period increase of the absolute amount of DNA occurs, indicating that further nuclei are appearing (since DNA is confined to the nuclei of cells, where it occurs in nearly constant concentration). Thus, fibres come to have increasing numbers of nuclei associated with them throughout the whole growing period. In the nervous system cytoplasm is added, nucleoprotein bodies appear, and axons and dendrites grow.

Thus, postnatal growth is, for at least some tissues, a period of development and enlargement of existing cells rather than the formation of new ones. Adipose tissue is perhaps an exception to this rule; the number of countable fat cells continues to increase till about the beginning of puberty; the rate of increase, however, gets continuously less. However, the fat cell can only be distinguished from a fibroblast when some fat has accumulated within it, and it is very possible that all the precursors of visible fat cells are, in fact, like muscle and nerve, laid down before birth. At any event, even in this tissue, the quantitatively more important change during postnatal growth is an increase in the size of each cell.

Fitting growth curves

Growth is in general an exceedingly regular process. The more carefully the measurements are taken — for example, with precautions to minimize the decrease in height that occurs during the day for postural reasons — the more regular does the succession of points in the graph become. Fig. 14.3 demonstrates the fit of a smooth mathematical curve to a series of

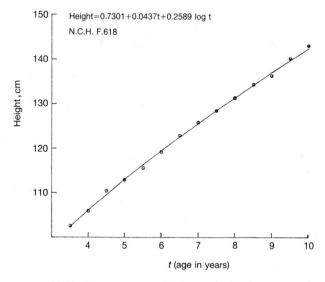

FIG. 14.3.   Curve of form $y = a + bt + c\ln t$ fitted to stature measurements taken at every 6 months by Whitehouse on a girl from age $3\frac{1}{2}$ to 10 years. Harpenden Growth Study. (From Israelsohn, in Tanner 1960.)

measurements all taken on a child by the same observer every 6 months from age $3\frac{1}{2}$ to 10 years. In a series of children followed this way deviations from the fitted curves were seldom more than 6 mm, and were distributed equally above and below the curve at all ages. Many children, however, show a regular seasonal variation, adding a 6-month rhythm about the general height curve, with an increased rate in the spring and a decreased rate in the autumn. This seasonal effect is large enough to make growth rates over periods of less than a full year unreliable for medical purposes; the average child in the UK grows about 3 times as fast in his fastest quarter of the year as in his slowest quarter.

Many attempts have been made at finding mathematical curves which fit human and animal growth data. Most have ended in disillusion or fantasy; disillusion because fresh data failed to conform to them, or fantasy, because the system held eventually so many parameters and was so complicated that it became quite impossible to interpret biologically. What is needed is a curve or curves with relatively few constants, each capable of being interpreted in a biologically meaningful manner. The fit to empirical data must, of course, be adequate, within the limits of measurement error. Part of the difficulty in reaching this goal is that most of the measurements to which curves are fitted are themselves biologically complex. Stature, for example, consists of leg and trunk lengths and head height, all of which have considerably different growth curves. Even with relatively homogeneous measurements such as tibia length or calf-muscle width, it is still not clear what purely biological assumptions should be made as the basis for the form of the curve. The assumption that all cells are continuously dividing leads to an exponential formulation such that the increment is proportional to the measurement ($dy/dt = ky$, or $y = e^{a+bt}$, where y is the measurement at age t, and k, a, and b are constants.) The contrary assumption that all cells are continuously adding or incorporating a given amount of non-dividing material leads to constant increment. The more complex and realistic assumptions that the proportion of cells either multiplying or incorporating new tissue decreases steadily with age and that the rates at which they do these things vary from one age period to another lead to a variety of further formulations.

However, fitting a curve to the individual values is the only way of extracting the maximum information about an individual's growth from the measurement data. This fact becomes increasingly inescapable as research progresses on such matters as the effects of illness on growth rate, or the genetics of growth patterns.

At the time of writing, there are two functions which are used more than others to describe growth in height, weight, or other dimensions. The simpler is the Preece-Baines curve, which is a curve with the shape of a Gothic S, like the better-known logistic. It models a system in which the velocity of height growth, say, at a given age, is a function of the distance yet to be traversed from present height to final mature height, but a function whose form itself changes as age progresses. In algebraic form we have

$$\frac{dh}{dt} = s(t).(h_A - h)$$

where h is height at time or age t, h_A is adult height, and $s(t)$ is a function of time. In the simplest of the Preece-Baines curves the explicit equation is

$$h = h_A - \frac{2(h_A - h_\Theta)}{\exp\{s_0(t-\Theta)\} + \exp\{s_1(t-\Theta)\}}$$

Here s_0 and s_1 are rate constants, and Θ is a time constant with h_Θ height at time Θ. There are thus five parameters to be estimated: h_A the final adult height; Θ, which in fact turns out to be very close to age at peak height velocity in the adolescent spurt; h_Θ, height at the moment of peak height velocity; and s_0 and s_1. The last two are less closely related to obvious points on the growth curve, but s_0 generally models prepubertal velocity and s_1 the velocity characterizing the adolescent spurt.

Fig. 14.4, taken from the original paper, shows how well the curve fits data on growth in height (given that all these measurements were taken

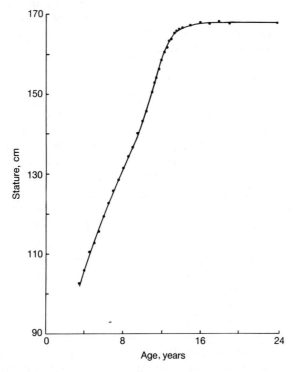

Fig. 14.4. Fit of Preece–Baines curve to measurements of height of a child in the Harpenden Growth Study. (From Preece and Baines 1978.)

by the same highly skilled observer). In general, the fit runs from age 2 years to adulthood; before 2 years this curve does not fit well.

The second curve, or set of curves, consists of three logistic functions, the second superimposed on top of the first, and the third superimposed on the second. This 'triple logistic' has nine parameters, and is thus both more complicated and more flexible than the Preece–Baines curve. It can be divided into three parts; the first section characterizes growth from birth to about 2 years, the second is of increasing importance from birth till the beginning of puberty, and the third characterizes growth at puberty. The temptation to regard these three sections as somehow real and each as under the control of a particular hormone or group of hormones is hard to resist, but any such reification would go considerably beyond the physiological facts. The triple logistic with its greater number of parameters can include the mid-growth spurt in the model, which the simpler Preece–Baines cannot do.

Both Preece–Baines and triple logistic have equations, and parameters which can be estimated. It is also possible, however, to fit non-parametric curves to growth data, which is in essence like drawing a line through the points by eye, though using a set of rules to do so. The most usual method is a system called cubic splines. Non-parametric curves are by their nature more flexible (though less elegant) than parametric curves, and they can be made with as many humps and troughs as the user wants in a particular case, wherein, as in graphic smoothing, lies the rub.

Longitudinal and cross-sectional data

Curves such as these are designed to be fitted to data on single individuals. Yearly averages derived from different children each measured once only in a mass-survey type of study do not, in general, give the same curve. The two sorts of investigation are distinguished as 'longitudinal' and 'cross-sectional'. In a cross-sectional study each child is measured once only, and all the children at age 8, for example, are different from all those at age 7. In a longitudinal study, on the other hand, each child is measured at each age and therefore all the children at age 8 are the same as those at age 7. A study may be longitudinal over any number of years. There are short-term longitudinal studies extending over a couple of years and full birth-to-maturity studies in which children may be examined once, twice, or more times every year from birth till 20 years. In practice it is always impossible to measure exactly the same group of children every year for a prolonged period; inevitably, some children leave the study and others, if desired, join it. A study in which this happens is called a 'mixed longitudinal' study, and special statistical techniques are needed to get the maximum information out of its data. In the past this has not been generally understood, with the truly appalling result that three-quarters and more of the useful information of mixed studies has been lost. One particular type of mixed study is that in which a number of relatively short-term longitudinal groups are overlapped; thus one might have groups of ages 0–6, 5–11, 10–16, and 15–20 years to cover the whole age range. However, problems arise at the 'joins' unless the sampling has been remarkably good.

Both cross-sectional and longitudinal studies have their uses, but they do not give the same information and cannot be handled in the same way. Cross-sectional surveys are obviously cheaper and quicker, and can include much larger numbers of children. They tell us a good deal about the distance curve of growth, and it is essential to have them as part of the basis for constructing standards for height and weight in a given community. However, they have one drawback: they can never reveal individual differences in rate of growth or in the timing of particular phases such as adolescence. It is these differences which chiefly throw light not only on the subtleties of the genetical control of growth, but also on the relationship of physical growth to educational achievement, psychological development, and social behaviour. Longitudinal studies are laborious and time-consuming, and they demand great perseverance on the part of those who make them and those who take part. Unless accompanied by cross-sectional surveys and animal experimentation they can sink over the years into sterile deserts of number-collecting. However, longitudinal studies are indispensable.

Cross-sectional data in some important respects can be misleading. Fig. 14.5 illustrates the effect on 'average' figures produced by individual differences in the time at which the adolescent spurt begins. The left half of

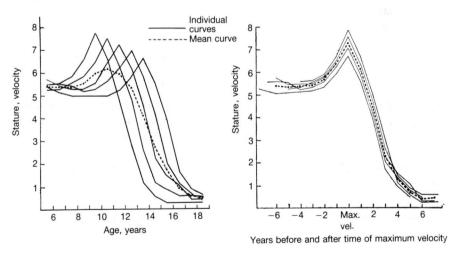

Fig. 14.5. Relationship between individual and mean velocities during the adolescent spurt. *Left*, the height curves are plotted against chronological age. *Right*, the height curves are plotted according to their time of maximum velocity. (From Tanner 1962.)

Fig. 14.5 shows a series of individual velocity curves from 6 years to 18 years, each individual starting his spurt at a different time. The average of these curves, obtained simply by treating the values cross-sectionally and adding them up at age 6, 7, etc., years and dividing by 5, is shown by the heavy interrupted line. It is obvious that the line in no way characterizes the 'average' velocity curve; on the contrary, it is a travesty of it. It smooths out the adolescent spurt, spreading it along the time axis. Averages at each age computed from cross-sectional studies inevitably do this and resemble the interrupted line; they fail to make clear the speed and intensity of the individual spurt. In the right half of Fig. 14.5 the same curves have been arranged so that their points of maximum velocity coincide; here, the average curve characterizes the group quite nicely. In passing from one diagram to the other the time scale has been altered, so that on the right side the curves are plotted, not against chronological age, but against a measure which arranges the children according to how far they have progressed along their course of development; in other words, according to their true developmental or physiological status. We shall return to consider this point at length in Chapter 21.

It is just this sort of problem that curve-fitting to individual data deals with so well. Suppose we have two children with adolescent spurts identical in form and intensity, but beginning at different times. If our fitted mathematical curve is well chosen, one or more of its parameters will characterize the slope of the spurt, another its peak intensity, and so on. These parameters will be equal for the two children because they are unrelated to time. A further parameter of the growth curve will refer to the time at which the spurt starts, and in this the children will differ, with values characterizing their chronological advancement or retardation.

In comparing different groups of children by means of curve-fitting one important point has to be borne in mind. To obtain the average value of a parameter a of the growth curve for a group of children it is necessary to ascertain the values of each individual's parameter a and then average these, if the equation of the curve is at all a complex one. A different and, for most applications, erroneous average will be obtained if the equation is fitted to the mean values of the *measurements* of the group at each age. The former, correct, curve for the group, is called the 'mean-constant curve' and cannot generally be reached by any route except fitting the curve to each individual's measurements.

Growth curves of different tissues and different parts of the body

Most skeletal and muscular dimensions follow approximately the growth curve described for height. So also do the dimensions of organs such as the liver, spleen, and kidneys. However, there are other tissues which have curves sufficiently different to need description. These are the brain and skull, the reproductive organs, the lymphoid tissue, and the subcutaneous fat.

In Fig. 14.6 these differences are shown in diagram form, using the size attained by various tissues as a percentage of the birth-to-maturity increment. Height and the majority of body measurements follow the 'general' curve. The reproductive organs, internal and external, follow a curve which is perhaps not very different in principle, though strikingly so in effect. Their prepubescent growth is slow and their growth at adolescence very rapid; they are less sensitive than the skeleton to one set of hormones and more sensitive to another.

The brain, together with the skull covering it and the eyes and ears, develops earlier than any other part of the body (see Chapter 22 for details). It thus has a characteristic postnatal growth curve. At birth it is

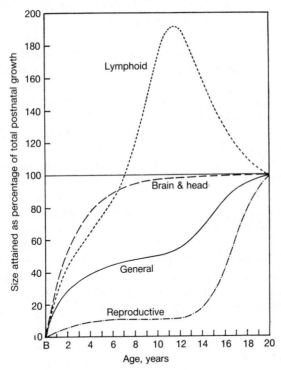

Fɪɢ. 14.6.   Growth curves of different parts and tissues of the body, showing the four chief types. All the curves are of size attained and plotted as percentage of total gain from birth to maturity (20 years) so that size at age 20 years is 100 on the vertical scale. *Lymphoid type:* thymus, lymph nodes, intestinal lymph masses. *Brain and head type:* brain and its parts, dura, spinal cord, optic apparatus, head dimensions. *General type:* body as a whole, external dimensions (except head), respiratory and digestive organs, kidneys, aortic and pulmonary trunks, musculature, blood-volume. *Reproductive type:* testis, ovary, epididymis, prostate, seminal vesicles, Fallopian tubes. (From Tanner 1962.)

already 25 per cent of its adult weight; at age 5 years, 90 per cent; at age 10 years about 95 per cent. Thus, if the brain has any adolescent spurt at all it is a very small one. A small but definite spurt occurs in head length and breadth, but all or most of this is due to thickening of the skull bones and the scalp together with development of the air sinuses. The face follows a curve midway between that of the top portion of the skull and the remainder of the skeleton. At birth it is nearer its mature dimensions than is body-length, but it has still a considerable adolescent spurt, which is greatest in the mandible. Thus, the head as a whole is more advanced than the remainder of the body, and the top part of it, that is, the eyes and brain, are more advanced than the lower portion, that is, the face and jaw.

The lymphoid tissue, of tonsils, adenoids, appendix, intestine, and spleen, has quite another growth curve (Fig. 14.6). It reaches its maximum amount before adolescence, and then, probably under the direct influence of the sex hormones, declines to its adult value.

The subcutaneous fat layer has also a curve of its own, and a somewhat complicated one. Its width can be measured either by X-rays or by specially designed calipers applied to a fold of fat pinched up from the underlying muscle. The distance and velocity curves of skinfolds taken on the back of the arm over the triceps muscle, and under the angle of the scapula, are shown in Fig. 14.7 (a) and (b). Subcutaneous fat begins to be laid down in the foetus at about 34 weeks and increases from then until birth, and from birth until about 9 months (in the average child; the peak may be reached as early as 6 months in some and as late as a year or 15 months in others). From 9 months, when the velocity is thus zero, the subcutaneous fat decreases, that is, has a negative velocity, until age 6 years to 8 years, when it begins to increase once again.

It must be noted that we have discussed the width of the fat layer; a decrease in this width does not necessarily imply a decrease in the cross-sectional area of fat. The fat is a ring around a musculo-skeletal centre which is itself increasing at all ages; if the cross-sectional fat area stayed constant the width of the ring would be reduced simply by enlargement of the musculo-skeletal core. However, calculations from measurements of fat on X-rays show that the cross-sectional area does in fact decrease

---

FIG. 14.7.   (a). Distance curve of subcutaneous tissue measured by Harpenden skinfold calipers over triceps and under scapula. Logarithmic transformation units. Data: *0–1 year*, pure longitudinal, 74 boys and 65 girls Brussels Child Study; *2–7 years* London Child Study Centre with pure longitudinal core 4–6 of 59 boys and 57 girls and actual mean increments subtracted or added to get means at 2,3, and 7; *5–16 years* London County Council, cross-sectional, 1000 to 1600 of each sex at each year of age from 5 to 14, 500 at 15, 250 at 16. (From Tanner 1962.) (b). Velocity curve of subcutaneous tissue as measured by skinfolds over triceps and under angle of scapula. Data as for (a), but with gains prior to 1 year smoothed. (From Tanner 1962.)

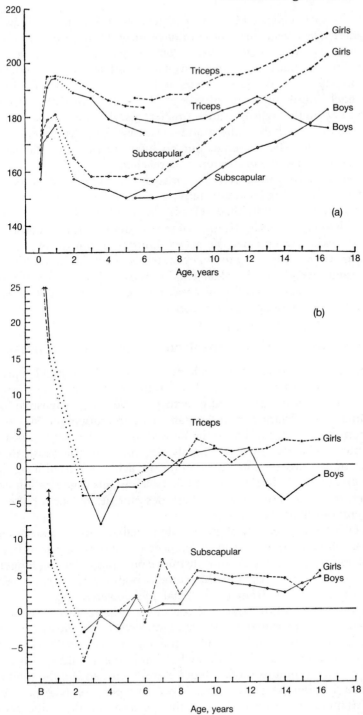

(a)

(b)

during these early childhood years. The decrease is less in girls than boys, so that after age 1 year girls come to have more fat than boys.

The increase from age 7 years or so occurs in both sexes, in measurements of both limb- and body-fat. At adolescence, however, the limb-fat in boys decreases [see 'triceps' in Figs. 14.7 (a) and (b)] and is not gained back until the age of about 20 years. In boys' trunk-fat ('subscapular' in the figure) a much smaller loss, if any at all, occurs; there is only a temporary halt to the gradual increase. In girls there is a slight halting of the limb-fat increase, but no loss and the trunk-fat shows a steady rise until the age of discretion is reached.

Because body-weight represents a mixture of these various tissues its curve of growth is often less informative than those of its component parts. In general, however, individual velocity curves of weight follow a similar course to the height curve. Though to some extent useful in following the health of a child, weight has the severe limitation that an increase may signify growth in bone and muscle, or merely an increase in fat. Similarly, failure to gain weight in the older child may signify little except a better attention to diet and exercise, whereas failure to gain height or muscle would call for immediate investigation.

Organization and disorganization of the growth process

We have seen that growth of a single dimension is a very regular process, and also that different tissues and different areas of the body grow at different rates. These differential growth velocities are responsible for the appearance of the characteristic human shape in embryogenesis, and for the change of shape from that of the baby to that of the adult in postnatal development. Clearly, they are highly organized in the sense that the initiation of one process must depend on the achievement of a particular stage in one or, more likely, several others. How this organization works, or rather in what terms we can best describe it, is one of the most fundamental problems of growth.

Differences in the velocity of growth of particular parts produce many of the foetal, childhood, and adult differences in morphology seen between different species or genera. Man is distinct from apes in having longer legs relative to body or arms; and this comes about by his greater relative velocity of leg growth from early foetal life onwards. Again, the joint between the head and the vertebral column is situated further forward in man than in other primates. At birth, however, this distinction is not present; all Anthropoidea have joints in about the same relative position. However, the postnatal growth of the head in monkeys and apes is greater in front of the occipital condyles than behind them, and so the joint shifts backwards. In man the growth rates of pre- and post-condylar parts of the head are more nearly equal, and the position of the joint remains unchanged. Differential growth rates are very often the mechanism of morphological evolution.

Canalization

Despite the importance of the problem, we know as yet very little about how these intricate growth patterns are organized. It is evident that there are regulative forces holding the processes of development in predetermined channels, for, in the long and complex process which intervenes between the primary chemical action of the genes and the finished adult form, there are many opportunities for slight deviations, slight discrepancies between chemical reactants, to occur and to get progressively multiplied. When a single egg divides to give identical twins, for example, it is unlikely that exactly equal amounts of cytoplasm go to each half. When the chemical substances produced by the genes go out to organize the cytoplasm it is unlikely, therefore, that exactly the same concentration of chemical reactants will be formed in the two organisms. During subsequent development these differences could become enormously magnified. Yet uniovular twins do, in fact, greatly resemble each other, to the degree where only precise measurements can distinguish them. Their similarity, not their difference, requires an explanation. Indeed, their characteristic similarity, at least in length, is not present at birth. Due to the predominance of the uterine effects, pairs of monozygotic twins actually differ slightly more at birth than do dizygotic twins. However, by 6 months the average difference between MZ pairs, in the best available study, was 1.3 cm, between DZ pairs 1.9; by 2 years the MZ pairs differed on average by only 1.1 cm, the DZ by 2.4 cm.

It is thought, in fact, that the processes of differentiation and growth are self-stabilizing or, to take another analogy, 'target-seeking'. Now that we are beginning to understand more about the dynamics of complex systems consisting of many interacting substances, we realize that the capacity to reach a similar final form by varying pathways is not such an exceptional property, nor one confined to living things. Many complex systems, particularly those (called 'open') which have a continuous interaction with other surrounding systems, even if composed of quite simple substances, show such internal regulation as a property consequent on their organization.

The power to stabilize and return to a predetermined growth curve after being pushed, so to speak, off trajectory persists throughout the whole period of growth and is seen in the response of young animals to illness or starvation. This property is called by Waddington 'canalization' or 'homeorhesis' (homeostasis being the maintenance of a static situation and homeorhesis the maintenance of a flowing or developing one). The unusually large velocity occurring during the process has been named 'catch-up' growth by Prader, Tanner, and von Harnack. An example is given in Fig. 14.8 (b); a similar catch-up occurs when, for example, a state of hypothyroidism is corrected or a cortisol-producing (hence growth-inhibiting) tumour is removed. The velocity during the initial period of catch-up may reach three times the normal for age. The allied term 'compensatory growth' is sometimes used by nutritionists to describe a similar phenomenon; however, that term was first applied to the quite different phenomenon of the replacement growth of organs or parts. Thus, when one kidney is removed the other hypertrophies and is said to be showing compensatory growth. Catch-up may be complete or incomplete; if the stress has been severe, and particularly if it has been applied early in the animal's life, then even though a catch-up velocity may be established for a while it may be insufficient to return the animal completely to its normal distance curve of growth. The mechanism of this regulation is almost totally obscure.

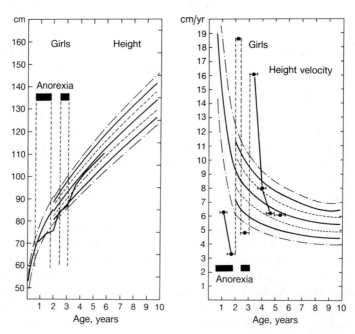

FIG. 14.8. Height and height velocity of girl with two periods of reduced food intake, each followed by catch-up growth. (Redrawn from Prader *et al.*, *J. Pediat.*, 1963.)

Growth gradients

One way in which the organization of growth shows itself is through the presence of maturity 'gradients'. One such is illustrated in Fig. 14.9. Taking the simpler, right-hand panel first, the percentage of the adult value at each age is plotted for foot length, calf length, and thigh length in boys. At all ages the foot is nearer its adult status than the calf, and the calf nearer than the thigh. A maturity gradient is said to exist in the leg, running from advanced maturity distally to delayed maturity proximally. In the left-hand panel of Fig. 14.9 the same gradient is illustrated in the upper limb, together with the fact that girls are more advanced in maturity at all ages than boys (see Chapter 15) without this affecting in any way the distal-proximal gradient.

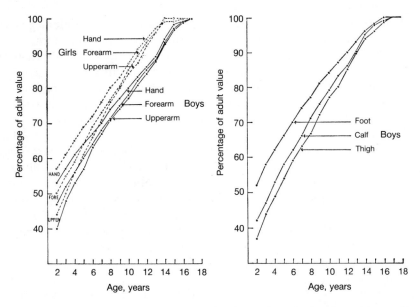

FIG. 14.9. Maturity gradients in upper and lower limbs. Length of segments of limbs plotted as percentage of adult value. Note hand nearer adult value than forearm, and forearm nearer than upper arm at all ages, independent of sex difference in maturity. Mixed longitudinal data. (From Tanner 1962.)

Many other gradients exist, some covering small areas only and operating for short periods, others covering whole systems and operating throughout the whole of growth. The head, for example, is at all ages in advance of the trunk, and the trunk in advance of the limbs. Within the trunk, however, this cephalocaudal gradient fails to be manifested; it is replaced by more complicated, smaller-area gradients. Gradients in the brain, clearly of the utmost importance for educational theory and practice, are described in Chapter 17.

The multitude of chemical reactions going on during differentiation and growth demands the greatest precision in the way one type of growth is linked to another. Thus, for normal acuity of vision to occur the growth of the lens of the eye has to be harmonized closely with the growth in depth of the eyeball. It is small wonder that the success of this co-ordination varies, and that most people are just a little long-sighted or short-sighted. Again, it would seem that many features of the face and skull are individually governed by genes which do not much influence other, nearby features. However, in general the parts of the face fuse to constitute an acceptable whole, and this is because the final growth stages are plastic, and in fitting together, for example, upper and lower jaws forces of mutual regulation come into play which do not reflect the original genetic curves of the discrete parts.

These regulative forces do not always succeed. If the original genetic forces begin by being too unbalanced, normal development cannot occur. For example, if one of the chromosomes is reduplicated so that an abnormal number and distribution of genes occurs in the fertilized egg, abnormalities occur which usually lead to abortion, but sometimes to viable offspring with abnormal mental and physical growth as in trisomy of the small chromosome 21, which gives rise to Down's syndrome.

Short of such disorders, however, it is clear that many individual differences in morphology, and probably in function too, arise through differential variations in the velocity of development of different structures. It is interesting that recently a number of psychologists have supposed that some of the many individual differences in personality structure may arise in a similar fashion. Furthermore, some psychological abnormalities, or culturally excessive deviations from average (analogous to an inconvenient degree of short-sightedness) are thought to arise from insufficient harmonization of the velocities with which various structures and functions develop. This could occur either for genetic reasons, the child carrying by chance a relatively disharmonic set of genes, or for environmental reasons, the development of one area of the personality having been speeded up by external forces, perhaps early in childhood, while another was relatively retarded. Though there is no certain proof that this occurs in man, a number of examples of disharmonious development affecting behaviour in animals are well known. Oedipus behaviour in the goose, for example, can be produced at will by mating a wild-strain gander and a domesticated goose. The domestic goose carries genes for early sexual maturation and in some of the young male offspring sexual maturity occurs before the mother-following response has disappeared. The young bird in consequence insists on copulating with its mother. Since the wild father's sexual activity arises only later in the spring he remains insensible to the drama.

Sensitive periods

Sensitive or critical periods are extreme examples of this linking of differential growth events. By 'sensitive period' is meant a certain stage of limited duration during which a particular influence from another area of the developing organism or from the environment evokes a particular response. The response may be beneficial, indeed essential, to normal development, or it may be pathological. An example of a normal sensitive period is given by the differentiation of the rat hypothalamus into male or female discussed below. During the first 5 days after birth the rat hypothalamus must receive the stimulus of testosterone if it is to become fixed as male; before this period testosterone has no effect and after the 5 days have passed the same is true. As in most sensitive periods the sensitivity is quantitative, rising gradually to a peak and then falling again.

A second example is furnished by the classical work of Hubel and Wiesel and their followers, who showed that a kitten must receive light during the first few weeks after its eyes have opened for the cells of the central nervous system subserving light reception to develop. When the 3 weeks have passed, the animal can no longer become sighted whatever its experience of light. Animals are born into 'expected' environments, where the events that have to occur in each sensitive period normally, of course, can be relied upon to do so.

Post-adolescent growth

Growth of the skeleton does not entirely cease at the end of the adolescent period. In man, unlike some other mammals such as the rat, the epiphyses of the long bones close completely and cannot afterwards be stimulated to grow again. However, the vertebral column continues to grow from age 20 to 30 years by apposition of bone to the tops and bottoms of the vertebral bodies. Thus, height increases by a small amount, on average 3–5 mm, during these years. From the age of 30 to 45 or 50 years it remains stationary, and then begins to decline. The timing suggests that androgenic hormones may be of importance in maintaining this growth, as they are in stimulating the vertebral column growth at adolescence (see Chapter 15).

For practical purposes, however, it is useful to have an age at which one may say that growth in stature has virtually ceased, that is, after which only some 2 per cent is added. Longitudinal records indicate that an average figure for this is currently about 17.5 years for boys and 16.0 years for girls, with a normal variation for different individuals of about 2 years either side of these averages.

Most head and face measurements continue to increase after adolescence steadily, though very slowly, to at least age 60 years. The increase from 20 to 60 years amounts to between 2 per cent and 4 per cent of the 20-year-old value.

The human growth curve as a Primate characteristic

The characteristic form of the human growth curve is shared by apes and monkeys. It is apparently a distinctive Primate characteristic, for neither rodents nor cattle have curves resembling it.

There are as yet few longitudinal series of linear measurements on species other than man, so that we use here curves for body-weight rather than length in comparing species. In Fig. 14.10 the weight velocity curve for the mouse is shown. There is little interval between weaning and puberty, and no visible adolescent spurt because there is no period of low velocity between birth and maturity. In terms of the maturation of its organs the mouse is born earlier in development than man. The peak velocity of its weight curve occurs at a time corresponding closely, by this organ maturation calendar, with birth in man, which is when the first peak of man's weight velocity occurs. The curve for the rat is similar to that of the mouse, except that the rat is born still earlier; the guinea-pig, born more mature, has its peak velocity actually at birth.

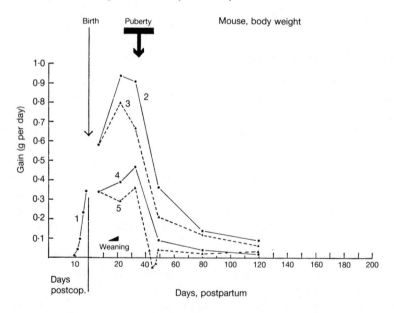

Fig. 14.10.   Weight velocity curve for the mouse. *Curve 1*, sexes combined, cross-sectional. *Curves 2 and 3*, males (18) and females (18), pure longitudinal, large strain bred by MacArthur. *Curves 4 and 5*, small MacArthur strain. Time of puberty from Engle and Rosasco, giving first oestrus in albinos at 37 days, standard deviation 5 days. (From Tanner 1962.)

In the chimpanzee, on the other hand, shown in Fig. 14.11, the curve is quite different, and resembles entirely that of man. The first peak velocity of weight must be shortly before or at birth, but this is followed by a gradual decrease of velocity during the long interval between weaning and puberty. At puberty a considerable adolescent spurt occurs, particularly in the male. The rhesus monkey has a similar curve, though with less time intervening between weaning and puberty. The magnitude of the adolescent spurt, and in particular the degree of sex dimorphism occurring during it, varies from species to species.

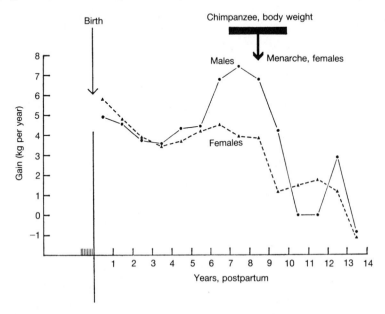

FIG. 14.11.   Weight-velocity curve for the chimpanzee. Menarche average 8.8 years, range 7.0 to 10.8 years. (From Tanner 1962.)

It seems, therefore, that the prolongation of the time between weaning and puberty, often with the acquisition of an adolescent growth spurt, is an evolutionary step taken by the Primates. The essential change seems to be a postponement of the time of puberty, for other mammals continue to grow for a good deal longer, relatively speaking, after sexual maturity has been reached. The immediate cause of the postponement appears to be traceable to a mechanism in the hypothalamus, for it is the brain which initiates the events of the adolescent spurt. The increased time necessary for the maturing of the primate brain has been sandwiched in between weaning and puberty, and the maturation of the hypothalamus has been put back until maturation of the associative areas of the cortex is well advanced.

This process has been carried successively farther in monkeys, apes, and man. At least some of the evolutionary reasons for it are not far to seek. It is probably advantageous for learning, and especially learning to co-operate in group or family life, to take place while the individual remains relatively docile and before he comes into sexual competition with adult males.

15 The adolescent growth spurt and developmental age

The adolescent growth spurt is a constant phenomenon and occurs in all children, though it varies in intensity and duration from one child to another. The peak velocity of growth in height averages about 10 cm a year in boys, and slightly less in girls. In boys the spurt takes place on the average between 12½ and 15½ years of age, and in girls some 2 years earlier.

The sex difference can be seen in Fig. 15.1, which shows the velocity curves for a group of boys who have their peak velocity between 14 and 15, and a group of girls with their peak between 12 and 13. These restricted groups have been taken so as to avoid as much as possible the time-spreading error referred to previously in Fig. 14.5. The difference in size between men and women is to a large degree due to differences in timing and intensity of the adolescent spurt; before it boys and girls differ only by some 2 per cent in height, but after it by an average of about 8 per cent.

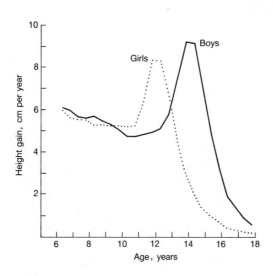

Fig. 15.1. Adolescent spurt in height growth for girls and boys. The curves are from subjects who have their peak velocities during the modal years 12–13 for girls, and 14–15 for boys. Actual mean increments, each plotted at centre of its ½-year period. (From Tanner 1962.)

The difference partly comes about because of the later occurrence of the male spurt, allowing an extra period for growth, even at the slow pre-pubertal velocity; and partly because of the greater intensity of the spurt itself. In absolute terms the adult sex difference is around 13 cm, of which 2 cm are due to prepubertal growth, 7 cm to the later occurrence of the male spurt, and 4 cm to the greater intensity of the spurt.

Practically all skeletal and muscular dimensions take part in the spurt, though not to an equal degree. Most of the spurt in height is due to trunk growth rather than growth of the legs. The muscles appear to have their spurt about 3 months after the height peak; and the weight peak velocity occurs about 6 months after the height peak.

The heart has a spurt in size no less than the other muscles, and other organs accelerate their growth also. Probably even the eye, the most advanced of any organ in maturity and thus the one with least growth still to undergo, has a slight spurt, to judge from the particularly rapid change towards myopia (short-sightedness) which occurs around this age. The degree of myopia increases continuously from age 6 or earlier till maturity, but this accelerated rate of change at puberty would be most simply accounted for by a fractionally greater spurt in axial than in vertical diameters.

It is not clear whether a spurt occurs in brain-growth. In the bones of the face there is a spurt, though a relatively slight one. Individual variability is sufficient so that in some children no detectable spurt occurs at all in some head and face measurements, including those of the pituitary fossa. In the average child, however, the jaw becomes longer in relation to the front part of the face, and also thicker and more projecting. The profile becomes straighter, the incisors of both jaws more upright, and the nose more projecting. All these changes are greater in boys than in girls.

Sex differences

Many of the sex differences of body size and shape seen in adults are the result of differential growth patterns at adolescence. The greater general size of the male has already been discussed. The greater relative widths of shoulders in the male and hips in the female are largely due to specific stimulation of cartilage cells, by androgens in the first instance and oestrogens in the second. The greater growth of the male muscles also results from androgen stimulation, as do some other physiological differences mentioned below.

Not all sex differences develop in this way. The greater length of the male legs relative to the trunk comes about as a consequence of the longer pre-pubescent period of male growth, since the legs are growing faster than the trunk during this particular time. Other sex differences begin still earlier. The male forearm is longer, relative to the upper arm or the height, than the female forearm; and this difference is already established at birth, and increases gradually throughout the whole growing period. It is probably caused by the laying down in early foetal life of slightly more tissue in this area in the male, or of slightly more active tissue. It occurs in some other primates, as well as in man.

A similar mechanism may be responsible for the sex difference in relative lengths of second and fourth fingers. The second finger is longer than the fourth more frequently in females than in males and this difference is also established before birth. The most striking of all the pre-puberty sex differences, however, is the earlier maturation of the female, discussed on p. 375.

Development of the reproductive system

The adolescent spurt in skeletal and muscular dimensions is closely related to the rapid development of the reproductive system which takes place at this time. In Fig. 15.2(b) the events of adolescence in the male are outlined diagrammatically. The solid areas marked 'penis' and 'testis' represent the period of accelerated growth of these organs, and the horizontal lines and rating numbers marked 'pubic hair' stand for its advent and development. The sequences and timings represent in each case the average value. To give an idea of the individual departures from this, figures for the range of ages at which the spurts for height, penis, and testis growth begin and end are inserted under the first and last points of the curves or bars. The acceleration of penis growth, for example, begins on average at about age $12\frac{1}{2}$ years, but sometimes occurs as early as $10\frac{1}{2}$ years, and sometimes as late as $14\frac{1}{2}$ years. There are thus a few boys who do not begin their spurts in height or penis development until the earliest maturers have entirely completed theirs. At age 13 and 14 there is an enormous variability in development amongst any group of boys, who range practically all the way from complete maturity to absolute pre-adolescence. The fact raises difficult social and educational problems and is itself a contributory factor to the psychological maladjustment sometimes seen in adolescence (see Chapter 17).

The *sequence* of events is much less variable than the age at which they take place. The first sign of puberty in boys is an accelerated growth in testes and scrotum. Slight growth of pubic hair may start at about the same time, but proceeds slowly until about the time the height and penis simultaneously accelerate, when it also grows faster. This is usually about a year after the first testicular acceleration. The testicular growth is mainly due to increase in size of the seminal tubules; the androgen-producing Leydig cells appear to develop more or less simultaneously.

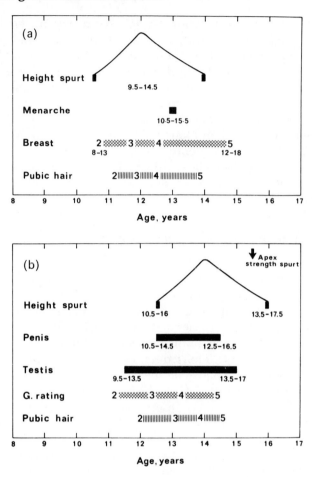

Fɪɢ. 15.2.   Diagram of sequence of events at adolescence in girls (a) and boys (b). An average child is represented; the range of ages within which each event charted may begin and end is given by the figures placed below its start and finish. (From Marshall and Tanner, *Archs Dis. Childh.* **45**, 1970.)

Axillary hair usually first appears about 2 years after the beginning of pubic hair growth, though there is sufficient individual variability so that in a very few children axillary hair actually precedes pubic hair in appearance. Circumanal hair, which arises independently of the spread of pubic hair down the perineum, appears shortly before axillary hair. In boys facial hair begins at about the same time as axillary hair. An increase in length and pigmentation occurs first in the hair at the corners of the upper lip, then spreads medially. Hair next appears on the upper part of the cheeks and in the midline just below the lower lip, and finally along the sides and lower border of the chin. The remainder of the body-hair appears from about the time of first axillary hair development until a considerable period after puberty. The ultimate amount of body-hair an individual develops seems to depend largely on heredity, though whether because of the kinds and amounts of hormones secreted or because of the reactivity of the end-organs is not known.

The enlargement of the larynx in boys occurs at about the time the penis growth is nearing completion. The voice change is a gradual one and is often not complete until adolescence is practically over. In boys at adolescence there are frequently some changes seen in the breast; the areola enlarges in diameter and darkens. In some boys — about a third of most groups studied — there is a distinct enlargement with projection of the areola and the presence of firm subareolar mammary tissue. This occurs about midway through adolescence and lasts from a year to 18 months, after which in the majority of boys the mound and tissue disappear spontaneously.

Sperm begin to appear in early morning urine samples on average a few months after peak height velocity; but in some boys sperm appear somewhat earlier than this. Whether they are fully functional during the first year or two after their appearance is doubtful.

A designation of how far a child has progressed through adolescence is frequently needed in clinical, anthropological, and educational work, and standards for rating the development of pubic hair, genitalia, and the breasts will be found in numerous texts, for example Tanner (1978).

A diagram of the events of adolescence in girls is given in Fig. 15.2 (a). As in boys, there is a large variation in the time at which the spurt begins, though the sequence of events is fairly constant. The appearance of the breast-bud is as a rule the first sign of puberty, though the appearance of pubic hair may sometimes precede it. The uterus and vagina develop simultaneously with the breast. Menarche (the first menstrual period) occurs almost invariably after the peak of the height spurt has been passed. In Fig. 15.3 is shown a Gompertz curve fitted to the growth in height of an individual girl, differentiated to give curves of velocity (above) and acceleration (below). The form of the acceleration curve is interesting and shows the gradual increasing acceleration, the change to sharp deceleration, and gradual reduction of the deceleration. The points marked SS and M stand for the first appearance of the breast-bud and menarche, respectively. It is striking how the one coincides with maximum acceleration and the other with maximum deceleration.

Menarche marks a definitive and probably mature stage of uterine development, but it does not usually signify the attainment of full reproductive function. The early menstrual cycles frequently occur without an ovum being shed; during the first year or two after menarche there is a period of relative infertility, characteristic of apes and monkeys as well as the human. In one study, 75 per cent of cycles during the first 2 years

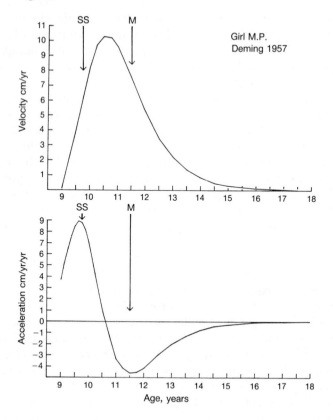

FIG. 15.3.   Velocity (*above*) and acceleration (*below*) curves of growth in stature of girl from age 9 years to 18 years. Calculated from data using first and second derivatives of fitted Gompertz curve. SS represents the first appearance of breast development and M the menarche. (From Israelsohn, in Tanner 1960.)

after menarche were anovulatory, and during the subsequent 2 years still 50 per cent. Two years later the figure was down to 25 per cent.

Changes in physiological function and motor development

Considerable changes in physiological function occur at the same time as the adolescent growth spurt. They are much more marked in boys than girls and serve to confer on the male his greater strength and physical endurance. Before adolescence boys are on average a little stronger than girls, there being more muscularly built or mesomorphic boys than girls in the population even then; but the difference is quite small. After adolescence boys are much stronger, chiefly by virtue of having larger muscles; their muscles probably produce no more force per gram tissue than the muscles of girls, though this is still disputed. Boys have larger hearts and lungs relative to their size, a greater capacity for carrying oxygen in the blood, and a greater power for neutralizing the chemical products of muscular exercise. In short, the male becomes at adolescence more adapted for the tasks of hunting, fighting, and manipulating all sorts of heavy objects, as is necessary in some forms of food-gathering.

In Fig. 15.4 are plotted (as distance curves) data for two strength tests taken from a group of boys and girls followed longitudinally through adolescence. Arm-pull refers to the movement of pulling apart clasped hands held up in front of the chest, the hands each holding a dynamometer handle; arm-thrust refers to the reverse movement, of pushing the hands together. Each individual test represents the best of three trials made in competition with a classmate of similar ability and against the individual's own figure of 6 months before. Only with such precautions can reliable maximal values be obtained. There is a considerable spurt in the boys from about 13 to 16 years. Little spurt can be seen in the girls' data, though figures for hand-grip taken from the same group show a slight acceleration at about 12 to 13½ years.

The male increase in the number of red blood cells at puberty, and consequently in the amount of haemoglobin in the blood, is shown in Fig. 15.5. No sex difference exists before adolescence; hence the combining of data from both sexes up to this age in the lower portion of the figure. The systolic blood-pressure rises throughout childhood, but this process accelerates in boys at adolescence; the heart-rate falls. The alveolar carbon-dioxide tension increases in boys and not in girls, giving rise to a sex difference in the partial pressure of carbon dioxide in arterial blood. Coincidentally, the alkali reserve rises in boys. Thus, the blood of an adult man can absorb during muscular exercise, without change of pH, greater quantities of lactic acid and other substances produced by the muscles than that of a woman — a necessity in view of the greater relative development of muscular bulk in the male. The efficiency of the response to exercise increases in several ways; the total ventilation required for each litre of oxygen utilized, for example, declines.

As a direct result of these anatomical and physiological changes the athletic ability of boys increases greatly at adolescence. The popular notion of a boy 'outgrowing his strength' at this time has little scientific support. It is true that the peak velocity of strength increase occurs a year or so after the peak velocity of most of the skeletal measurements, so that a short period may exist when the adolescent, having completed his skeletal growth, still does not have the strength of a young adult of the same body-size and shape. However, this is a temporary phase; considered absolutely, power, athletic skill, and physical endurance all increase progressively and rapidly throughout adolescence.

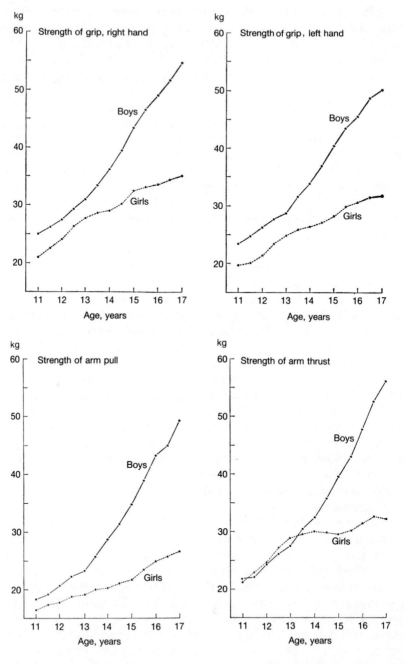

FIG. 15.4.   Strength of hand-grip, arm-pull, and arm-thrust from age 11 years to 17 years. Mixed longitudinal data, 65–93 boys and 66–93 girls in each group, from Jones (1949). (From *Motor performance and growth*, University of California Press).

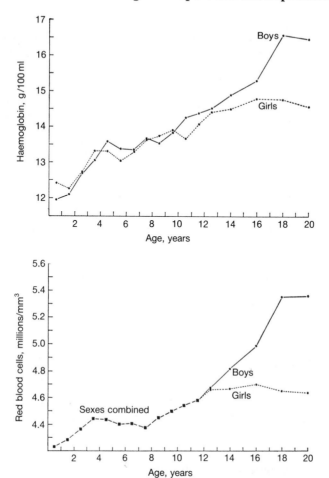

Fig. 15.5.   Change in blood haemoglobin (measured by Van Slyke manometric O_2 capacity) and number of circulating red blood corpuscles during childhood, showing the development of the sex difference at adolescence. Distance curves. Mixed longitudinal data reported cross-sectionally, from Mugrage and Andresen (1936, 1938). *Am. J. Dis. Child.* (From Tanner 1962.)

Developmental age and the concept of physiological maturity

Though all the events of adolescence described above usually occur together, linked in a rather uniform sequence, the *age* at which they happen varies greatly from one child to another. From a file of photographs of normally developing boys aged exactly 14 years it is easy to select three examples which illustrate this. One boy is small, with childish muscles and no development of reproductive organs or body-hair; he could be mistaken for a 12-year-old. Another is practically a grown man, with broad shoulders, strong muscles, adult genitalia, and a bass voice. The third boy is in a stage intermediate between these two. It is manifestly ridiculous to consider all three as equally grown-up physically, or, since much behaviour at this age is conditioned by physical status, in their social relationships. The statement that a boy is 14 years old is in most contexts hopelessly vague; all depends, morphologically, physiologically, and sociologically, on whether he is pre-adolescent, mid-adolescent, or post-adolescent.

Evidently some designation of physical maturity other than chronological age is needed, and in this instance the obvious one would be the degree of development of the reproductive system. However, the same differences in tempo of growth, as its first describer, Franz Boas, called it, occur at all ages, though less spectacularly than at adolescence. Thus, we need a measure of developmental age or physiological maturity applicable throughout the whole period of growth. Three possible measures exist at present; skeletal maturity, dental maturity, and shape age.

Skeletal maturity

The most commonly used indicator of physiological maturity is the degree of development of the skeleton as shown by radiography. Each bone begins as a primary centre of ossification, passes through various stages of enlargement and shaping of the ossified area, acquires in some cases one or more epiphyses, that is, other centres where ossification begins independently of the main centre, and finally reaches adult form when these epiphyses fuse with the main body of the bone. All these changes can be seen easily in a radiograph, which distinguishes the ossified area — whose calcium content renders it opaque to the X-rays — from the areas of cartilage where ossification has not yet begun. The sequence of changes of shape through which each of the bone centres and epiphyses pass is constant from one person to another and skeletal maturity, or bone age, as it is often called, is judged both from the number of centres present and the stage of development of each.

In theory any or all parts of the skeleton could be used to give an assessment of skeletal maturity, but in practice the hand and wrist is the most convenient area and the one generally used. A radiograph of the hand is easily done without any radiation being delivered to other parts of the body, it requires only a minute dose of X-rays, and it demands only the minimum of X-ray equipment, such as a dental or a portable machine. Finally, the hand is an area where a large number of bones and epiphyses are developing. The left hand is used, placed flat on an X-ray film with the palm down and the tube placed 76 cm above the knuckle of the middle finger.

The figure for skeletal maturity is derived by comparing the given radiograph with a set of standards. There are two ways in which this may be done. In the older 'atlas' method one matches the given radiograph successively with standards representing age 5, age 6, and so on, and sees with which age standard it most nearly coincides. The more recently developed method is to establish a series of standard stages through which each bone passes, and to match each bone of the given radiograph with these stages. Each bone is thus given a score, corresponding to the stage reached, and the whole radiograph scores a total of so many maturity points. This score is then compared with the range of scores of the standard group at the same age and a percentile status is then given to the child in skeletal maturity, that is, the percentage of normal children with lower scores at that age, perhaps 80 per cent, is read off. (The child would be in this case at the 80th percentile.) A skeletal age may also be assigned, this being simply the age at which the given score lies at the fiftieth percentile.

Dental maturity

Dental maturity can be obtained by counting the number of teeth erupted and relating this to standard figures in much the same way as skeletal maturity. The deciduous dentition erupts from about 6 months to 2 years and can be used as a measure of physiological maturity during this period. The permanent or second dentition provides a measure from about 6 years to 13 years. From 2 to 6 years and from 13 years onwards little information is obtainable from the teeth by simple counting, but recently new measures of dental maturity have been suggested which use the stages of calcification of teeth as seen in jaw X-rays in just the same way as the skeletal maturity assessment uses the stages of wrist ossification.

Shape age

As the child grows older his shape changes because of the differential growth velocities described in Chapter 14. In principle, the degree of shape change achieved could be made a measure, and a practically very convenient one, of developmental maturity. However, a difficulty enters here which does not arise in the skeletal and dental measures. The shape change is useful only if its measure is completely independent of the final shape reached (as skeletal and dental ages are, the final state being identical in everybody). This is the trouble which besets the 'height age' and 'weight age' once much used by paediatricians, and also the mental age or IQ used by psychologists. The height age of a given child is the age at which the average child achieves the height of the given one. Suppose the given child is tall for his age; this may be *either* because he is advanced developmentally (which is what we are trying to measure) *or* because he is simply going to be a taller-than-average adult and is already exhibiting the fact.

For shape age to be effective, therefore, a combination of body measurements must be found which would change with age, but independently of final size and shape. This is a mathematically complex and difficult proposition, but not an impossible one. Shape age is at present a research problem, not a practical method for use.

Relationships between different measures of maturity

The data of Fig. 15.2 must not be allowed to obscure the fact that children vary a great deal, both in the rapidity with which they pass through the various stages of puberty and also in the closeness with which the various events, including skeletal maturity, are linked together. At one extreme one may find a perfectly healthy girl who has not menstruated although she has reached adult status in breasts and pubic hair, and is 2 years past her peak velocity of height. Her bone age, however, will be retarded like her menarche, for skeletal maturity and menarche are quite closely related. The standard deviation of chronological age at menarche is approximately 1 year; the SD of bone age at menarche, however, is only 0.7 years; thus the normal limits in the UK currently are 11.0–15.0 years in terms of chronological age but 11.5–14.5 in terms of bone age. Bone age is not related to the age at which breasts first develop and only slightly to the age of pubic hair appearance. Pubic hair and breasts develop their sequences with considerable independence as do genital growth and pubic hair in boys. Differences between individuals in the extent to which maturational events are tightly or loosely linked together presumably reflect differences in the organization of the hypothalamic–hypophysial system.

As Fig 15.6 shows, children tend to be consistently advanced or retarded during their whole growth period, or at any rate after about age 3 years. In the figure three groups of girls are plotted separately; those with an early, those with a middling, and those with a late menarche. The early menarche girls are skeletally advanced not only at adolescence, but at all ages back to 7 years; the late menarche girls have a skeletal age which is consistently retarded. The points M1, M2, and M3 represent the average age of menarche in each group.

At all ages from 6 to 13, children who are advanced skeletally have on average more erupted teeth than those who are skeletally retarded. Likewise, those who have an early adolescence erupt their teeth earlier, as illustrated in Fig. 15.7. But the relationship of dental to skeletal maturity is not a very close one, as the figure also implies: even with only three maturity groups in each sex a certain amount of crossing of the lines takes place.

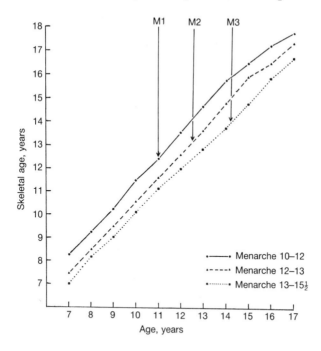

Fɪɢ. 15.6.   Relationship of skeletal maturity and age at menarche. Skeletal development ages (Todd Standards) for early-, average-, and late menarche groups of girls, from age 7 to maturity. M1,M2,M3, average time of menarche for each group. Mixed longitudinal data. (From Tanner 1962.)

This relative independence of teeth and general bodily development is not altogether surprising. The teeth are part of the head-end of the organism, and we have already seen in Chapter 14 how the growth of the head is advanced over the rest of the body and how for this reason its curve differs somewhat from the general growth curve.

Evidently there is some general factor of bodily maturity throughout growth, creating a tendency for a child to be advanced or retarded as a whole; in his skeletal ossification, in the percentage attained of his eventual size, in his permanent dentition, doubtless in his physiological reactions, probably also in his intelligence test score, as described below, and perhaps in other psychological reactions also. Set under this general tendency are groups of more limited maturities, which vary independently of it and of each other. The teeth constitute two of these limited areas (primary and secondary dentition being largely independent of each other), the ossification centres another, and probably the brain at least one more. Some of the mechanisms behind these relationships can be dimly seen; in children who lack adequate thyroid gland secretion; for example, tooth eruption, skeletal development, and brain organization are all

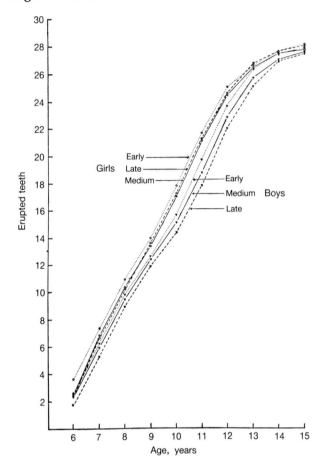

F_{IG}. 15.7. Total number of erupted teeth at each age for early-, medium-, and late-maturing girls and boys. Maturity groups defined by age at peak height velocity. Mixed longitudinal data, reported longitudinally. (From Tanner 1962.)

retarded; whereas in children with precocious puberty, whether due to a brain disorder or a disease of the adrenal gland, there is advancement of skeletal and genital maturity without any corresponding effect upon the teeth or, as far as we can tell, upon the progression of organization in the brain.

The percentage of adult height attained at a given age is quite closely related to skeletal maturity from about age 7 years onwards. Regression equations are available for predicting the adult height of a child from height, chronological age, and bone age. In the case of girls some improvement can be made by including information as to whether or not menarche has occurred, and in both sexes a further allowance can be made for parents' height.

Sex difference in developmental age

Girls are on the average ahead of boys in skeletal maturity from birth to adulthood, and also in dental maturity during the whole of the permanent dentition eruption (though not, curiously, in primary dentition). It would seem, therefore, that the sex difference lies in the general maturity factor (as well as in various more detailed specific factors), which prompts the question as to whether it may not exist in intelligence tests and social responses also.

The skeletal age difference begins during foetal life, the male retardation being ultimately traceable to the Y-chromosome. Children with the abnormal chromosome constitution XXY (Klinefelter's syndrome) have a skeletal maturity indistinguishable from the normal XY male, and children with the chromosome constitution XO (Turner's syndrome) have skeletal maturities closely approximating to the normal XX up till puberty. In what manner these genes work we cannot say. Possibly, the slowing up of male maturation may begin as early as the differentiation of testis or ovary in the second intra-uterine month and represent some basic difference in developmental timing. More probably it may be due to the secretion shortly after this time of sex-specific hormones by the foetal gonads or adrenals. The curious feature here is that all male-specific hormones so far known produce advancement rather than retardation of bone maturity.

At birth, boys are about 4 weeks behind girls in skeletal age, and from then till adulthood they remain about 80 per cent of the skeletal age of girls of the same chronological age. It is for this reason that girls reach adolescence and their final mature size some 2 years before boys. The percentage difference in dental age is not so great, the boys being about 95 per cent of the dental age of girls of the same chronological age.

This sex difference in maturity is not confined to man; it occurs in apes, monkeys, and rats, and may well be characteristic of all or most mammals. Its full biological significance is not at present obvious.

Physical maturation, mental ability, and emotional development

There is considerable evidence that intellectual and emotional advancement is to some extent linked to advancement in skeletal maturity. This may be most simply construed, at least so far as intellectual development goes, as evidence that the brain is affected by the general factor of development tempo, in the same manner as the teeth. Thus, those advanced in physical development do better in mental tests than those retarded in physical development. This subject is further discussed in Chapter 17.

There is little doubt that being an early or later maturer has considerable repercussions on emotional development and social behaviour, particularly at adolescence. These problems are also discussed in Chapter 17. Clearly, the occurrence of tempo differences in human development has profound implications for educational theory and practice.

16 Hormonal, genetic, and environmental factors controlling growth

The endocrine glands are of great importance in the control of growth and development, being one of the chief agents for translating the instructions of the genes into the reality of the adult form, at the pace and with the result permitted by the available environment.

Prenatal period

Genes on the Y-chromosome cause the previously undifferentiated gonad to become a recognizable testis at the ninth week of foetal age, reckoned post-menstrually (or seventh week post-fertilization). Whether this is the result of hormonal action is at present uncertain. At the eleventh post-menstrual week Leydig cells appear in the testis and by the twelfth week they secrete testosterone or an allied substance, probably under the influence of chorionic gonadotrophin, which reaches a peak in the mother's urine at this time (where its presence is used as the standard test for pregnancy). The testicular hormone causes the previously undifferentiated external genitalia to form a penis and serotum. In the female, it seems that differentiation of the ovary and external genitalia proceeds more passively. In the absence of the Y-chromosome, nothing happens at the ninth week and at about the tenth post-menstrual week the gonad turns into an ovary. The external genitalia become female at around the fourteenth week, apparently without hormonal intervention.

There is another aspect of this sexual differentiation, so far studied only in animals, but of much importance in man in principle and perhaps in practice too. In the rat the Leydig-cell secretion acts on the brain as well as on the external genitalia. In all mammals investigated, endocrinological and to a large extent behavioural maleness is dependent on the structure of the hypothalamus. If a female rat pituitary is grafted into an adult male whose own pituitary has been removed, then when vascular connections with the hypothalamus have been established the pituitary will secrete gonadotrophic hormones in a male, not a female cycle. The converse is also true.

In the rat, differentiation of the hypothalamus is caused by testosterone secreted by the Leydig cells during the first 2 or 3 days after birth. This is a true sensitive period. Testosterone given a few days before birth will not cause brain differentiation, nor will testosterone given later than 5 days after birth to a rat whose testes were removed at birth. The message has to reach the hypothalamus at exactly the right time. A single injection of female sex hormone on the fifth day after birth will stop the proper male differentiation, and a single injection of testosterone into a female on the fifth day will produce the 'androgen-sterilized female', a rat without female reproductive cycles when it becomes adult. It is known that some areas of the brain selectively take up testosterone. These must include areas concerned in sexual behaviour as well as in control of gonadotrophin releaser. Female rats given testosterone neonatally do not show any female sexual behaviour when adult, even though ovariectomized and given oestrogen-progesterone replacement therapy so that their sex-hormone state is that of a normal female. If ovariectomized and given testosterone, however, they behave as males.

It is already clear that the rat is not an exception among mammals in this respect. To what extent and with what timing an analogous situation holds in man is not yet known. Birth in the rat corresponds probably to about the sixteenth to eighteenth post-menstrual week in man, though since man seems to develop Leydig cells relatively earlier than other mammals investigated, perhaps we should think in terms of about the fourteenth to sixteenth week. In fact, this would correspond well with foetal testosterone secretion. Whether this work has significance for human sexual behaviour patterns is not at present known.

The prenatal role of other endocrine glands is somewhat uncertain. Maternal oestrogen passes across the placenta and causes the uterus of newborn girls to be temporarily enlarged at birth. Thyroid hormone is necessary for the normal development of the brain, and is secreted by the foetal gland. The adrenal gland has a special zone which is well developed at birth and regresses soon afterwards; its significance and its cause, however, are still matters of debate.

Postnatal period

The most important hormone controlling growth from birth up to adolescence is somatotrophin or growth hormone. This is a polypeptide secreted by the pituitary and showing a greater degree of species (or rather order) specificity than other pituitary hormones. Thus, only human or monkey hormone has a growth-stimulating effect in man.

Though growth hormone is present in the foetus it is not necessary for foetal growth. From birth onwards, however, it is essential if a normal rate of growth is to occur. By the age of 2, children with isolated growth-hormone deficiency are recognizably smaller than normal (though in fact they seldom are recognized till age 5 or later when they go to school, or when younger siblings come to surpass them in height).

Growth hormone does not itself cause the epipbyses to grow. It causes growth by stimulating an increase in cartilage-generating cells and by making them secrete another hormone called Somatomedin C, or Insulin-like Growth-Factor I. It stimulates the liver also to produce Somatomedin C. Somatomedin C is also a peptide but of smaller size than growth hormone. The administration of growth hormone to a person who lacks it causes growth of muscle with increased incorporation of amino acids into tissues to form protein. It also causes diminution of the amount of adipose tissue, shifting the metabolic balance from the laying down of fat to the laying down of protein. Thus, children who lack the hormone are fat as well as small.

The secretion of growth hormone, like that of other pituitary hormones, is controlled by the hypothalamus. There are two hypothalamic hormones concerned, a stimulator called growth-hormone releasing factor (GRF) and an inhibitor called somatostatin. Both are relatively small peptides, of known structure, and have been synthesized. The exact way in which they interact is not yet entirely clear.

Growth hormone is secreted in pulses throughout the 24 hours of the day, not continuously. Exercise, anxiety, and sleep regularly cause secretion, but other factors are uncertain. Under normal circumstances some six or eight pulses occur each 24 hours. The amplitude, and perhaps the frequency of pulses increases at puberty, contributing to the adolescent growth spurt. It is at present not clear whether shortness and tallness within the normal range are caused by differences in amounts of somatomedin but much more probably in amounts or characteristics of receptors in the cartilage cells.

Thyroid hormone plays a vital role throughout the whole of growth. The activity of the thyroid, judged by the basal metabolic rate, decreases gradually from birth to adolescence, at which time it probably increases, or at least falls less rapidly, for a year or so. So far as rate of growth in size is concerned, the action of the thyroid is permissive and not controlling. In hypothyroidism growth is delayed; skeletal maturity, dental maturity, and growth of the brain are all affected.

Though clearly the normal mechanism controlling the rate of skeletal maturation must be hormonal, the balance of hormones is not yet clear. Lack of thyroid hormone and lack of growth hormone both cause retardation; sex hormones and adrenal androgens cause advance. Small quantities of sex hormones and adrenal androgens circulate in the blood before adolescence, but what part variations in their amount play in controlling tempo of growth is quite unknown.

Puberty

At adolescence a relatively new phase of growth occurs in which hormones from the gonads combine with growth hormone to produce the adolescent spurt. It seems that a full spurt is dependent on both sets of hormones being present; boys with growth hormone deficiency have a spurt only reaching about half the normal peak velocity.

Two out of the three major groups of hormones produced by the adrenal circulate in the blood at relatively unchanged levels from birth onwards; these are cortisol and aldosterone, the latter being the hormone which maintains within acceptable limits the concentrations of electrolytes in the tissue fluid.

The third group of adrenal hormones, the androgens, appears in quantity only in mid-childhood, at the time of the mid-growth spurt. Androgens increase gradually from about age 7 until puberty begins, when their rate of increase about doubles. What part they play in the body's economy is uncertain; it seems likely to concern muscular function. However, testosterone is the major cause of the increase in size and strength of the male muscles at adolescence, and of the increase in number of red blood-cells.

The sequence of endocrine events at puberty is fairly clear, though rather complicated. The sequence is initiated by events in the hypothalamus. Before puberty the pituitary contains gonadotrophins, or can manufacture them, but does not release them to the general circulation because it is not stimulated by the hypothalamus to do so. Gonadotrophin release is caused by a releasing substance, luteinizing hormone releasing hormone (LHRH), an octapeptide which is synthesized in the cells of the arcuate nucleus of the hypothalamus and reaches the pituitary via the hypophysial-portal system of blood-vessels. It is the hypothalamus that carries the information as to maturity, not the pituitary. At the 'correct' stage of bodily maturity the hypothalamus matures in some way and LHRH is released.

The way in which this happens has a general importance for the clarification of developmental mechanisms. In the neonatal period there is a feedback system already established and in operation, whereby the levels of circulating testosterone (in boys) and oestrogen (in girls) inhibit the hypothalamic neurons which secrete LHRH and thus reduce the initially high gonadotrophin level. Thereafter, throughout childhood, this system seems to go into cold storage; the arcuate nucleus goes to sleep, apart from a very occasional burst. Then, at puberty, something wakes up the arcuate, LHRH is secreted, gonadotrophins are released, at first only at night and subsequently in bursts throughout the 24 hours. The gonads respond by secreting sex steroids and these now re-establish the original feed-back system. Just what causes the awakening of the arcuate nucleus remains a mystery. Clearly, we are dealing with some sort of internal clock, but one dependent on the passage of numerous prior events in the organism and not simply dependent on chronological nor even wholly on developmental time.

The reason for the increase of adrenal androgens at puberty is less clear. Since cortisol continues to be secreted at pre-adolescent rates it is unlikely that adrenocorticotrophic hormone secretion is increased at puberty. It seems likely that there is a specifically adrenal-androgen-stimulating pituitary hormone, not yet isolated.

Much else remains obscure. The cause of the pre-adolescent increase in fat is unknown, though its timing seems to coincide with the fast increase of adrenal androgen. Though we are beginning to understand the delicate linkage of the hormonal events of adolescence, detailed knowledge will have to wait upon longitudinal studies conducted with the more sensitive chemical and biological methods that have recently become available.

The interaction of heredity and environment in controlling growth rate

Many factors that affect the rate of development are known. Some are hereditary in origin and act by hastening or retarding physiological maturation from an early age. Others, such as dietary restriction, season of the year, or severe psychological stress, originate in the environment and simply affect the rate of growth at the time they are acting. Others again, such as socio-economic class, reflect a complicated mixture of hereditary and environmental influences.

The height, weight, or body-build of a child or an adult always represents the resultant of both the genetical and environmental forces, together with their interaction. It is a long way from the possession of certain genes to the acquisition of a height of 2 m. In modern genetics it is a truism that any particular gene depends for its expression firstly on the internal environment created by all the other genes, and secondly on the external environment. Furthermore, the interaction of genes and environment may not be additive. That is to say, bettering the nutrition by a fixed amount may not produce a 10 per cent increase in height in all persons irrespective of their genetical constitutions; instead a 12 per cent rise may occur in the genetically tall and an 8 per cent rise in the genetically short. This type of interaction is called 'multiplicative'. In general, a particular environment may prove highly suitable for a child with certain genes and highly unsuitable for a child with others. Thus, it is very difficult to specify quantitatively the relative importance of heredity and the environment in controlling growth and physique under any given circumstances; the particular circumstances must always be made clear. In general, the nearer optimal the environment the more the genes have a chance to show their potential actions, but this is an overall statement only and undoubtedly many more subtle and specific interactions occur, especially in growth and differentiation.

Genetic factors, however, are clearly of immense importance. The fundamental plan of growth is laid down very early, in the comparative safety of the uterus. An immature limb bone removed from a foetal or newborn mouse and implanted under the skin of the back of an adult mouse of the same inbred strain (which therefore produces no antibodies to it) will continue to develop until it closely resembles a normal adult bone. Furthermore, the cartilage scaffolding of the bone, removed at the stage preceding actual bone formation, will do the same. Thus, the structure of the adult bone in all its essentials is implicit in the cartilage model of months before. The later action of the bone's environment, represented by the muscles pulling on it and the joints connecting it to other bones, seems to be limited to the making of finishing touches.

Genetics of growth

The genetical control of tempo of growth is manifested most simply in the inheritance of age at menarche. Identical twin sisters reach menarche an average of 2 months apart; non-identical twin sisters an average of 10 months apart. The correlation coefficient between age at menarche of mother and daughter is about 0.4, only slightly lower than similar correlations for height. These are indications that a high proportion of the variability of age at menarche in populations living under European conditions is due to genetical causes. The inheritance of age at menarche is probably transmitted as much by the father as by the mother, and is due not to a single gene, but to many genes each of small effect. This is the same pattern of inheritance as that shown by height and other body measurements.

This genetical control operates throughout the whole period of growth; skeletal maturity shows a close correspondence at all ages in identical twins. The time of eruption of the teeth, both deciduous and permanent, and also the sequence in which teeth calcify and erupt, is largely determined by heredity. Genes controlling growth range all the way from those affecting rate of growth of the whole body, probably through endocrine mechanisms, to those bringing about a highly localized growth gradient causing one tooth to erupt before another, or one ossification centre in the wrist to appear before another.

Not all genes are active at birth. Some express themselves only in the physiological surroundings provided by the later years of growth; their effect is said to be 'age-limited'. This is the probable explanation of the curve described by the correlations between measurements of a child at successive ages and his or her measurements as an adult, which have been obtained by long-term longitudinal studies (see Fig. 16.1). The correlation of length at birth with adult height is very low, since birth length reflects

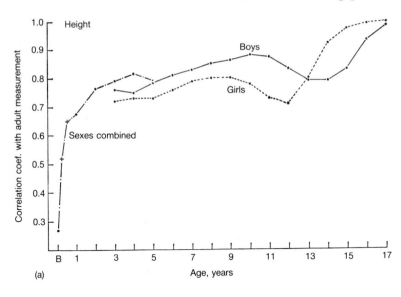

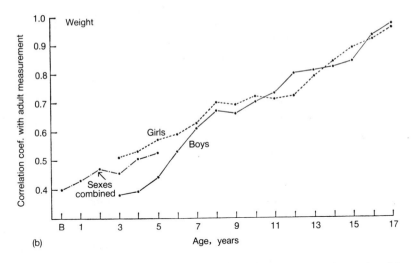

FIG. 16.1. Correlations between adult height and weight and heights and weights of same individuals as children. Sexes-combined lines (0–5) from 124 individuals of a study in Aberdeen with + points from Bayley. Boys' and girls' lines (3–17) from 66 boys and 70 girls of California Guidance Study. All data pure longitudinal. (From Tanner 1962.)

uterine conditions and not the child's genotype. The child's genes increasingly make themselves felt and the correlation rises steeply during the first 2 years; but after this only a small rise occurs until adolescence. It seems likely that the magnitude as well as the time of the spurt is genetically controlled, perhaps by genes causing the secretion of large or small amounts of androgenic hormones. Such genes may produce no effect until the moment when androgen secretion begins. Certainly, there is a considerable degree of independence between growth before and growth at adolescence.

Race and ecological conditions

There are racial differences in rate and pattern of growth, leading to the differences seen in adult build. Some of these are clearly genetically determined, while others depend perhaps on climatic differences and certainly on nutritional ones. We must suppose that in each of the major populations of the world the growth of its members was gradually adjusted, by means of selection, to the environmental conditions in which they evolved. We should be able to see the remnants of this process in modern populations — the remnants only, because relatively recent migrations have much altered the distributions of peoples, so that many no longer live in the areas in which they evolved. There is, in fact, a quite close positive relation between the linearity of peoples, as judged by their adult weight for height, and the average annual temperature of where they live. Differences in size must be sharply differentiated from differences in shape, for the former are relatively easily affected by malnutrition and the latter are not. A European who is starved throughout childhood ends up a small adult, but his skeletal shape is little affected, though he will be lacking in fat and, if the malnutrition has been severe and prolonged enough, lacking also in muscle.

Height-for-age curves of the best-documented groups of European, African (in the sense of origin), and Asian peoples each in comparable and well-off circumstances, thus under similar, presumed near-optimal, nutritional conditions, show little if any difference between Negroes and Europeans; but the well-off Chinese and Japanese are shorter and clearly finish their growth earlier. In contrast, in groups of genetically similar populations under very different environmental circumstances, the gross restriction of growth in the malnourished is plain to see.

Contrary to popular belief, climate has little direct effect on rate of growth. The average age of menarche in relatively well-nourished Nigerian schoolgirls has been reported as 14.3 years, and that of Eskimo girls as 14.4 years. Burmese and Assamese girls living under excellent nutritional and medical circumstances, but with a hot-weather temperature of 45°C have an average age of menarche of about 13.2 years, a figure practically identical with the average in Europe at the same time. Nutritional effects on menarche (see below) are so marked that they overwhelm possible climatic ones, which, if existent at all, are relatively minor.

Some differences in shape between populations seem mostly to be due to genetic causes. Children between 6 and 11 years in American Indian tribes in Arizona are heavier for their height than the local white children, despite being in worse economic circumstances. One could conceivably explain this by a differential effect of malnutrition, the diet causing a stunting of growth in length but an excess of growth in breadth of bone and muscle. But data on African negroes, and particularly Nilotics, shows that they, by contrast, are lighter at all ages for their height than are whites. Either the character of the malnutrition is totally different in the two areas, or, far more probably, genetic differences are involved.

Certainly genetic differences are the cause of the negroes, in West African, East Africa, and the USA, being ahead of the white in skeletal maturity at birth and for the first year or two. This is associated with advancement in motor behaviour, and earlier passing of the milestones such as sitting-up and crawling. The advancement, at least in Africa, disappears by about the third year, either partially or wholly, because of inadequate nutrition. Well-off negro girls in the USA remain in advance of whites throughout the whole growth period. The permanent teeth also erupt earlier in negroes than whites, by an average of a year. The teeth-buds are laid down early in life and their growth is more resistant to malnutrition and disease than is the skeleton, probably because it is less affected by hormonal alterations.

Season of year

In most data from industrialized countries in temperate areas a well-marked seasonal effect on growth velocity can be seen. Growth in height is on average fastest in spring and growth in weight fastest in the autumn. The average velocity of height from March to May is about twice that from September to October in most of the older western European data.

Individual children differ surprisingly, however, both in the time when their seasonal trend reaches its peak, and in the degree to which they show a seasonal trend at all: in a considerable number little evidence of any seasonal effect is seen. These differences may reflect individual variation in endocrine reactivity.

Nutrition

Malnutrition delays growth, as is shown from the effects of famine associated with war. In Fig. 16.2 the heights and weights of schoolchildren in Stuttgart are plotted at each year of age from 1911 to 1953. There is a uniform increase at all ages in both measurements from 1920 to 1940 (see secular trend discussion below), but in the latter years of the Second World War this trend is sharply reversed.

Children have great recuperative powers, provided the adverse conditions are not carried too far or continued too long. During a short period of malnutrition the organism slows up its growth and waits for better times. When they arrive growth takes place unusually fast until the

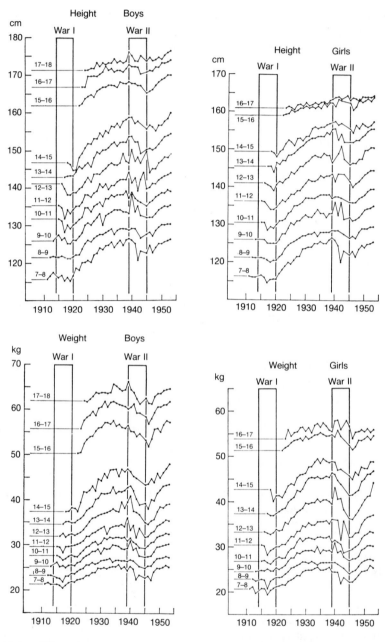

Fɪɢ. 16.2.   Effect of malnutrition on growth in height and weight. Heights and weights of Stuttgart schoolchildren (7–8 years to 14–15 years, Volkschule; 15–16 years upwards, Oberschule) from 1911 to 1953. Lines connect points for children of same age, and express secular trend and effect of war conditions. (From Tanner 1962.)

genetically determined growth curve is reached or approached once more, and subsequently followed. During this 'catch-up' phase, weight and height and skeletal development seem to catch up at approximately the same rate. In cattle, alternation of periods of good feeding and underfeeding may alter the final shape and tissue composition according to the timing of the periods, the fastest-growing tissues suffering most during malnutrition. There is little evidence that anything similar occurs in malnutrition in man.

Girls appear to be better buffered than boys against the effects of malnutrition or illness. They are less easily thrown off their growth curves, perhaps because the two X-chromosomes provide better regulatory forces than one X- and the small Y-chromosome.

Psychological disturbance

That adverse psychological conditions might cause a degree of retardation in growth is a thought that comes readily to mind. In recent years it has been clearly established that in certain children under emotional stress the growth hormone secretion is inhibited and they come closely to resemble cases of idiopathic growth-hormone deficiency. However, when taken out of the stressful conditions they begin to secrete growth hormone again and have the usual rapid catch-up growth.

A similar thing may happen under less extreme circumstances and account for smaller variations in individuals' growth, though good evidence on this is naturally hard to come by. However, one experimental investigation by Widdowson, is clearly (one might almost say providentially) controlled.

In studying the effect of increased rations on orphanage children living on the poor diet available in Germany in 1948 Widdowson had the rare opportunity of observing the change brought about by replacement of one sister-in-charge by another. The design of the experiment was to give orphanage B a food supplement after a 6 months' control period and to compare the growth of the children there with those in orphanage A, which was not to be supplemented. As shown in Fig. 16.3, however, the result was just the reverse of that expected; though the B children actually gained more weight than the A children during the first, unsupplemented, 6 months, they gained less during the second 6 months, despite actually taking in a measured 20 per cent more calories. The reason appeared to be that at precisely the 6-month mark a certain sister had been transferred from A to become head of B. She ruled the children of B with a rod of iron and frequently chose mealtimes to administer to individual children public and often unjustified rebukes, which upset all present. An exception was the group of eight favourites (represented by the curve with squares in the figure) whom she brought with her from orphanage A. These eight always gained more weight than the others, and on being supplemented in B gained still faster. The effect on height was less than that on weight, but of the same nature. 'Better', quotes Widdowson, 'a dinner of herbs where love is than a stalled ox and hatred therewith'.

Possibly similar factors may explain in part some of the observations made on gains in height and weight in boarding-school schoolchildren during term-time as opposed to holidays.

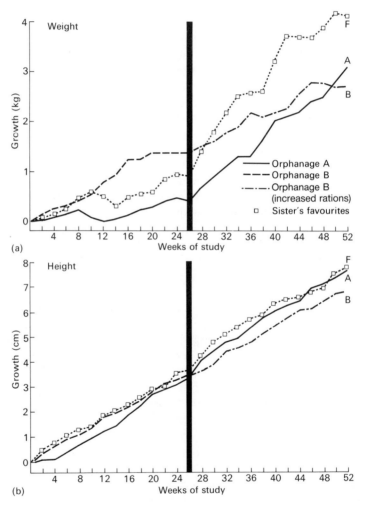

FIG. 16.3.　Influence of sister-in-charge S on growth in weight (a) and height (b) of orphanage children. Orphanage B diet supplemented at time indicated by vertical bar, but sister simultaneously transferred to B from A. Note magnitude of growth follows presence or absence of sister, not amount of rations. The curves (F) with squares are for 8 favourites of sister, transferred with her to B from A. (From Tanner 1962.)

Socio-economic class: size of family

Children from different socio-economic levels differ in average body-size at all ages, the upper groups always being larger. In most studies socio-economic status has been defined according to the father's occupation, though in recent years it is becoming clear that in many countries this does not distinguish people's living standards or life-style as well as formerly; an index reflecting housing conditions is becoming a necessary adjunct, as is some measure of the child-centredness of the family budget.

The difference in height between children of the professional and managerial classes and those of unskilled labourers is currently about 2 cm at 3 years, rising to 5 cm at adolescence. In weight the difference is relatively less since the lower socio-economic class children have a greater weight for height.

In Fig. 16.4 the heights of a national sample of 7-year-old children from all over Great Britain (those born in one week of 1958) are plotted in relation to socio-economic class and numbers of children in the family. The tendency of the better-off children to be taller is visible in families of all sizes.

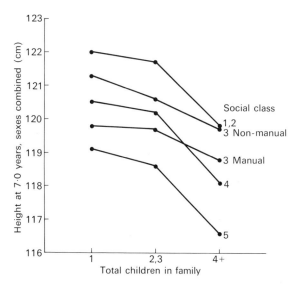

Fig. 16.4.   Differences in height of 7-year-old children according to occupation of father ('social class') and number of siblings in family. Sexes pooled. (From Goldstein, *Hum. Biol.* **43**, 1971.)

Some of the height difference is due to earlier maturation of the well-off classes, but most persists into adulthood, where the social class difference I/II to V is currently (1980s) about 3 cm in men and 2 cm in women. There is a difference in age at menarche of 2–3 months between daughters of the managerial class and those of unskilled workmen; permanent tooth eruption occurs earlier in the more favoured groups by about the same margin, when all the teeth are averaged.

The causes of this socio-economic differential are probably multiple. Nutrition is almost certainly one, and with it all the habits of regular meals, sleep, exercise, and general organization that distinguish, from this point of view, a good home from a bad one. Home conditions are more related to the growth differences than are the economic conditions of the families, and home conditions reflect to a considerable degree the intelligence and personality of the parents. Minor illnesses such as measles, influenza, and even antibiotic-treated middle-ear infection or pneumonia cause no discernible retardation of growth in the great majority of well-nourished children, but they may have some effect on relatively ill-cared-for ones. Possibly the greater incidence of such illnesses in the worse-off and more socially disorganized families contributes to their reduction in growth rate, though this has not yet been certainly established. Smoking may also play a part. Babies whose mothers smoked during pregnancy average some 100 g and 1 cm smaller than others at birth, and the height deficit, though small, is apparently persistent throughout the whole of childhood. The socio-economic size differential has been getting somewhat less during the last 50 years, as social conditions have improved. In Sweden and Norway it has now disappeared. In all other countries investigated however, including the United States and the United Kingdom, it still persists, being seemingly dependent now more upon home conditions and parents' education than upon simple income.

It is perhaps not altogether surprising therefore that more intelligent children (at least by tests of ability) are at all ages taller than less intelligent children from the same occupational background. This association probably represents a complex mixture of environmental and genetical effects, the one reinforcing the other. There is evidence that the height differential between social classes in the adult population is kept in being by a system of social mobility which, perhaps rather curiously, produces an average movement of tall persons upward and short persons downwards.

Secular trend

During the last 100 years there has been a very striking tendency for children to become progressively larger at all ages. The magnitude of this trend is considerable and quite dwarfs the differences between socio-economic classes. In Fig. 16.5 are plotted the heights and weights of Swedish schoolchildren measured in 1883, 1938, and 1965–71. At all ages

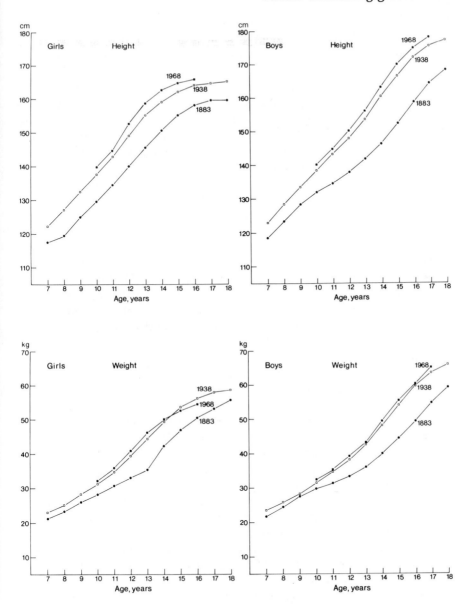

FIG. 16.5. Secular trend in growth of height in Swedish children 1883–1968. (From Ljung, B., Bergsten-Brucefors, A. and Lindgren, G. *Ann. hum. Biol.* **1**, 245, 1974.)

from 7 years onwards the 1938 children are larger than their 1883 counterparts and the 1965–71 children taller still.

British, Scandinavian, German, Polish, and North American data all give secular trends of very similar magnitude. The average gain between

1880 and 1950 is about 1 cm in height and 0.5 kg in weight per decade at ages 5–7 years. It increases to about 2.5 cm and 7 kg per decade during adolescence and decreases to a figure of about 1 cm per decade for the fully grown adult. The rather scanty pre-school-age data indicate that the trend starts at birth and relative to absolute size is probably actually greater between 2 years and 5 years than subsequently. It seems that this trend is still continuing in European countries, for example Holland. In Japan there has been an especially marked trend, which has now almost stopped. Japanese data show clearly that the trend is entirely accounted for by an increase in leg length; there is none in sitting height. Consequently, Japanese today have practically the same trunk–leg proportions as Europeans, though they are both earlier-maturing and a little shorter. In the better-off classes in most industrialized countries the trend has virtually ceased; these children seem to be attaining their full genetic potential.

It is not clear when the present trend started, though an astonishing series of Norwegian growth data stretching back to 1741 indicate that little gain in adult height took place from 1760 to 1830, a gain of about 0.3 cm per decade took place from 1830 to 1875, and a gain of about 0.6 cm per decade from 1875 to the present day. Danish data stretching back to 1815 show a similar lack of gain till about 1845. In Fig. 16.6 figures for the growth of boys in England are plotted from 1833 to the present. The secular trend has overridden the social-class differences and, though these still exist, the average boy of today is taller at all ages than the upper-class boy of 1878.

This trend in children's size is due both to earlier maturation, culminating in final adult height being reached earlier now than formerly, and also to this adult height having itself increased. The secular trend at completion of stature is about 1 cm per decade or approximately 2.5 cm per generation in most European data.

The acceleration of growth is shown in the marked secular trend of the age of menarche, shown in Fig. 16.7. The trend is very similar in all the series of data, and over the whole period for which records are available. Menarche occurred earlier by about 4 months per decade in Western Europe over the period 1830–1960. Recently, however, there is evidence the trend has stopped in Oslo and in London and, since 1970, in Holland and Hungary. Most European countries now have an average age of menarche between 12.8 and 13.2, with Italy a little earlier. The figure for whites in the USA is currently 12.8 years of age. It seems likely that in several countries maximum rate of growth is being reached. Whether such a maximal rate of growth is medically, let alone socially, desirable, is anybody's guess.

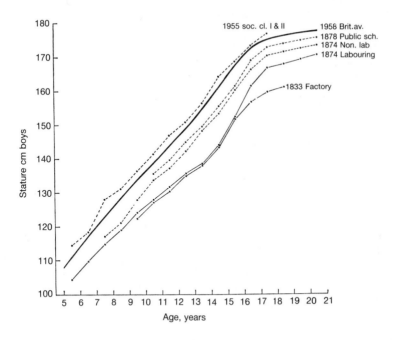

FIG. 16.6.   Height of English boys, 1833–1958, to show secular trend. 1833 factory boys; 1874 labouring and non-labouring classes; 1978 public school (upper classes); 1955 social class I and II from Birmingham Survey; 1958 British average. (From Tanner 1962.)

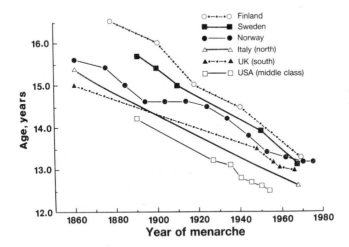

FIG. 16.7.   Secular trend in age at menarche 1860–1980. Values are plotted at year in which menarche took place. (From Tanner 1978.)

Nobody knows for certain why the secular trend has occurred. Better nutrition and generally improved environmental circumstances are usually given the credit, and with considerable reason. However, the increase is by no means confined to the less-well-off classes, and if nutrition is the cause then it must be a change in the balance of the diet or of the intake of certain essential factors, and not just an increase in calories. It has been suggested that the trend of adult stature may be genetical in origin and caused by the progressive breaking down of genetical isolates, that is, of the tendency for marriages to be contracted between members of the same village community. An increasing degree of out-marriage has certainly been occurring ever since the introduction of the bicycle. But for this to cause an increase in height demands that the many genes governing height act so that on average the children of a tall and a short parent would not be exactly halfway between the parents in height, but a little taller. To date there is no really solid evidence that height genes do fulfil this condition though a little presumptive evidence that they may. The idea that taller people now survive to have more offspring than formerly, perhaps due to the suppression of bacterial infections, is entirely unable to account for the speed of the trend, even if it were true, which is improbable.

Growth as a mirror of the condition of society

As we have seen, children's growth responds with reliability and precision to the circumstances of their lives; nutritional, ecological, even psychosocial. Economic historians have recently used the heights of children and adults to throw light on the much-debated question: did the early industrial revolution produce a fall in the living standards of the working class?

There is data which shows that the heights of vagabond London boys aged 13–16 fell during the period 1780–1800, then rose some 3 inches between 1800 and 1830, reflecting changes in the conditions of the poor at this time. The economics of slavery in America and the West Indies have been studied in the same way.

Nearer home, it has been said that one of the best measures of the classlessness of society is the height of its children. Not the absolute height of all the population, of course, so much as differences in heights between different social, ethnic, occupational, country- and town-dwelling groups. In a number of countries, especially Holland, Cuba, the United States, and, more recently, the UK, national surveillance systems have been set up to monitor the growth of children from time to time. In the UK continuous monitoring of samples of preschool and primary school children has been in operation since 1972, producing data, for example, on the secular trend of the decade 1972–1982, the association of unemployment of the father with heights and weights of his children, and the interaction of the effects of poverty and disease on growth.

17 Physical and psychological development

Clearly, the study of growth and development qualifies as one of the two or three basic sciences on which educational theory and practice must ultimately rest. Early educators had close links with human biology, and some of the pioneers were themselves doctors or anthropologists. However, during the first half of this century educators paid little attention to the facts of physical growth, perhaps because the rapid rise of educational psychology temporarily filled the whole horizon. This state of affairs has now clearly ended, and teachers increasingly demand expert and comprehensive data from human biology on the growth of the brain, the occurrence of critical periods and of stages of growth, the relationship of intelligence-test results to rate of maturation, the effects of early or late physical maturation upon emotional stability, the secular trend, and so forth.

Many of the answers are all too sketchy, for the study of growth and development has been a neglected field. However, we do have a considerable array of facts and a number of principles derived from them which are of importance in education. Chief amongst these are the questions of growth gradients in the brain; individual variability in tempo of growth, particularly at adolescence; and the relationship between tempo of intellectual and physical development.

Growth of the brain

From early foetal life onwards the brain, in terms of its gross weight, is nearer to its adult value than any other organ of the body, except perhaps the eye. In this sense it develops earlier than the rest of the body (see Fig. 14.6, p.349). At birth it is about 25 per cent of its adult weight, at 6 months nearly 50 per cent, at $2\frac{1}{2}$ years about 75 per cent, at 5 years 90 per cent, and at 10 years 95 per cent. This contrasts with the weight of the whole body, which at birth is about 5 per cent of the young adult weight and at 10 years about 50 per cent.

Different parts of the brain grow at different rates, and reach their maximum velocities at different times. Figure 17.1 illustrates this for the prenatal period in the same manner as Fig. 14.9 (p.355) which showed the growth gradients in upper and lower limbs. In Fig. 17.1 the percentages of

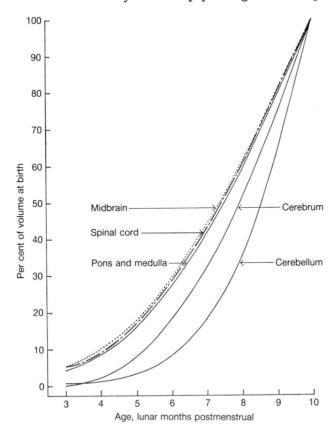

FIG. 17.1.   Percentage of their volume at birth reached at earlier months by parts of the brain and spinal cord. Cerebrum includes hemispheres, corpus stratium, and diencephalon. (From Tanner 1961.)

the value at birth are plotted for the weights of the cerebrum (including the corpus callosum, basal ganglia, and diencephalon, with the thalamus and hypothalamus), the cerebellum, the midbrain, the pons and medulla, and the spinal cord. The midbrain and the spinal cord are the most advanced at all ages from 3 foetal months to birth, and the pons and medulla are next. The cerebrum is less advanced, but still much ahead of the cerebellum. Though data on the postnatal growth of these parts are mostly lacking it is clear that these relationships would be essentially unchanged if the plots were made in terms of percentage of adult value; that is, no counter-gradients appear, so far as is known, during postnatal period.

   The maximum velocity of growth is reached first by the spinal cord, midbrain, and pons at about 2 post-menstrual months, then by the cerebral hemispheres at about 3 post-menstrual months, and finally by the cerebellum at about 6 post-menstrual months. Thus, the brain shows regular growth maturity gradients in just the same way as other parts of the body.

Cerebral cortex development

As yet we know all too little about the growth of the brain and the development of its organization. Anatomical studies of brain structure are immensely laborious and few workers have had the courage, persistence, and technical support needed to carry out morphological analyses of the brains of children at different ages. Physiological studies, such as the pattern of electroencephalogram change with age, are still in their infancy.

Most of our knowledge of the development of brain structure is due to the devoted studies of Conel and Rabinowicz, who have published analyses of the cerebral cortex at 6, 7 and 8 foetal months, birth, 3, 6, and 15 months, and 2, 4, 6 and 8 years.

The cerebral cortex is identifiable at about 8 post-menstrual weeks; thereafter, it increases in width and by about 26 weeks it has developed the typical structure of 6 somewhat indeterminate layers of nerve-cells with a layer of fibres on the inside. The layers do not mature simultaneously; the cells of the fifth layer are most advanced up to birth, followed in order by those of the sixth, third, fourth, and second. All the nerve-cells present in the adult are thought to be formed during the first 15–18 post-menstrual weeks, except perhaps for some in the cerebellum, which may appear a little later. Thereafter, axons and dendrites grow, nucleoprotein appears in the cytoplasm and the cells increase in size, and axons acquire varying amounts of myelin as sheaths, but no new nerve cells are formed. Neuroglia, the cells of the supporting connective tissue, continue to appear for considerably longer; after the early period of development they outnumber the neurons, and eventually contribute some 90 per cent of the cells present in the brain.

From these changes a series of criteria for maturation of parts of the cortex can be obtained, just as criteria for skeletal maturity can be obtained from the changes in appearance of the ossification centres of the hand and wrist. Conel uses nine criteria, amongst which are the number of neurons per unit tissue, size of neurons, condition of Nissl substance and neurofibrils, length of axons, and degree of myelination.

Two clear gradients of development occur, the first concerning the order in which general areas of the brain develop and the second the order in which bodily localizations advance within the areas. The leading part of the cortex is the primary motor area of the pre-central gyrus (see Fig. 17.2); next comes the primary sensory area of the post-central gyrus; then the primary visual area in the occipital lobe; then the primary auditory

Lateral view

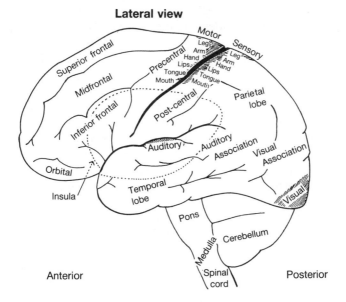

Medial View

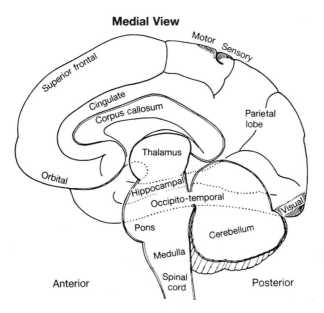

FIG. 17.2. Lateral and medial views of the brain, to show divisions of cerebral cortex and areas of localization of function. (*Above*) Lateral view of cortex; (*below*) medial view of cortex and subcortial structures. (From Tanner 1961.)

area in the temporal lobe. All the association areas lag behind their primary stations. Gradually development spreads out, as it were, from the primary areas; thus, in the frontal lobe the parts immediately in front of the motor cortex develop next and the tip of the lobe last. The gyri on the medial surface of the hemisphere and in the insula are in general last to develop.

Within the motor area the nerve-cells controlling movements of the arms and upper trunk develop ahead of those controlling the leg. The same is true in the sensory area. This corresponds, of course, to the greater maturity of the arm relative to the leg in bodily development and also to the infant's greater capacity to control his arms.

At birth the cortex is very little developed and its appearance does not suggest that much, if any, cortical function is possible. By 1 month the histological appearance of the primary motor area of upper limb and trunk suggests that it may be functioning, and by 3 months all the primary areas are relatively mature, correlating with the child's vision and hearing, though the association areas subserving the interpretive functions are not yet mature. By 6 months some fibres reaching the cortex from lower down have become myelinated, though few association fibres within the cortex are mature. Between 6 months and 2 years much further development takes place, and the primary sensory area catches up the primary motor area. However, many areas are still quite immature, most notably the hippocampal and cingulate gyri, and the insula.

During the period from birth to 4 years, and presumably for some time after, there is a continuous increase in the number and size of dendrites in all layers of the cortex, and in the number and complexity both of exogenous fibres from lower in the brain, and in association fibres within and between cortical areas. The 'connectivity' (i.e. the probability of one cell influencing others through its connections with them) increases, and this is clearly of paramount importance to the exercise of the more complicated brain functions.

It is clear from the studies on myelination by Yakovlev and his colleagues that the brain goes on developing in the same sequential fashion at least till adolescence and perhaps into adult life. Myelination of nerve-fibres is only one sign of maturity, and fibres can and perhaps sometimes do conduct impulses before they are myelinated. But the information from myelin studies agrees well with Conel's information on nerve-cell appearances where the two overlap. As a rule the fibres carrying impulses to specific cortical areas myelinate at the same time as those carrying impulses away from these areas to the periphery: thus, maturation occurs in arcs or functional units rather than in geographical areas.

A number of tracts have not completed their myelination even 3 or 4 years after birth. The fibres which link the cerebellum to the cerebral cortex and which are necessary to the fine control of voluntary movement only begin to myelinate after birth, and do not have their full complement of myelin till about age 4 years. The reticular formation, a part of the brain especially developed in primates and man and concerned with the maintenance of attention and consciousness, continues to myelinate at least until puberty and perhaps beyond. Myelination is similarly prolonged in parts of the forebrain near the midline. Yakovlev suggests that this is related to the protracted development of behavioural patterns concerned with metabolic, visceral, and hormonal activities during reproductive life.

Throughout brain growth from early foetal life the appearance of function is closely related to maturation in structure. Fibres of the sound-receiving system (the 'acoustic analyser') begin to myelinate as early as the sixth foetal month, but they complete the process very gradually, continuing until the fourth year. In contrast, the fibres of the light-receiving system or 'optic analyser' begin to myelinate only just before birth, but then complete the process very rapidly. Yakovlev points out that in foetal life the sounds of the functioning of maternal viscera are the chief sensory stimuli, apart from anti-gravity sensation. They are evidently not perceived at a cortical level; but at a subcortical one the analyser is working. After birth, however, visual stimuli rapidly come to predominate, for man is primarily a visual animal. These signals are very soon admitted to the cortex; the cortical end of the optic analyser myelinates in the first few months after birth. The cortical end of the acoustic analyser, on the other hand, myelinates slowly, in a tempo probably linked with the development of language.

There is clearly no reason to suppose that the link between maturation of structure and appearance of function suddenly ceases at age 6 or 10 or 13 years. On the contrary, there is every reason to believe that the higher intellectual abilities also appear only when maturation of certain structures or cell assemblies, widespread in location throughout the cortex, is complete. Dendrites, even millions of them, occupy little space, and very considerable increases in connectivity could occur within the limits of a total weight increase of a few per cent. The stages of mental functioning described by Piaget and others have many of the characteristics of developing brain or body structures and the emergence of one stage after another is very likely dependent on (i.e. limited by) progressive maturation and organization of the cortex.

To what extent environmental stimulation can influence brain maturation or organization is not clear. Cajal and Hebb supposed that use of a cell actually increases its connectivity, and there is some experimental evidence to support this view.

Many aspects of brain function seem quite unaffected by variations in environmental input within the range of what we consider normal environments. Thus children born before the normal 40 weeks' gestation period develop in most neurological aspects quite in parallel with children of the same post-fertilization age growing in the uterus. Pre-term babies become able to stand and walk no sooner by being exposed to the stimuli of the outside environment longer. This is not to say that maturation of the brain is not affected by any outside conditions. Certain states, such as severe malnutrition or the presence of toxic substances, can affect normal growth. To what extent the sort of malnutrition encountered in some areas of underdeveloped countries can retard or prevent brain maturation is a much disputed and at present unresolved issue. Much confusion has been caused by experimenters who failed to realize that starvation of a rat immediately after birth corresponded, auxologically speaking, to starvation of the human mid-term foetus and not to the human infant. Most of the rat work showing permanent effects of such starvation relates only to children born small for gestational age due to disease of the placenta. A classic follow-up study of children severely starved in Holland in 1944–5, while they were foetuses, newborns, etc., has been made by Susser and Stein (Stein *et al.* 1975). All the male children were measured as to height and mental ability on entry to the Dutch army aged 18. In neither measurement did they differ from 18-year-olds who were not starved. The evidence at present is that the great potential for catch-up ensures full restoration of height and probably mental development even after an episode of severe malnutrition, provided that in the rehabilitation period conditions are good. Often this last proviso is not met in developing countries. Susser and Stein sum up the present state of knowledge in this area admirably when they write 'We believe we must accept that poor *prenatal* nutrition cannot be considered a factor in the social distribution of mental competence among surviving adults in industrial societies. This is not to exclude it as a possible factor in combination with poor *postnatal* nutrition, especially in preindustrial societies' (1975).

Effects of the tempo of growth

Individual differences in tempo of growth have been described in Chapter 15, and it was pointed out there that they had important social and educational effects, particularly at adolescence. These can be conveniently discussed first in relation to intellectual, and secondly in relation to emotional, development.

Intellectual ability

There is good evidence that, in the European and North American school systems, children who are physically advanced towards maturity score on average slightly higher in most tests of mental ability than children of the same age who are physically less mature. The difference is not great, but it is consistent and it occurs at all ages that have been studied, going back as far as 6½ years. Thus, in age-linked examinations physically fast-maturing children have a significantly better chance than slow-maturing children.

It is also true that physically large children score higher in IQ tests than small ones, at all ages from 6 years on. In a random sample of all Scottish 11-year-old children, comprising 6940 pupils, the correlation between height and score in the Moray House Group Test was 0.25 ± 0.01, allowing for the effect of age difference from 11.0 to 11.9 years. An approximate conversion of these test scores to Terman–Merrill IQ leads to an average increase of 0.67 points for each centimetre of stature. A similar correlation has been found in London children. The effects can be very significant for individual children. In 10-year-old girls there was a 9-point difference in IQ between those whose height was above the 75th percentile and those whose height was below the 15th. This is two-thirds of the standard deviation of the test score.

It was usually thought that the relationship between test score and height would disappear in adulthood. If the correlations represented only the effects of co-advancement both of mental ability and physical growth, this might be expected to happen. There is indeed no difference in height between early- and late-maturing boys when both have finished growing. However, it is now clear that, curiously, at least part of the height–IQ correlation persists in adults. It is not clear in what proportion genetical and environmental factors are responsible for this; differential social mobility is probably the main factor involved.

Emotional development

There is little doubt that being an early or late maturer has repercussions on behaviour, and in some children these repercussions may be considerable. The world of the small boy is one where physical prowess brings prestige as well as success, and where the body is very much an instrument of the person. Boys who are advanced in their development, not only at puberty, but before as well, are more likely than others to be the leaders. Indeed, this is reinforced by the fact that muscular powerful boys on average mature earlier than others and have an early adolescent growth spurt. Conversely, it is the unathletic, lanky boy, unable perhaps to hold his own in the pre-adolescent rough and tumble, who on average has a late adolescence.

At a much deeper level the late developer at adolescence sometimes begins to have doubts about whether he will ever develop his body properly and whether he will be as well endowed sexually as those others he has seen developing around him. Much of the anxiety about sex is, of course, at an unconscious level, and much proceeds from sources that are considerably more complex and deep-rooted than these. Yet even here the events — or lack of events — of adolescence may act as a trigger to reverberate fears accumulated deep in the mind during the early years of life.

The early maturers perhaps appear to have things all their own way. It is indeed true that most studies of the later personalities of children whose growth history is known do show the early maturers as more stable, more sociable, less neurotic, and more successful in society, at least in the USA (though this may be because on average they are less linear in build and more muscular). However, the early maturers have their difficulties also. Though some glory in their new possessions, others are embarrassed by them. The girl whose breasts are beginning to develop may slouch instead of standing erect when asked to recite in front of a class; and the adolescent boy may have similar embarrassments. The early maturer, too, has a longer period of frustration of his sex drive, and his drive towards independence and the establishment of vocational orientation, factors which all writers on adolescence agree are major elements in the disorientation that some young men and women experience at this time.

Such are some of the social problems brought about by the great variability in tempo of growth. Practically all this variability is biological in origin; there are no social steps by which we can significantly reduce it. It therefore behoves us to fit our educational system, in theory and in practice, to these biological facts, matching the biological variability with an equal degree of social flexibility.

18 Analysis and classification of physique

Introduction

The study of human constitution is an attempt to answer the question 'In what ways do men consistently differ from one another, and how do these differences come to exist?' 'Consistently' is used here in its ordinary sense; it is those aspects of structure, function, or behaviour which do not change much in a single individual from day to day, or even from year to year, that are referred to as constitutional traits.

In any group of people studied the total variability of a character observed repeatedly over a period of time can be divided into the between-individual variability and the within-individual variability, by which latter is meant the changes which occur with time in a single individual. These sources of variability can be distinguished and estimated separately by refinements of the statistical technique of analysis of variance, provided the experiment has been correctly designed. It is those traits for which between-individual variation is high in relation to within-individual variation that are the concern of constitutional study. Characteristics which alter in immediate response to the environment are the ones excluded from consideration. Traits showing only the individually consistent and steady changes characteristic of growth, maturation, and senescence, may be just as much constitutionally controlled as those which do not change at all. Also, it must be remembered that manner of reacting to environmental change may be a constitutional characteristic. In some physiological traits, resting values show greater within-individual variation than values taken under maximal stress; for example, an individual's resting heart-rate may be more variable and less constitutionally controlled than his heart-rate upon maximal exertion, where constitutional factors assume greater importance. Perhaps similar considerations hold for some psychological traits. An individual's constitution embraces characteristics of morphology, physiology, and psychology; the historical trisection has no theoretical place in constitutional study, though in the techniques of investigation it necessarily serves as the present-day framework.

No explicit distinction between heredity and environment enters into the definition of constitution, because the study of constitutional differences is older than the science of genetics. Obviously, genetical factors are of prime importance in this field. Yet many constitutional traits have a very complex genetical background, and some, it is reasonable to suppose, may be due more to the ineradicable effects of early experience than to genetical influences. Thus, the study of constitution covers a wider field than human genetics in the strict sense. Nowadays, however, genetics embraces more and more the study of epigenesis, that is, the development of adult characteristics through the many interactions of growth, and in this way constitutional and genetical studies are becoming increasingly indistinguishable.

In medicine, the chief branch of applied human biology, constitutional diseases are assuming more and more importance as the bacterial disorders are successfully brought under control. The aim here was memorably stated in 1881 by Beneke, a pathologist who was amongst the first to apply really scientific methods in constitutional work. 'The different constitutions', he wrote, 'and the different resistances conditioned by them form only the soil in which certain diseases develop when the individual is subjected to certain stresses. The importance of this point of view for general hygiene and therapy goes without saying. It is in our power to lead the different constitutions happily through the dangers of life if we recognize them correctly and if we understand rightly their physiological differences.' This we are still far from doing.

In morphological features the within-individual changes are relatively small; stature changes little from week to week, and not very much from year to year, except in the periods of childhood and senescence. Some physiological characters, such as blood-groups or the ability to taste PTC (phenyl thiocarbamide), change even less, and these are the characteristics classically used in human genetics. However, most physiological characteristics change more, even the stable ones such as excretion of ketosteroids or creatinine, or basal body temperature. Behavioural characters are still harder to pin down to constancy, though certain aspects of reacting to circumstances, and certain traits of character and temperament persist throughout life and seem clearly to be constitutional in the sense given above. Morphology offers the clearest and simplest hunting ground in constitutional research at present, even if not fundamentally the most important. Accordingly, in this section constitutional differences in physique, or body-build, will claim most of our attention; physiological differences linked with them will be described in Chapter 19 and behavioural patterns, also alleged to be related to physique, will be discussed there also (see p.427).

There are great differences in bodily form amongst humans, and these differences occur in all degrees of detail, from the general size and shape of the individual to the curve of the eyelid or the form of the particular finger. These variations are certainly not recent in origin; they represent a form of quantitative balanced polymorphism, if we may somewhat stretch the meaning of that word. Other animals show a similar polymorphism in build, though whether to a greater or lesser degree than man it is impossible to say, since our main knowledge comes from artificially maintained and hence not comparable stocks.

Presumably, the same forces that maintain single-locus polymorphisms such as the blood-groups, also maintain the quantitative polymorphism of build. Different builds have presumably different advantages in different circumstances and at different times. Selection may favour first one build than another. There is evidence, for example, that the long, thin build of the Nilotic Negro has arisen through the very definite advantage such a build confers in maintaining physiological function in a hot environment (see section on ecology). As in the case of the blood-groups, certain diseases select against one build more than against others; tuberculosis, for example, caused more deaths in long, thin people than in short, stocky ones (see below), and coronary heart disease nowadays causes more deaths in fat and muscular people than in lean, bony ones. Sometimes it may be the build that is selected for or against as such (as in the case of Nilotes); more usually perhaps it is physiological characteristics associated with build which are selected. Clearly, there are considerable physiological differences between persons of differing builds; but as yet we know almost nothing about them.

The sex difference in physique is the most striking example of this quantitative balance polymorphism. Even before puberty there are differences, not, probably, between boys and girls of the same physique (or somatotype, see below), but between the incidence of different physiques, or somatotypes, in either sex. The more muscular physiques occur in smaller numbers amongst the girls. [In Sheldon's system (see p.409), it may be that before puberty a girl 4–4–2 differs very little from a boy 4–4–2, but there are few girl 4–4–2s, and no girl 2–7–1s, born.] The advantage of the mechanism is obvious; it reconciles the demand for a wide variety of physiques in the population, on the one hand, and for specialization of physique for various tasks on the other. Primate reproduction requires a relatively less mobile female who for considerable periods cannot be relied upon in fighting and in rapid changes of territory. Selection might thus produce an equilibrium such that the predominance of the more highy mobile and powerful physiques appeared in the males and the predominance of less mobile, but in other ways more advantageous physiques in the females. The process could not go too far without a loss of physical variability occurring. The chromosomal and embryological mechanisms of the equilibrium are not known; there is some slight evidence, however, from the study of XXY and XO individuals that genes on the X chromosome may somehow inhibit the development of large muscles. Individuals with the karyotype XYY have a tendency to be tall, but are not more muscular than other people.

The classification of physique by external body form

There have been numerous attempts to classify the varieties of physique in ways which will relate to physiological function, habitual behaviour, or susceptibility to disease. It is, of course, perfectly easy to take a series of anthropometric measurements on an individual and compare each of these figures with the range of values exhibited by a standard group of the same age and sex. The given individual can then be described as being at the 70th percentile for height, 75th for sitting height, 65th for weight, 80th for head breadth, 50th for calf circumference, and so on. (Since most anthropometric measurements are distributed in a Gaussian, or Normal, curve, with the exception of measurements by caliper or X-ray of subcutaneous fat thickness, which are log-normally distributed, the individual can alternatively be said to be $+0.5$ standard deviations above the mean for height, etc.)

One difficulty with such a description is its inefficiency. Practically all body measurements are positively correlated one with another, so that the amount of new information becomes less in each subsequent measurement. When height and leg length have been measured, arm length is practically known and its measurement tells us very little. Thus, a careful selection of measurements has to be made with some particular end in view. This may be to define a few basic components of physique which are uncorrelated (see below), or to estimate the amounts of different tissues, such as fat, muscle, and bone, in the body.

The second difficulty about this simple description is that it fails to communicate much information about body shape, even if the information does in fact lie locked in the figures. This is a real difficulty in many circumstances, particularly in medical work. One wants to be able to describe a person's physique so that one's audience can form a mental picture of his size, shape, and composition. This is very hard to do from a recital of a multitude of measurements, though it is not altogether impossible with practice.

The human body varies in a thousand ways, even in externals alone, and each classification of physique must necessarily be based on the selection of some characters and the ignoring of others. Classifications stand or fall by the way in which they link up with other areas of human biology and illuminate problems of growth, evolution, physiology, disease, or behaviour. Judged on this criterion there are four classifications by external form which merit description. Three are associated with those of the names of their originators: Viola, an Italian physician; Kretschmer, a German psychiatrist; Sheldon, an American psychologist; and the fourth results from the application of the statistical technique known as factor analysis to multiple anthropometric measurements.

Viola

Viola's classification, dating from the beginning of this century and now largely fallen into disuse, was the first to be based on a comprehensive system of body measurements. For general purposes, ten measurements were used. These were combined together in a rather empirical way (modern biometry being at that time in its infancy) to give four indexes. In each of the indexes the position of the individual relative to the standard group of the same age and sex was recorded and the person classified as 'longitype', 'brachitype', 'normotype', or 'mixed-type'. The longitypes had long limbs relative to their trunk volume, a large thorax relative to their abdomen, and large transverse diameters relative to antero-posterior ones. Brachitypes were the reverse, normotypes in between and mixed types those whose four indexes failed to agree amongst themselves, one placing the individual in one category and another elsewhere (mixed-types were thus what we would call 'dysplastics' nowadays).

An enormous amount of work was done using this system, and several hundred papers published covering morphological analyses, differential susceptibility to disease, and, to a lesser extent, physiological and psychological relationships. Viola's work has been unjustly neglected, and some parts of it have still considerable value.

Kretschmer

Kretschmer's system is better known because it was translated into English and made its way into the psychiatric textbooks. It was much less objective than Viola's, however, relying entirely on anthroposcopic inspection. Kretschmer described and illustrated three types, the 'pyknic', the 'leptosome', and the 'athletic'. The pyknic was broad, round and fat, sturdy and stocky; the leptosome long, thin, and linear; and the athletic heavily muscled with large thorax and shoulders, and narrow hips.

This system is now entirely outmoded, for it suffered from a fatal error (partly, but only partly, shared by Viola's). It supposed people were really classifiable into separate discrete types, with only a few unfortunates left out in the cold in between. This assumption — widespread up to about 1930 — involved the later practitioners of the system in hopeless difficulty, for honest classifiers simply had to admit that most people fell in between the established, and obviously fairly extreme, types.

Sheldon

Though Sheldon's system has some relation to Kretschmer's, being a three-way rather than a two-way classification, it starts from the outset with the idea now universally accepted, that there are no discrete 'types', but only continuously distributed 'components' of physique. 'The concept of types' Sheldon wrote in 1940, 'has been useful in the study of personality, but, like the poles supporting a clothes-line, it provides only end suspension for distributive classifications. As the line becomes filled, the notion of types recedes and finally vanishes altogether, perhaps submerged under a smooth distribution. The path of progress is from the notion of dichotomies to the concept of variation along dimensional axes.'

Sheldon's choice of three components sprang from his initial observational technique. He began by taking nude standardized photographs showing front, side, and rear views of some 4000 college students. Disregarding the attribute of largeness or general body size, which his classification ignores, he sorted for extremes of body shape, and he found three. These are illustrated in Fig. 18.1. Each extreme represented the end of the distribution of a component. Every individual was then assigned a place in each component. This was done anthroposcopically, using a rating scale of 1 to 7 with equal-sized intervals between the numerals (i.e. the man rated 3 for one particular component appears to be as much more so than one rated 2, as the 2 is more than the 1). Thus, the first extreme example was rated 7-1-1, the second extreme 1-7-1, the third 1-1-7: the components were named 'endomorphy', 'mesomorphy', and 'ectomorphy', respectively, on a theory, not generally accepted, of their genesis from embryonic germ layers. The whole system is known as 'somatotyping' — a word strictly reserved for this system and not to be used as synonymous with any effort at classifying physique. The set of three numerals is a person's 'somatotype'.

The components are best described by reference to their extreme manifestations. The extreme in endomorphy (7-1-1) approaches the spherical as nearly as is humanly possible; he has a round head, a large fat abdomen predominating over his thorax, and weak, floppy, penguin-like arms and legs, with much fat in the upper arm and thigh, but slender wrists and ankles. Relative to his general size he has a large liver, spleen, and, it is said, gut; large lungs and a heart shaped differently from those of the other extreme physiques. He has a great deal of subcutaneous fat and might be simply called a fat man were it not that his whole body, including his thoracic and pelvic skeleton, is greater in the antero-posterior than in transverse direction. When starved, he becomes, in Sheldon's phrase, simply a starved endomorph, not somebody high in ectomorphy or mesomorphy. It seems that fatness is related to this build more or less inevitably, and it is thought that the amount of weight put on as a person gets older is fairly directly related to his rating in endomorphy. Though the tide can be delayed by dieting and exercise it continually threatens, in a way that it fails to do in persons high in ectomorphy. It may be that persons high in endomorphy have more fat cells than persons low in this component, just as those high in mesomorphy have, presumably, more muscle-cells. However, little is known about the quantitative human histology either of fat or muscle cells, and this is hypothesis, not fact.

The extreme in mesomorphy is the classical Hercules. In him muscle and bone predominate. He has a cubical, massive head, broad shoulders and chest, and heavily muscled arms and legs, with the distal segments strong in relation to the proximal. Relative to his size his heart muscle is large. He has a minimal amount of subcutaneous fat and the antero-posterior diameters of his body are small.

The extreme in ectomorphy is the linear man; he has a thin, peaked face with a receding chin and high forehead, a thin, narrow chest and abdomen, a narrow heart, and spindly arms and legs. He has neither much muscle nor much subcutaneous fat, but, relative to his size, a large skin area and a large nervous system.

Naturally, the vast majority of people are not extremes like these, but have a moderate amount of each component. Thus, the common somatotypes are the 3-4-4, 4-3-3, or 3-5-2. A method of plotting somatotypes on a plane diagram is shown on p.413 (Fig. 18.1).

All parts of the body may not agree in the extent to which they show these characteristics, and the difference beween the somatotypes of different regions is called 'dysplasia'. Dysplasia is in theory probably of great importance, but in practice in Sheldon's system it is hard to assess consistently since its assessment is made by assigning regional somatotypes to head, arms, legs, trunk, and thorax separately, and this involves more error than somatotyping the whole body.

Somatotyping is carried out anthroposcopically, by inspection of photographs, which for this purpose have to be well made, with the subject standing in a rigidly standardized pose, rotated on a turntable between pictures, and placed a long distance from the camera to avoid the nearer parts of his body appearing significantly larger on the photograph than parts further away. The technique has been much developed, and pictures are now taken from which bodily dimensions can be accurately measured with special apparatus, a technique known as photogrammetric anthropometry.

In somatotyping, however, such measurements are not used. (In Sheldon's original publication, tables of photographic measurements are given and a system for reaching the somatotype through their application is described, but this sytem does not work and has never, in fact, been used.) The somatotype is assigned by inspection of the photograph, but its comparison, if necessary, with photographs of known somatotypes, and by reference to tables of height/$\sqrt[3]{}$ weight for each somatotype at each age, published in Sheldon's *Atlas*. Though this sounds and is a subjective procedure, trained somatotypers, who nowadays use half-unit ratings to give thirteen instead of seven points for each component, agree with each other to half a unit in 90 per cent of ratings. In only 10 per cent of cases should there be a difference of a whole unit, and very seldom a difference of more than that. The technique is not a difficult one to acquire, though

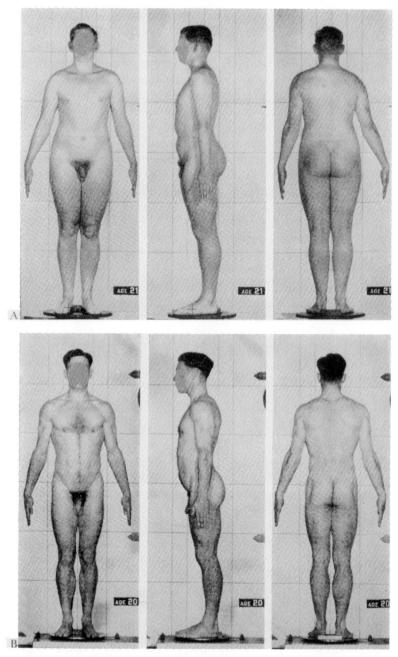

FIG. 18.1. Persons representing the extremes of the three Sheldon somatotype components, together with a person of average physique. A. extreme in endomorphy, somatotype 6 – 3 – 2; B. extreme mesomorphy, somatotype $1\frac{1}{2}$ – 7 – 2; C. extreme in ectomorphy, somatotype $1\frac{1}{2}$ – 2 – $6\frac{1}{2}$; D. average physique, somatotype 3 – 4 – 4.

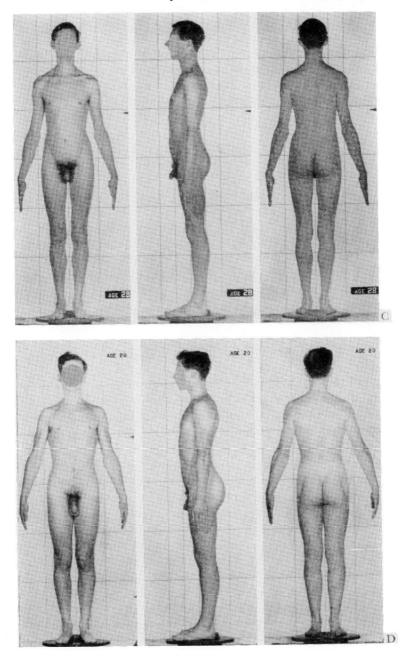

like others, it needs a month or two of training, and spells of practice from time to time.

A number of criticisms can be levelled at somatotyping as a system. First amongst them is that the components are not independent; they are oblique, not orthogonal, in the language of multivariate analysis. They are negatively intercorrelated so that a high rating in one precludes to some extent high ratings in the others. Thus, there are 2–3–5s, 3–3–5s, and 4–3–5s, but no 1–3–5s, 5–3–5s, 6–3–5s, or 7–3–5s. Also there are 6–4–1s and 4–4–4s, but no 7–7–1s, 5–5–5s, 3–3–3s, or 1–1–1s. The lack of independence makes for biometrical difficulties in handling somatotype correlations with other variables and, more seriously, creates difficulties in thinking, for it is remarkably hard not to think of the components as if they were independent or orthogonal. Orthogonal components are certainly more convenient; however, if the somatotype components truly have their roots in physiological and genetical mechanisms then the decision between oblique and orthogonal components depends on the biological, not the statistical, situation. It may be that embryological development does in fact produce variation of this negatively interrelated sort. Recent data on the amounts of fat, muscle, and bone in various parts of the body, discussed below, seem mercifully to indicate, however, that at least these tissues develop along quite independent directions.

The independence of widths of bone and muscle (see below) is a more serious criticism of the concept of mesomorphy as at present defined, since it includes both wide bones and large muscles. However, it is the size of the muscles that is chiefly used in the assessment, and dropping the bone width from the description would make little difference in practice.

The subjectivity of somatotyping is naturally a source of criticism, although in one way it is a strength and not a weakness. Sheldon defines the somatotype as the best guess the investigator can make at the subject's 'morphogenotype'. The somatotype, by definition, remains constant throughout life; the appearance of the body changes, and so do the body's measurements, but not the somatotype. Hence, different measurement criteria have to be used for somatotyping at different ages. Again, the effects of disease, or the muscular hypertrophy caused by weight-training, change body contours, but not the somatotype. Hence, the somatotyper confronted with a picture has to take into account, so far as he is able, age, disease, and muscular exercise. This is not always as hard as it sounds; to an experienced eye the effects of weight-lifting are usually obvious enough, and a weight loss of more than 4 or 5 kg is usually revealed by wrinkles and folds in the skin clearly visible in the photograph. The eye can see and make allowances where the caliper is blind.

Though the assignment of a somatotype is theoretically best done only after following a subject for several years, in practice the physique shown in a state of normal nutrition at age 20–25 years is usually taken as the basis of the definitive estimate. Somatotyping children is different, since the necessary longitudinal studies demonstrating what each child turns into at age 20 are very rare. However, the recent publications of Tanner and Whitehouse's *Atlas of children's growth* (1982) gives many serial pictures of children of different somatotypes followed until adult. Trying to assign the correct somatotype to a child without looking at his adult picture, Tanner found that in general he could get within 1 point on the scale from age 8 onwards, but that before that age, for reasons that were not clear to him, his error was much greater. Similar results were found in a different series of children by Walker and Tanner (1980).

Somatotyping women also raises difficulties at present since no published atlas exists; the same general criteria apply as in somatotyping men, however, and this leads to an excess of high endomorphs in women as compared with men, and a total lack of degrees 6 and 7 in mesomorphy.

There have been a number of variations of somatotyping suggested since Sheldon's work first appeared, but none has much to recommend it. Parnell promulgated a system dependent on body measurements only, designed to approximate the somatotype. It does this less effectively, however, than the standard multiple regressions of somatotype on body measurements given by Damon and his colleagues. Heath and Carter have suggested a variant in which ratings above 7 are admitted, with the object of relating the components more linearly to measurements such as height, weight, skinfolds, and limb-bone widths. The advantages of these methods are not obvious; the Sheldon system as conceived in 1940 and illustrated in the 1954 atlas is coherent; whatever its theoretical background it is intensely practical and communicates its information very readily. To know that a man is a large 3–2–5 is sufficient to recognize him at once amongst a hundred others.

Factor analysis of physique

The factor analysis classification results from the purely quantitative, measuremental approach to human physique — an approach not opposed to but complementary with that of photographs and subjective assessments. Factor analysis is a branch of multivariate statistical technique used for reducing a large number of measurements, all of which are intercorrelated, to a smaller number of factors, which account for most of the variability defined by the original measurements. The factors may be themselves intercorrelated, or 'oblique', or they may be independent or 'orthogonal', the choice being made by the investigator. The starting-point is a table of correlation coefficients for a group of, say, a dozen or more body measurements. The finish is a series of, say, three orthogonal factors A, B, and C which together perhaps account for some 80 per cent of the variability of the dozen measurements. Factor analysis enables one to escape from the otherwise well-deserved charge, in physical anthropology, of measuring the same thing over and over again.

There are several alternative techniques of factor analysis, but the application of the one most used gives rise to a general factor of gross size, followed by several group factors, which can be themselves subdivided into smaller groups. The results of analysing various tables of correlations are quite consistent and define the series of factors outlined in Table 18.1. The true situation is probably not so simple as this implies, and additionally should include factors for head, hand, and feet size and shape.

Factors are essentially statistics, just like standard deviations. They cannot *a priori* be equated with genetical or physiological mechanisms. However, in the analysis of shape they will probably only be useful if it turns out that they do represent fairly directly physiological mechanisms, and link up straightforwardly with the facts of growth. It seems quite hopeful that they may do so, at least if the original measurements to be factored are chosen with biological considerations clearly in mind. The multivariate statistical approach to the classification and understanding of physical differences is capable of much further development. One of the hopeful lines which this development is taking is described in the next chapter.

The analysis of physique by tissue components

The classifications of physique described above are all based on external body-form. An alternative approach is to classify by means of quantitative differences in tissue structure, e.g. by the relative amounts of fat, muscle, and bone possessed by different individuals. The two approaches are neither independent nor mutually exclusive. Some aspects of external form, for example overall size, may be practically independent of structure. However, shape and structure are inevitably somewhat related; a fat man, try as he may, cannot be anything but rounded; a heavily muscled man cannot appear fragile and linear. For each separate tissue shape and structure are more independent; a man with a given percentage of bone may have long limbs and a short trunk, or vice versa.

The total amount of fat and water in the body can now be estimated by chemical and physical means with an accuracy of about ± 15 per cent and the amount of muscle approximated in favourable circumstances by the estimation of the radioactive potassium ^{40}K content. Such methods, however, only lead to statements about the over-all amount of a certain constituent of the body. They cannot tell us where the fat, for example, is, nor how, except in quantity, it is contributing to the external bodily contours.

A bridge between the two approaches has been made by the use of X-rays for measuring the widths of fat, muscle, and bone in the limbs and by the introduction of a special caliper for measuring the thickness of

TABLE 18.1

Summary of factor-analysis classification of physique in terms of orthogonal subdivided group factors

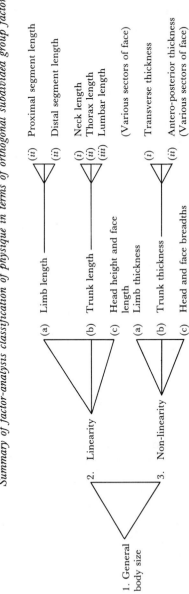

Slightly modified from Tanner (1953). Growth and constitution. In *Anthropology today* (ed. A. L. Kroeber). Chicago University Press.

subcutaneous body fat. This is done by pinching up a fold of skin and fat away from the underlying muscle and applying the caliper to this skin-fold. At certain sites in the body this can be done readily and accurately.

The sites most simply X-rayed are the upper arm, calf and thigh. The dose of X-rays is extremely small and delivered at low kilovoltage; leaded clothing is used to shield the gonads completely. In the radiographs of the upper arm the widths of the subcutaneous fat, the muscle, and the humerus are measured halfway between the acromion and the top of the radius. In the thigh, the widths of fat, muscle and femur are taken at a point one-third of the leg-length up from the bottom of the femoral condyles. In the calf the widths of fat, muscle, and tibia (fibula omitted) are taken at the maximum total diameter.

When these X-ray measurements are intercorrelated a clear pattern emerges (Table 18.2) (analysis of calculated cross-sectional limb areas yields identical results). The three measurements of each tissue correlate positively one with another, but are virtually independent of measurements of the other tissues. Thus muscle width in the calf correlates with muscle width in the arm and thigh, but not with bone or fat width in the calf or elsewhere. In Table 18.2 the averages of the three fat–fat, three

TABLE 18.2

Average intercorrelations of measurements by X-ray of widths of subcutaneous fat, muscle, and bone (humerus, femur, tibia) in upper arm, thigh, and calf. 166 young women (upper rows) and 125 young men (lower rows)

	Fat	Muscle	Bone
Fat	0.66	—	—
	0.75	—	—
Muscle	0.09	0.43	—
	0.08	0.49	—
Bone	0.09	0.13	0.37
	0.07	0.09	0.48

(After Tanner, Healy, and Whitehouse, unpublished.)

muscle–muscle, three bone–bone, and corresponding cross-correlations are given. The within-tissue correlations, shown in the diagonal, are high for fat and moderate for muscle and bone; the between-tissue correlations, including those between muscle and bone, are virtually zero. There are, therefore, three orthogonal tissue-component factors, representing fat, muscle, and limb-bone width, respectively. Further measurements show that these limb-bone widths represent width of medullary cavity irrespective of width of cortical bone, which relates somewhat to the amount of muscle but not at all to the width of the marrow cavity.

This type of analysis can be pushed further by including other measurements of the skeleton taken by ordinary anthropometric methods, of fat taken by the skinfold caliper, and of external body contours measured by photogrammetry. Correlations can be calculated (using the logarithms of the fat measurements, since these are distributed log-normally) and the matrix factor-analysed. As more measurements are added the situation becomes naturally somewhat more complex, but the three prime orthogonal components persist. In Table 18.3 an analysis of this sort is shown. The columns of figures represent the saturations of each measurement in the factors; these saturations vary from 0 to 1, and are interpreted in a similar fashion to correlation coefficients.

Fat

All measurements of subcutaneous fat correlate very highly with each other; thus a fat factor is defined. In Table 18.3 this is factor (I), with very high saturations in arm, calf, and thigh fat, and low saturations in the other measurements. Other data show that in both sexes there is a subsidiary factor for fat on the trunk versus fat on the limbs. Systematic regional variations, that is, fat arms versus fat legs, or fat chest versus fat abdomen, seem to be less in evidence, but a good deal of purely local variation remains, this meaning that one person may have a particularly large fat layer over the calves, another a particularly large layer over the upper arms, and so on. Such single-measurement factors are known as 'specifics'. It is the trunk-versus-limbs factor and the specific local factors which give the characteristic shape of fat covering to an individual, whether fat or lean. We may suppose that these 'fat-shape' factors reflect the numbers of fat-cells present at the various sites and that the over-all fat factor represents the degree to which all these cells are filled with fat. The amount of internal abdominal fat, however, seems to be largely independent of the amount of subcutaneous fat, at least in young adults. Because of this, the estimation of total body fat by the score in subcutaneous fat factor, or by such an approximation to it as the sum of four log-transformed skinfolds has a fairly large error.

In the factor analysis of Table 18.3 it will be seen that, of the skeletal measurements, only sitting height and bicondylar femur breadth (width of knee) have consistent saturations in the fat factor (I). The knee-width saturation comes about for purely technical reasons; some fat is unavoidably included in the measurement itself. However, the negative saturation of sitting height definitely implies that in these young adults the capacity for becoming fat is greater in short-bodied individuals than in long-bodied ones. The embryological and physiological reasons for this are not known. Adherents of somatotyping say that the degree of endomorphy can be gauged from the skeletal shape, at least in theory and with some difficulty; and they adduce, in fact, a short trunk and large antero-posterior

TABLE 18.3

Factor analysis (maximum likelihood) of measurements on 166 young women (left-hand columns) and 125 young men (right-hand columns). Saturation of measurements in first five rotated orthogonal components

	I (women)	I (men)	II (women)	II (men)	III (women)	III (men)	IV (women)	IV (men)	V (women)	V (men)
Fat width										
arm	0.81	0.92	0.09	0.05	0.09	0.10	0.02	0.01	−0.09	−0.11
calf	0.83	0.82	−0.01	0.03	0.03	−0.07	0.00	−0.15	−0.02	0.11
thigh	0.86	0.87	0.01	−0.05	−0.10	−0.06	−0.02	0.01	0.08	0.03
Muscle width										
arm	0.13	0.04	0.52	0.70	0.06	−0.04	0.10	0.15	0.05	0.02
calf	0.06	0.10	0.65	0.67	−0.01	0.10	−0.22	0.11	−0.12	0.05
thigh	−0.04	0.07	0.72	0.84	0.10	0.30	0.26	0.12	0.01	0.07
Bone width										
arm	−0.05	−0.08	0.12	0.09	0.28	0.28	0.56	0.51	−0.01	−0.03
calf	0.05	−0.12	0.03	−0.09	0.45	0.37	0.75	0.51	0.06	−0.03
thigh	0.10	−0.06	−0.11	−0.01	0.47	0.37	0.57	0.51	0.06	0.04
Sitting height	−0.34	−0.10	0.21	0.17	0.69	0.72	0.11	0.00	−0.23	0.02
Leg-length (subischial)	−0.05	0.02	0.03	−0.28	0.66	0.77	−0.10	0.21	0.48	0.55
Biacromial diameter	−0.17	0.02	0.26	−0.16	0.63	0.75	0.07	0.07	0.22	0.11
Bi-iliac diameter	0.02	0.04	0.18	−0.07	0.46	0.54	0.03	0.29	−0.17	0.03
Bicondylar humerus	0.17	0.03	0.35	0.23	0.49	0.55	0.24	0.41	−0.00	−0.12
Biconylar femur	0.40	0.31	0.36	0.30	0.53	0.53	0.22	0.28	−0.14	−0.02
Percentage of total variance accounted for	16.4	16.2	11.0	13.0	17.2	20.1	9.7	8.2	2.8	2.4
Average saturations										
fat	*0.83*	*0.87*	*0.03*	*0.01*	*0.01*	*−0.01*	*0.00*	*−0.04*	*−0.01*	*0.01*
muscle	*0.05*	*0.07*	*0.63*	*0.74*	*0.05*	*0.09*	*0.05*	*0.13*	*−0.02*	*0.05*
all skeletal measurements	*0.03*	*0.01*	*0.01*	*0.02*	*0.52*	*0.54*	*0.27*	*0.31*	*0.03*	*−0.05*
limb-bone breadth	*0.02*	*−0.09*	*0.16*	*0.00*	*0.40*	*0.34*	*0.63*	*0.51*	*0.04*	*−0.01*
leg-length	*−0.05*	*0.02*	*0.03*	*−0.28*	*(0.66)*	*(0.77)*	*−0.10*	*0.21*	*0.48*	*0.55*

(From Tanner, Healy, and Whitehouse, unpublished.)

skeletal diameters as indicators of this component. Possibly endomorphy, as they conceive it, measures the number of fat-cells; and fatness the amount of fat per cell.

Muscle

The correlations between muscle widths in different parts of the body are not as high as those between fat measurements, but are still very appreciable. The muscle factor (Table 18.3, component II) can be reasonably estimated by simply adding the three muscle widths in the limb X-rays. Photogrammetry makes it clear that this is a factor measuring muscle all over the body, including the trunk and back. There is little evidence of important regional factors, such as arm muscle versus leg muscle, but large specifics certainly exist.

Skeleton

The skeleton is far and away more complicated. If we describe it in terms of orthogonal factors, however, the result shown in Table 18.3 (factors III, IV, and V) is at least reasonable. One of the great difficulties with factor analysis is that it does not lead to a unique solution; it is possible, even maintaining orthogonality, to find other components which would equally well describe the original correlations between measurements. One must be guided by biological considerations; to provide components different from the fat and muscle ones would be absurd, but the skeletal configuration is more subjective and more debatable.

In Table 18.3 the skeleton is described by way of a skeletal frame size factor (III), in which all skeletal measurements saturate; a limb-bone width factor (IV), which represents the widths of the shafts of the long bones; and a limb-length factor (V), which represents the length of limbs as opposed to trunk.

Further analyses show that the limb-bone width is unrelated to the width of the vertebral bodies, so that slenderness of limb bones is independent of slenderness of spine. Furthermore, trunk length may be split into several surprisingly independent components; head and neck length, thorax length (top of sternum to umbilicus), and pelvic length (umbilicus to ischial tuberosities). Measured this way, thorax and pelvic lengths are entirely independent. Bi-iliac diameter is similarly largely independent of other skeletal measurements. There is evidently a 'rear-end' or 'hindquarter' factor in human physique, both in males and females. A similar factor has been described in cows. The neck length, on the other hand, is little related to thorax length, but more so to limb length. Long arms, long legs, and a long neck go together. The size of the hands and feet, related one to the other, are largely independent of all the factors so far considered, and so also is head-size and the shape of the individual features of the face.

The cause of these patterns into which body measurements fall must be sought in differential growth rates. Many components similar to those in man are found in animals, and experimental work using inbred strains is helping to elucidate their genesis. An example of the sort of explanation that must be looked for may be given from Table 18.3. In men there is a negative saturation of leg-length in the muscle component (− 0.28). We know that muscular males are in general early maturers (see above) and that early maturers have a shorter period during which the legs are growing faster than the trunk, so that they will end up relatively short-legged.

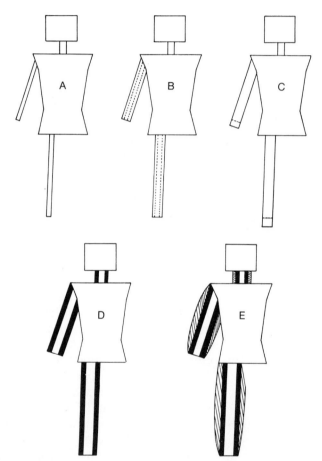

Fig. 18.2.   Diagram of shape successively defined by 5 orthogonal factors ((A), (B), and (C) skeletal size and shape, (D) and (E) soft tissue). (A) Individual defined by a score in skeletal frame-size factor (other skeletal factors all having average values.) (B) Individual now given a higher-than-average score in limb-bone width factor. (C) Individual additionally given a higher-than-average score in limb-bone length factor. (D) Individual additionally given a score in muscle factor. (E) Individual additionally given a score in fat factor. (From Tanner, Healy, and Whitehouse, unpublished.)

The first simple results of tissue-component analysis are summed up in the diagrams of Fig. 18.2. We can postulate a set of orthogonal or independent factors so that we reconstruct the major elements of an individual's physique (shorn of head, hands, feet, and various interesting soft parts) in terms of scores in:

(a)   skeletal frame size (Fig. 18.2 A);
(b)   limb-bone width (Fig. 18.2 B, where figure has larger than average limb-bone width);
(c)   limb-bone length (Fig. 18.2 C, where figure has larger than average limb-bone length also);
(d)   muscle width (Fig. 18.2 D);
(e)   fat thickness (Fig. 18.2 E).

Factors (b) + (c) together define the amount of extra limb size over and above that average demanded by the score in factor (a), general skeletal size; and factor (b)/(c) defines the shape of the limbs, that is, their linearity in relation to their size. Naturally, this scheme leaves much out of account, specifically the shape of the trunk. It is, however, a start; all the factors can be measured objectively and at least some make embryological sense. These components apply equally to men and women, though in some women score higher on average than men, and in others lower. Sex differences, discussed in the previous chapters on growth, require yet further factors for their definition.

19 Physique and its relationship to function, disease, and behaviour

Function

Remarkably little is yet known about relationships between body-build and physiological function. It seems fairly certain that such relationships must exist, at least in endocrine function and in metabolism. It is unlikely, for example, that persons high in the muscular component of physique have exactly the same amount of certain endocrine secretions as those low in it, or that persons at the opposite ends of the scale of endomorphy have metabolisms which are indistinguishable. It seems, unlikely, also, that differences in the structure and the habitual function of the central nervous system are entirely independent of differences in body-build, but this is as yet a completely unexplored field.

The older literature on constitution contains a good many allegations about relationships between endocrine function and physique. There is even one system of classification, by 'biotypes', which is based upon these alleged relationships. Nearly all these suggestions are quite lacking in scientific support, and founded on the flimsiest clinical analysis. Physique may indeed relate to endocrine activity, or it may relate to amounts of the tissue receptors which translate the endocrine message into information regulating the cells themselves. Only very recently have tests of endocrine secretion reached a point where habitual individual differences may be characterized; tests of amount of receptor substance are in their infancy.

Only physiological functions which are relatively stable from week to week or year to year can be expected to relate to build, and so the first search must be for measurable functions characterized by a low within-individual variability and a high between-individual one. By using a suitable statistical design involving measuring the individuals twice or more over a period of time, it is possible to separate the stable and the fluctuating components and to calculate the relationship between the stable values for each individual and various indexes of his build.

The results of one such study are shown in Table 19.1. Urinary 17-ketosteroid and 17-ketogenic steroid excretions are measures of two quite different aspects of adrenal function. The 17-ketosteroids represent the end-products of adrenal androgen secretion, together with those of testosterone from the testis. The 17-ketogenic steroids are the end-products of adrenal corticoid secretion, that is, of the hormones controlling, amongst other things, carbohydrate metabolism. Both groups of substances are excreted in relatively constant amounts from week to week by young individuals living under ordinary circumstances. Creatinine is a substance known to be derived wholly or mostly from the muscle in the body, and is also excreted at a relatively constant rate.

TABLE 19.1

Correlation coefficients between stable ('between-person') values of daily urinary excretions of chemical substances and widths of fat, muscle, and bone in the limbs, and height and weight

	Fat	Muscle	Bone	Height	Weight
17-ketosteroids	0.10	0.39†	0.23	0.05	0.31‡
17-ketogenic steroids	− 0.12	0.10	0.49†	0.29‡	0.30‡
Creatinine	0.16	0.77‡	0.26	0.17	0.76‡

† Significant at 5 per cent.   ‡ Significant at 1 per cent.
(From Tanner, Healy, Whitehouse, and Edgson 1959.)

The 17-ketosteroid excretion is significantly related to the amount of muscle in the body; the multiple correlation with muscle width and body surface (to give an approximate measure of total muscle bulk) is 0.56. The 17-ketogenic excretion in this data, rather surprisingly, is related to the width of the shafts of the bones, or more precisely to the width of the marrow cavities. It is also related, much more than is 17-ketosteroid excretion, to the overall size of the body. Creatinine excretion, as anticipated, relates closely to muscle bulk.

These results are still a long way from revealing differences in rates of secretion of hormones by the adrenal itself, for the amounts of steroids excreted in the urine are influenced by tissue utilization, liver conjugation, and the amount of each individual steroid cleared from the blood by the kidney. However, we can be confident that ultimately research along these general lines will reveal the true manner in which the emergence and maintenance of each physique is dependent on hormonal agents and their receptors.

There are few confirmed data on other relationships. A persistent tradition has it that linear, ectomorphic individuals have higher habitual rates of thyroid secretion than others, but no studies with modern methods have been reported. A number of studies have shown that systolic and diastolic blood-pressure, as measured by the cuff method, are higher in broadly built persons than in linear ones; but there is some doubt as to whether this represents merely the effect of a large arm-circumference producing a systematic measuring error, or whether it would also occur in intra-arterial measurements. Blood-volume has been shown to be higher, relative to body-weight, in persons high in mesomorphy and lower in those, either ectomorphs or endomorphs, who are low in mesomorphy. It is said that the reaction time is less the higher the degree of ectomorphy, but this requires confirmation.

Disease

There is no question that persons of particular builds are more susceptible to certain diseases. In the case of at least one disease, pulmonary tuberculosis, it has been shown by studies of physique made before the disease started that the build differences are not caused by the disease, but genuinely precede it.

Pulmonary tuberculosis is a disease of ectomorphic builds, or at least of persons with a low weight for their height. In one longitudinal study it has been shown that the initial infection with the tubercle bacillus occurs irrespective of physique, but that a spread of the germ to cause clinical disease several years later occurs much more frequently in tall people of relatively low weight than in others, despite all living under similar circumstances. Evidently, the linear builds supply a more favourable soil, though in what way we do not know. One difficulty in this and in other disease studies is that investigators have often been content to characterize build simply by height and weight, thus failing to differentiate fat from muscle and heavy from light bones. Consequently, all chance of a physiological analysis of the situation is lost.

Coronary thrombosis is also certainly related to physique, the incidence being higher in persons of high weight for height. The details of this are still not clear, but it seems that both above-average fat and above-average muscle contribute to the susceptibility. In principle, either or both of these relationships may be of physiological or of sociological origin. It may be that mesomorphic men put themselves more than others into positions where they are at increased risk of developing coronary thrombosis, perhaps because of increased stress. There is some data to indicate that a special susceptibility may characterize mesomorphic men who fail to exercise their muscles, while not affecting less muscled but equally lazy persons. A physiological mechanism for such a differential susceptibility might reside in utilization of androgenic hormones in exercise, or something similar.

Diabetes is another disease related to build. Clinically, two forms are distinguished. One appears early in life and requires much insulin for its control; the other appears in middle age and requires little insulin. The physiological background of the two forms is believed to differ; the build of the patients certainly does. Sufferers from the late-onset form are more endomorphic than sufferers from the early-onset form and have a fat pattern in which trunk fat predominates over limb fat. In the old Italian literature most cancers are said to occur more frequently in the brachitype than the longitype; and more modern studies by Damon and his colleagues tend to bear this out, at least for breast cancer.

Amongst mental disorders there is a marked affinity with build. Schizophrenics are usually high in ectomorphy; manic-depressives low in ectomorphy and high in both other Sheldon components; paranoids are high in mesomorphy. All these results have been repeatedly confirmed by various techniques and can be regarded as proven, but again no mechanisms are known. It has been said that neurotic breakdowns occur most frequently in those of more than ordinary degrees of dysplasia, but this needs further confirmation. It is also believed that the type of symptoms shown in neurotic breakdown is associated with build; hysterical and depressed patients tend to be high in mesomorphy and endomorphy, anxious patients high in ectomorphy.

Thus, many empirical relationships have been described, but few causal physiological or sociological analyses performed. It seems most likely that in many of these cases physique acts as an expression of the gene complex affecting the penetrance or expressivity of more specific genes predisposing to particular disorders. The field is one wide open to research.

Behaviour

Practically all those who have concerned themselves with human physique have been interested also in the relation between body-build and behaviour. The older authors took such a relationship for granted and felt no need for justifying statistics. More recently, research has been concentrated on finding out what elements of behaviour remain more or less consistent and unaltered throughout life, whether these are related to physical differences, and, if so, to what differences and how closely.

Intellectual capacity in the sense of ability to score highly on IQ tests remains relatively constant, but is very slightly, if at all, related to build. The relationship with build is greater for temperament, by which is meant the deepest layer of personality, wherein lie habits of behaviour ranging from such traits as 'indiscriminate amiability' right down into obviously physique-conditioned aptitudes, such as love of physical adventure and relaxation in posture. Kretschmer described originally two types of temperament, the 'cyclothyme' and the 'schizothyme', which corresponded more or less to Jung's 'extravert' and 'introvert'; he related these closely to the types of build, the cyclothyme being pyknic, the schizothyme leptosomic. Later he also ascribed a particular temperament to the athletic.

Sheldon carried the same work much further and used, for his description of temperament, components rather than types, in precisely the same way as he had for physique. He began by rating some 30 subjects on 50 traits chosen to define the persistent aspects of behaviour as obtained from a 50-hour clinical history. The traits fell into three clusters, and from this start, by adding other traits gradually, Sheldon produced three components of temperament, each defined by 20 traits. Each trait is rated on a 7-point scale, and the average of the 20 ratings gives the score in each component. A selection of traits is given in Table 19.2. The temperamental components are called 'viscerotonia', 'somatotonia', and 'cerebrotonia', and, when the scores in these of 200 subjects were related to their somatotype components, Sheldon found correlations of the order of 0.8 between viscerotonia and endomorphy, somatotonia and mesomorphy, and cerebrotonia and ectomorphy.

TABLE 19.2

A selection of traits from Sheldon's scale of temperament

Trait no.	Viscerotonia	Somatotonia	Cerebrotonia
1	Relaxation in posture and movement	Assertiveness of posture and movement	Restraint in posture and movement, tightness
2	Love of physical comfort	Love of physical adventure	Inhibited social address
3	Greed for affection and approval	Psychological callousness	Secretiveness of feeling, emotional restraint
4	Smooth, easy communication of feeling; extraversion of viscerotonia	Horizontal mental cleavage; extraversion of somatotonia	Vertical mental cleavage; introversion
5	Relaxation and sociophilia under alcohol	Assertiveness and aggression under alcohol	Resistance to alcohol and to other depressant drugs
6	Need of people when troubled	Need of action when troubled	Need of solitude when troubled
7	Orientation toward childhood and family relationships	Orientation toward goals and activities of youth	Orientation toward later periods of life

This is a somewhat closer, or at any rate more precise, relation than has previously been described, and the study has been criticized, though more on the grounds of incredulity than of conflicting evidence. The chief problems calling for attention seem to be two. First is the extent to which different raters of Sheldon's temperamental traits agree when each has acquired the necessary prolonged life-history of the subject, and the extent to which the ratings of a subject differ at different times in his life. The genetical background of the temperamental components also needs to be examined. Scales constructed on traits of this sort are always liable to petrify what is really plastic. Second, the work needs confirmation, with particular efforts to avoid any possible 'halo' effect, that is, with the somatotyper and the psychologist separate, and the psychologist working either from another man's notes or from a recording of another man's interviews, so that he never sees the subject's physique at all. One such study has been done, and, though it covered only a portion of the field, relating dominant cerebrotonia to dominant ectomorphy, in this portion it confirmed Sheldon's findings.

There are two theories as to how the relationships between physique and temperament may come about. One ascribes them to the actions of pleiotropic or of closely linked genes, the other to early conditioning. The first maintains that genes concerned in body shape also have effects on the brain and endocrine structure and that these dictate temperament: alternatively, it maintains that there exist separate but closely linked genes for each. The other view is, for example, that the highly mesomorphic child, on mixing with other children, finds by chance that he can knock down the others and continues thereafter to do so because of his success and because it comes to be expected of him. The two theories are not, of course, mutually exclusive; probably conditioning supports and reinforces an originally genetic tendency.

Persons of different physique do, on average, choose different careers, and this must reflect at least to some degree a temperamental inclination to lead one sort of life rather than another. In Fig. 19.1 are shown the somatotype distributions of a sample, believed relatively representative, of Oxford University undergraduates; an entire entering class of officer cadets at the Royal Military Academy, Sandhurst; and a class of student teachers of physical education at Loughborough Training College, an institution specializing in the training of physical educators. The cadets are significantly more mesomorphic than the Oxford students and the physical educationists are still more mesomorphic than the cadets. (The method of representing somatotype distributions on a plane surface is visually useful, but tests of significance have to be applied, of course, to the figures themselves.)

This is a simple, and perhaps expected, result. One scarcely enters physical education unless good at games, or the regular army without some proclivity for rough and tumble. However, more subtle choices of career have also been described: men on the research side in a factory, at all wage levels, were more linear than those on production; long-distance transport drivers and aircraft pilots were more mesomorphic than the general population; students at Harvard University enrolled in the faculties comprising natural and social sciences were more 'masculine' in body-build than those in arts, letters, and philosophy; students of engineering, medicine, and dentistry were more mesomorphic and less ectomorphic than students of physics and chemistry; youths convicted of delinquent behaviour were decidedly more mesomorphic than the average.

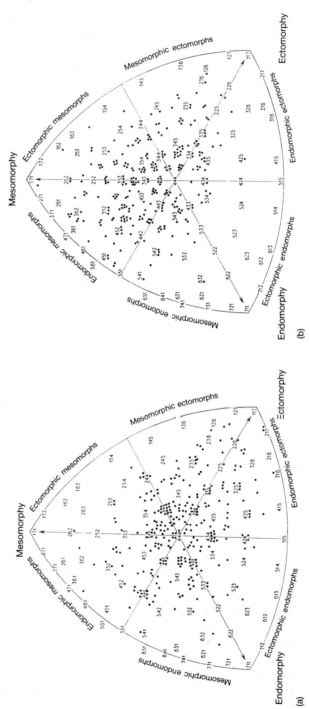

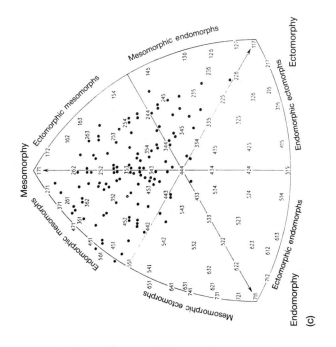

Fig. 19.1.   Somatotype distribution of samples of (a) Oxford University students, (b) officer cadets of Royal Military Academy, Sandhurst, and (c) student teachers of physical education at Loughborough Training College.

All these results point in a rather consistent direction. The fact that both army officers and juvenile delinquents are comparable in being above average in mesomorphy should occasion no surprise. For both careers some of the expectations are the same — power, risk, action, crisis; it is the *object* towards which this energy is directed which is different. It may well be that somatotonic energy and drive is an inborn characteristic of the same nature as the instinctual responses described by the ethologists. Somatotonia may be as much an instinctual drive as mating behaviour, but with objects more easily manipulated by the culture.

Most direct studies of the physique-behaviour relationship — and they are neither numerous nor particularly well conceived — lead to a lower degree of association than one would expect from Sheldon's work. One should perhaps maintain an open mind upon the whole question of this association, though in face of experience with friends and with patients it is difficult to do so. Everyday experience certainly inclines one to believe in the existence of some degree of relationship of the same general sort as that described by Kretschmer and Sheldon.

Certainly, some such belief, even if erroneous, is deeply embedded in our culture. One cannot imagine Laurel and Hardy with shapes the same, but behaviour transposed. One cannot imagine Falstaff and Mephistopheles, Scrooge and Pickwick each in each other's physical shoes. One cannot imagine a portrait of Christ that looks like Buddha. Evidently we have a stereotyped image linking physique and behaviour in our minds and this stereotype has persisted unchanged since the Middle Ages or before. It may be a collective cultural fantasy, but it seems more likely to spring from the firm ground of human biology.

Suggestions for further reading (Part III)

Growth

Bielicki, T. and Charzewski, J. (1983). Body height and upward social mobility. *Ann. Hum. Biol.* **10**, 403–8.

Brundtland, G. H., Liestøl K., and Walløe, L. (1980). Height, weight and menarcheal age of Oslo schoolchildren during the last 60 years. *Ann. Hum. Biol.* **7**, 307–22.

Fogel, R. W. and Engerman, S. L. (eds) (1982). Trends in nutrition, labor welfare and labor productivity: the uses of data on height. *Social Sci. Hist.* **6**, 395–581.

Goss, R. J. (1986). Modes of growth and regeneration: mechanisms, regulation, distribution. In *Human growth*, 2nd edn (eds F. Falkner and J. M. Tanner) Vol I, pp. 3–26. Plenum Publishing Corporation, New York.

GREENE, L. A. and JOHNSTON, F. J. (eds.) (1980). *Social and biological predictors of nutritional status, physical growth and neurological development.* Academic Press, New York.

LE GROS CLARK, W. E. and MEDAWAR, P. B. (eds) (1945). *Essays on growth and form presented to D'Arcy Wentworth Thompson.* Clarendon Press, Oxford.

LINDGREN, G. (1976). Height, weight and menarche in Swedish urban school children in relation to socio-economic and regional factors. *Ann. Hum. Biol.* **3**, 501–28.

MARSHALL, W. A. and TANNER, J. M. (1986). Puberty. In *Human growth,* 2nd edn (eds F. Falkner and J. M. Tanner) Vol 2, pp. 171–210. Plenum Publishing Corporation, New York.

MUELLER, W. H. (1986). The genetics of size and shape in children and adults. In *Human growth,* 2nd edn (eds F. Falkner and J. M. Tanner) Vol 3, pp. 145–68. Plenum Publishing Corporation, New York.

RONA, R. J. and CHINN, S. (1984). The National Study of Health and Growth: nutritional surveillance of primary school children from 1972 to 1981 with special reference to unemployment and social class. *Ann. Human. Biol.* **11**, 17–28.

STEIN, Z., SUSSER, M., SAENGER, C., and MAROLLA, F. (1975). *Famine and human development.* Oxford University Press, Oxford.

TANNER, J. M. (1962). *Growth at adolescence: with a general consideration of the effects of hereditary and environmental factors upon growth and maturation from birth to maturity* (2nd edn). Blackwell Scientific Publications, Oxford.

—— (1963). Regulation of growth in size of mammals. *Nature, Lond.* **199**, 845–50.

—— (1978). *Foetus into man: physical growth from conception to maturity.* Open Books, London, and Harvard University Press, Cambridge, Mass.

—— (1981). *A history of the study of human growth.* Cambridge University Press, Cambridge.

—— (1983). Physical growth and development. In *Textbook of paediatrics* 3rd edn (eds J. O. Forfar and G. C. Arneil), pp. 278–330. Churchill Livingstone, Edinburgh.

—— (1986). Growth as a target-seeking function: catch-up and catch-down growth in Man. In: *Human growth,* 2nd edn (eds F. Falkner and J. M. Tanner) Vol. 1, pp. 167–80. Plenum Publishing Corporation, New York.

—— HUGHES, P. C. R., and WHITEHOUSE, R. H. (1981). Radiographically determined widths of bone, muscle and fat in the upper arm and calf from age 3–18 years. *Ann. Hum. Biol.* **8**, 495–518.

—— and INHELDER, B. (eds) (1956–60). *Discussions on child development, Vols I–IV. A consideration of the biological, psychological and cultural approaches to the understanding of human development and behaviour.* Tavistock Publications, London.

THOMPSON, D'ARCY WENTWORTH (1942). *Growth and form* (revised edn) Cambridge University Press, Cambridge.

WADDINGTON, C. H. (1957). *The strategy of the genes*. Allen and Unwin, London.

WIERMAN, M. E. and CROWLEY, W. F. JR (1986). Neuroendocrine control of the onset of puberty. In *Human growth*, 2nd edn (eds F. Falkner and J. M. Tanner) Vol 2, pp. 225–42. Plenum Publishing Corporation, New York.

Physique

BROZEK, J. (ed.) (1965). *Human body composition*. Symposia of the Society for the Study of Human Biology, 7. Pergamon Press, London.

DAMON, A., BLEIBTIEN, H. K., ELLIOT, O., and GILES, E. (1962). Predicting somatotype from body measurements. *Am. J. Phys. Anthropol.* **20**, 461–73.

HEATH, B. H. and CARTER, L. (1967). A modified somatotype method. *Am. J. Phys. Anthropol.* **27**, 57–74.

PETERSEN, G. (1967). *Atlas for somatotyping children*. Royal Vangorcum, Assen Netherlands.

ROSENBAUM, S., SKINNER, R. K., KNIGHT, I. B., and GARROW, J. S. (1985). A survey of heights and weights of adults in Great Britain, 1980. *Ann. Hum. Biol.* **12**, 115–28.

SCHREIDER, E. (1960). *La biométrie*. Presses Universitaires, Paris.

SHELDON, W. H., DUPERTUIS, C. W. and McDERMOTT, E. (1954). *Atlas of men*. Harpers, New York.

—— HARTL, E. M., and McDERMOTT, E. (1949). *The varieties of delinquent youth*. Harpers, New York.

—— and STEVENS, S. S. (1942). *The varieties of temperament*. Harpers, New York.

—— and TUCKER, W. B. (1940). *The varieties of human physique*. Harpers, New York.

TANNER, J. M. (1956). Physique, character and disease: a contemporary appraisal. *Lancet* **2**, 635–7.

—— (1963). The regulation of human growth. *Child Devel.* **34**, 817–47.

—— (1964). *The physique of the Olympic athlete*. George Allen and Unwin, London.

—— HEALY, M. J. R., WHITEHOUSE, R. H., and EDGSON, A. C. (1959). The relation of body build to the excretion of 17-ketosteroids in healthy young men. *J. Endocrinol.* **19**, 87–101.

—— and WHITEHOUSE, R. H. (1982). *Atlas of children's growth: normal variation and growth disorders*. Academic Press, London.

VIOLA, G. (1935). Critères d'appréciation de la valeur physique, morphologique et fonctionelle des individus. *Biotypologie* **3**, 93.

—— (1936). Il mio metodo di valutazione della costituzione individuale. *Endocrinologia Patologia costituzionale* **12**, 387.

WALKER, R. N. and TANNER, J. M. (1980). Prediction of adult Sheldon somatotypes I and II from ratings and measurements at childhood ages. *Ann. Hum. Biol.* **7**, 213-4.

Part IV

Human adaptability

P.T. BAKER

20   Human ecology and human adaptability

Introduction

Over the long period of time from the first appearance of the genus *Homo* to the emergence of the species *sapiens* frequent adaptations must have occurred to allow survival in the changing environment. During this time hominoids also appear to have spread beyond their natural range requiring still further adaptations. Finally, in the form of *Homo sapiens*, people adapted to almost the total range of terrestrial environments.

While *Homo sapiens* had the same fundamental problems of adjustment to the natural environment as other species, it is clearly unusual in the way it has in the past and continues at present to spread far beyond the original habitat in which it evolved. Much of this ability to inhabit apparently hostile habitats is related to an ability to manipulate the environment into one suitable for human survival. An ability to modify the natural environment is not unique to people, as shown by such common examples as birds' nests and mammals' burrows, which modify the micro-environments they live in. What is unusual in the adaptive responses of our species is, first, the vast amount of learned information which can be passed from generation to generation without genetic encoding and, second, the constantly expanding nature of that base. Along with this accumulating information, we pass on a modified physical and natural environment, which in some aspects enhances the survival of the next generation. This process of transmission, often called a population's culture, forms the basic mechanism whereby populations have adapted to diverse environments.

The basic biological flexibility of human populations called *human adaptability* also contributes to the adjustment process. The term *adaptability* is often interpreted to encompass only those responses in the phenotype, which are produced by the action of the environment upon a given gene system. The generic term for this ability of a genetic system to produce a viable phenotype in variable environments is *genetic plasticity*. At the same time, the term 'human adaptability' is often applied only to those responses which improve some function of the organism or population in a specific environment. These definitions satisfy a need to distinguish those specific genetic traits which adapt a given individual or population to its environmental niche from the general genetic structure of the species which allows all populations to adjust to a variety of environments. In practice, however, such distinctions are generally impossible at our present state of knowledge. The study of human adaptability, therefore, involves biological responses, which include biochemical, physiological, and behavioural components. They may be present in a population as a consequence of population-specific and/or species-universal genetic adaptations.

Human ecology

In order to place the study of human adaptability in the broader context, which encompasses all of the mechanisms by which human populations adjust to their environment, one must first examine an interdisciplinary field of study called 'human ecology'. While the study of how human groups adjust to their environment is very old in all societies, the term 'human ecology' was not used extensively to define this body of information and studies until the 1960s. Perhaps for this reason there remains very little agreement on the definition of terms or upon topics of study which should be included under the name. Instead, there have developed a series of sub-disciplines, which are each called 'human ecology', but share only a limited overlap in perspective and terminology. These differences may be traced primarily to the initial discipline or disciplines from which the human ecology sub-disciplines arose.

Cultural ecology

Within the social sciences Amos Hawley, a sociologist, recognized in the 1930s that the studies of animal ecologists on spatial grouping and habitat use within a species might offer interesting insights into human behaviour, particularly in cities. He proposed that such studies be called 'human ecology'. This methodological approach to studying human social behaviour is now frequently called 'social ecology'.

By the early 1950s, Julian Steward, an anthropologist, became convinced that the natural environment through its effect on human subsistence behaviour had a strong effect on the kinds of social and political structures which developed in societies utilizing comparable natural habitats. As his propositions and evidence were integrated with the views and studies of other archaeologists and cultural anthropologists, a field variously termed 'cultural' or 'human' ecology developed. Within this approach the emphasis is on the regularities of human behaviour, social structures, and values which develop or evolve in response to particular natural environmental niches or situations. It is assumed that the mechanisms are universal to our species and the need to obtain essential natural resources, such as food, the driving force. The term adaptation in this schema usually refers to a response which appears to enhance the ability of the human society to utilize its natural habitat.

While most studies in cultural ecology have been focused on subsistence-related behaviour in traditional human societies, many studies, using historical records, have considered how changing environmental resources, population sizes, and external intrusions of other societies affect cultural evolution. Archaeologists, in particular, have continued to be concerned with how these changes relate to the diachronic changes often observed in social, spatial, and political organization.

Evolutionary ecology

Beginning in the 1970s a group of anthropologists developed an alternative ecological approach to the study of human behaviour based on an evolutionary ecology model developed for other species. This approach is termed 'evolutionary ecology' and focuses on individual human behaviour. As with cultural ecology, there is the assumption of a common genetic capacity throughout the species which is being investigated. The term adaptation, although rarely used, appears to refer to innate behaviour, which increases the efficiency of resource utilization. It is assumed that people individually and, as a consequence in aggregate, will tend to optimize their exploitation of the natural environment meaning that their behaviour will tend to become the most efficient (in terms of time and energy) in obtaining needed resources from the environment. If populations do tend to optimize, then it is argued that observed differences in many of the behaviour patterns of human groups may be seen as a direct consequence of attempts to extract the maximum benefits from various kinds of environments.

In order to test the validity of this approach, the behaviours of human groups are examined in relation to optimizing predictive mathematical models often previously validated on other animal species. As Smith (1983), one of the proponents of this approach, notes, the basic assumptions and methods of evolutionary ecology are closely tied to formal economic theory, which also assumes attempts to reap maximum benefits, but they differ in that evolutionary ecology is based on the assumptions of natural selection, while economic theory assumes only that maximization behaviour is common to all human endeavours. Proponents of evolutionary ecology believe it can help clarify human behaviours, including (i) why people use specific behaviours for obtaining food; (ii) why they develop various mating practices; (iii) why they distribute themselves over space in different manners; and (iv) why they differ in such interrelated characteristics as population dynamics and community structure.

Proponents such as Winterhalder (1980) have developed an extensive set of hypotheses based on evolutionary ecology theory and have suggested methods for testing some of the models developed from the hypotheses. Perhaps because of the enormous amount of observer time and data collection required, most of the research undertaken to date in human evolutionary ecology has focused on behaviour related to food acquisition and use. The models being tested comprise optimal foraging theory and have been shown to reasonably predict behaviour in a variety of other animal species.

Biological human ecology

Within the disciplines of physical anthropology, environmental physiology, and, to a lesser extent, epidemiology, nutrition, and human genetics, there developed by the 1960s a common interest which was often called 'human ecology'. The content and research objectives of the group associated with this form of human ecology were best presented by J. S. Weiner in the earlier editions of this book. In contrast to the societal focus of the cultural ecologists and the behavioural focus of the evolutionary ecologists, the emphasis in this type of human ecology was on how the natural and cultural environment affected the biological characteristics of given human populations.

Despite a common background in the biological sciences, these individuals from several disciplines concerned with human biology approached human ecology with somewhat disparate sets of assumptions and objectives. All accepted the assumption that evolutionary selection processes had produced the human species and that the processes had produced a set of genetic characteristics which adapted our evolving species to their environment. The human physiologists had based their previous research on the implicit assumption that the physiological responses of individuals to an environmental characteristic were representative of the species response. The epidemiologists also often assumed that the response of any group of people to a given infectious disease was representative of the response of any given group to the same disease in the same environment. Thus, how Englishmen or Americans responded biologically to an aspect of the natural environment was often taken as an indication of how all populations would respond. While these assumptions remained the basis of considerable research, the belief that all populations within the species would respond the same was challenged.

The traditional definitions and theoretical postulates of physiology became and remain very important in this type of human ecology. They are also widely accepted in the studies subsumed under the title 'human adaptability' in this section. Within this framework, the individual organism contains a complex set of neurological, physiological, and behavioural mechanisms which function to keep it within viable limits. These mechanisms contribute to *homeostatic* processes. For example, as the air temperature surrounding the individual becomes low enough to lower the core body temperature below a neutral state, peripheral vasoconstriction begins reducing heat loss to the surface. As the air temperature becomes colder and core body temperature tends to fall, such behaviours as huddling, which reduces heat loss, and then shivering, which increases heat production, begin to accelerate to the limits of the system's ability. If these various mechanisms of response cannot adequately maintain the core body temperature, a critical level is reached leading to a loss of consciousness and death. Comparable mechanisms of vasodilation followed by sweating are found if the environment tends to drive up body core temperature. Again at a critical level of stress, the system fails and death follows. The entire set of responses are viewed as attempts by the organism to maintain near optimal or hormeostatic core temperatures. Many homeostatic systems have been described in all living matter and, for people, even social systems have been so described. The homeostatic concept is derived from principles in physics, and, thus, terminology developed, which also has its origin in the physical sciences.

Figure 20.1 exemplifies both the principles and the terminology. In the illustration, two springs, representing mechanisms to maintain, homeostasis, hold a pointer between them. This represents the neutral state. The two springs are not under tension, the pointer rests at zero, indicating no *strain* on the springs, and the system is in energetic equilibrium. The effect of a force pushing the pointer either way creates a

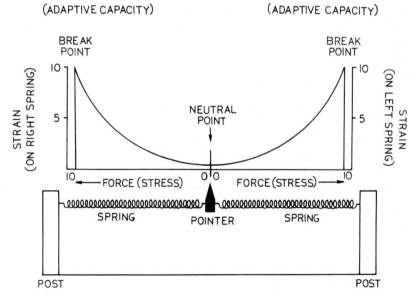

Fig. 20.1. Dynamics of homeostatic systems: a comparison of terminology and behaviour of a physical system with physiological homeostatic systems.

stress on the system and results in a strain on the spring or homeostatic system, which attempts to keep the pointer centred. Of course, if the stress exceeds the flexibility of the spring (*adaptive capacity*), it breaks and the system no longer functions. Within this analogy an *adaptation* is anything which reduces the stress or improves the strain tolerance (adaptive capacity) of the system.

This conceptual model has proven reasonably acceptable to other individuals within biological human ecology. However, many human population geneticists and physical anthropologists do not believe that the homeostatic mechanisms are identical within and between populations. Having been trained in disciplines which emphasize the genetic diversity of individuals and groups, they doubt that most individuals or even groups would respond to identical environments in an identical manner. Based on the synthetic theory of evolution developed from a combination of Darwinian theory and genetics, these disciplines use the term adaptation almost synonymously with *genetic fitness*.

From a synthesis of these viewpoints, a more comprehensive approach to examining how people have adapted to their particular environments has developed. This paradigm emphasizes that a population's adaptive response to an environmental stress likely to involve several mechanisms, including learned responses, species-wide physiological adjustment capabilities (*acclimatization*), and population-specific genetic characteristics, which are the result of natural selection.

An example of an individual and population-specific genetic adaptation is provided by haemoglobin S. The presence of a single allele for haemoglobin S improves the survival chances of a child exposed to falciparum malaria. It may also enhance the reproduction performance of female carriers. The gene in a single dose therefore may be viewed as an adaptation in the physiological sense to an environment containing falciparum malaria even though it exists only in some individuals and some groups. It should also be noted that, for many physical anthropologists and population geneticists, only this kind of known genetic difference with evidence of enhanced reproduction would be termed an adaptation. In biological human ecology, such instances are generally referred to as *genetic adaptations*.

In most studies of how human populations are adapted to the natural and social environment, it has been impossible to distinguish between the specific genetic adaptations and species-wide genetic adaptability or plasticity which enhance functional capacities. A few specific genetic adaptations are known and several adaptability responses such as the ability to acclimatize to heat appear in almost all individuals. However, most other biological adaptations to environmental stresses are not so easily classified. The population variation in skin melanin is a good example of the more complex adaptability responses our species shows to an environmental stress.

It is well known that the damage of ultraviolet rays to the skin is inversely proportional to the amount of melanin in the dermis since the melanin screens ultraviolet penetration. On the other hand, in some nutritional environments where sources of vitamin D are scarce, lack of ultraviolet penetration may lead to undesirable health results (childhood rickets), since with enough penetration people produce adequate amounts of vitamin D without dietary sources. Given these effects of skin colour, dark skin colour in the high radiation areas is adaptive in the general sense, while light skin colour is adaptive in low radiation areas with certain diets. Melanin density in the skin was, in general, inversely related to the amount of ultraviolet radiation prior to the migrations of recent centuries. Since individual skin colour variation appears to be closely tied to the specific genetic structure of the individual, this appears to be a good example of an individual and population-specific genetic adaptation. Yet, it must be remembered that suntanning, which increases the melanin content of the skin, is nearly a universal response within our species, occurring even in individuals with the darkest of natural skin colour. Furthermore, if the population has an adequate dietary intake of vitamin D, heavy pigmentation does not adversely affect health in a low radiation environment. In this instance, as perhaps in many others not yet understood, the adaptive response of any given population to the evironmental stresses may involve genetic differences, genetic plasticity, and behavioural responses in various combinations.

Human adaptability

Attempts to found a unified body of theory and focus of research under the title 'human ecology' have encountered many difficulties as a consequence of the varied assumptions and goals of the various disciplines. Despite these differences, there is an agreement that an adequate understanding of how we have adapted, adjusted, and coped with our natural and social environments through time will be the key to explaining why we function and behave the way we do. In fact, barring chance and the supernatural, no other cause can be suggested.

This commonality of perspective, along with increasingly productive research efforts, has led to some sharing of ideas among the component groups in human ecology and may in the long term lead to a unified approach. At the present time, the presentation of the knowledge gained is perhaps best presented in terms of the sub-disciplines of human ecology, which have been described.

For several reasons, the sub-discipline of human ecology, which was devoted to studying how the environment affected the morphological and physiological characteristics of people, was renamed human adaptability. Although this may be a somewhat improper use of the term, it is probably the best currently available. It is not assumed that all the biological adaptations made are a result of genetic plasticity nor is it assumed that the causal relationships are all one-way. That is, some of the biological adjustments or adaptations people made to their natural and social environments have also modified how they adjusted to subsequent environments. The adjustments we have made to improve our adaptations to a given environment have produced a new environment to which we, in turn, adapt in an ongoing process of new stress and new adaptation.

Research design and evidence in human adaptability

The causes of individual and group differences in morphology, physiological abilities, and behaviour have intrigued scholars in most complex societies. During the development of Western cultural systems, scholars in the Judaeo-Christian tradition ascribed major morphological differences between groups to man's fall from grace and God's wrath. The Greeks, on the other hand, sought more secular explanations suggesting, for example, that the skin colour differences between Europeans and Africans might be caused by differences in the intensity of the sun.

As late as the 1940s, the apparently inherent differences in the morphological characteristics of the peoples found in the large land masses of the world in the sixteenth century were still often treated as if they had no cause or had been accidental in origin. This undefined process had presumably created biological subdivisions of early *Homo sapiens* which later spread to various geographical regions. Subsequent population genetic research did, indeed, show that many of these morphologically distinct groups, taxonomically referred to as races, also differed in the frequencies of identifiable genes. However, population genetics theory provided little explanation for the genetic and morphological differences between the groups other than showing that at least some of the gene frequency differences might have arisen through chance factors. While the biological variation in human groups remained unexplained, the study of group differences in other species used the Greek approach of relating variability in morphological characteristics to variation in an aspect of the physical or biological environment. The associations found were often called rules by the German scientists who described them.

Coon, Garn, and Birdsell in a small speculative book published in 1949 suggested that the 'racial' variation might also have developed as a consequence of variation in the environment. Among the first demonstrations that some of the 'rules' for other species applied to our own, were the correlation analyses performed by Roberts and Schreider.

These researchers showed that average body size and weight relative to height of human groups was, in previous times, significantly correlated to measures of air temperature.

This statistical technique along with others remains a common method for showing that an environmental characteristic is associated with a genetic, morphological, or physiological response to an environmental characteristic. However, the student of the subject must, as the investigators did, interpret the causal meaning of such associations very cautiously. Not only may A cause B in a correlation, but B may cause A. Furthermore, the association between A and B may be entirely the product of the associations of A and B to an unmeasured variable C. For example, in many countries between 1930 and 1950 the yearly *per capita* consumption of cigarettes smoked rose. At the same time life expectancy rose. Thus, a correlation of cigarette consumption per year and life expectancy shows a highly positive association. Of course, very few individuals now accept this as evidence that increasing cigarette consumption will increase life expectancy, but comparable analyses of associations have been accepted as evidence of cause in the analysis of less familiar topics. While evidence of association does not in and of itself demonstrate a causal relationship between an environmental and human biological trait, it is a useful tool of evidence in several regards. As in the example of skin colour and ultraviolet radiation, the very strong correlation between this environmental and human characteristic strengthens the hypotheses that the inherited differences in the skin colours of human populations are related to the ultraviolet-blocking effects of melanin. Correlations can also provide the basis for a hypothesis to be tested by more rigorous methods.

Carefully designed experiments provide the most reliable and reproducible results of the research methods available. Thus, the monitoring of skin colour changes with a controlled exposure to ultraviolet radiation has shown that tanning occurs in most individuals irrespective of untanned skin colour or population of origin. From this evidence one can be rather certain that it is a part of human adaptive capacity. When this finding is combined with the fact that the tanned individual shows less dermal damage than the same untanned individual, tanning can then be described as a species-wide adaptation to high solar radiation.

While the experimental research designs provide the basic information on how individuals and populations differ in their responses to the stresses of the natural and socio-cultural environment, they cannot be used to determine directly how or if specific environmental stresses affect such incremental processes as physical growth or the development of many degenerative diseases. To examine whether exposure to the stressfully low oxygen pressure encountered in the high mountains affects the morphological characteristic or oxygen consumption ability of a mouse population, one only need raise two groups of mice from a common genetic strain under identical conditions except for environmental oxygen pressure and compare the end products. While such experiments are, of course, not conducted on people, it does happen that populations and individuals live naturally under a wide variety of environmental circumstances. It is, therefore, sometimes possible to find individuals or groups whose living conditions approximate the conditions required by an experimental design. Such instances, generally termed *natural experiments*, have been extensively examined to provide basic evidence on how populations respond to various environmental characteristics.

An example of how such natural experiments may be exploited is illustrated in Fig. 20.2. Haas (1976), who developed this design, had reason to believe that the growth of children from birth to age 2 was slowed by the environmental effects of high altitude. He also realized that nutritional factors, a genetic difference between populations, and medical care could all affect growth rates. Based on his knowledge of the populations of Southern Peru, he designed a study comparing the growth of infants in the 12 groups identified in the boxes. Comparisons of these infant groups would then indicate whether the high altitude environment *per se* affected growth. In practice he was able to obtain only four of the high altitude groups and four of the lowland ones, but this was sufficient to demonstrate that within this region altitude was the most important factor in slowing infant growth.

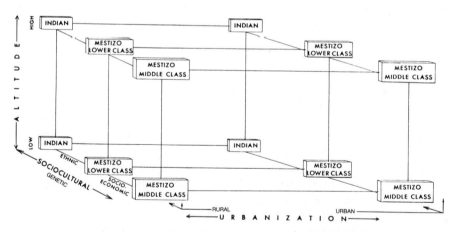

FIG. 20.2. A *natural experiment* research design developed to examine how altitude nutrition, health care, and ethnic characteristics interact to regulate growth in Peruvian infants. (From Haas, in Baker and Little 1976.)

Natural experiments are always to some extent flawed research designs, since it is impossible to find groups of individuals living under identical circumstances except for a single environmental characteristic such as the partial pressure of oxygen. For this reason, causal conclusions from such studies must be drawn with due caution.

Since none of the individual methods available for human adaptability research are adequate in themselves to examine some of the complex problems generated, most proofs are based on a combination of the methods mentioned. Among the most important of these proofs is again evidence gained by examining a population in its natural environment. For example, controlled experiments show conclusively that, in cold air and water temperatures, increased body fat aids in the maintenance of body core temperature, but it was only by the study of English Channel swimmers and the study of Andean natives sleeping in their traditional houses that it was conclusively demonstrated that the normal variation in body fatness significantly affects the core temperatures of some people in their normal activities.

In the subsequent chapters of this section the adaptive responses of human populations to specific stresses are explored. It should, however, be remembered that it is unlikely that most of these responses arose in response to a single environmental stress. It appears that one of the trends of evolution, which characterizes the more complex animals, is the development of response systems which serve to keep individuals and population units functional through frequently changing stresses. Our research methods, thus, often remain quite inadequate for determining either why most specific genetic traits arose or what are the limits of our species adaptability.

21 The physical environment

The potential stressors

Although modern technology permits travel into space, people are able to survive the experience only by surrounding themselves with a micro-environment which resembles the temperatures and atmosphere on earth. Even so, it appears that without gravity our stay in space remains limited. All species, including *Homo sapiens*, must be adapted to their physical environment. In order to function the basic organic building block, the cell, must be kept within a set temperature range, prevented from excessive dehydration, and protected from overdoses of some kinds of radiation. Most cells also require oxygen in the surrounding environment in order to be metabolically active.

While some orders of mammals evolved species capable of satisfying these adaptive requirements, in almost all terrestrial environments, most primate forms were limited to lowland tropical and warm temperate environments. Neither fossil forms nor current species have been found where climatic conditions included winter temperature substantially below freezing. The coldest climates where non-human primates have been found in recent times are those in central Japan and North Africa. *Homo*, of course, became an exception to this generalization, but from present evidence we must conclude that even this intrusion into cold climates did not occur until the human line had evolved into a form such as *Homo erectus* or more certainly the Neanderthal form of *Homo sapiens*.

The evolutionary record still does not permit the full delineation of the physical environment in which hominid forms evolved, but it appears that the early forms all developed in dry tropical savannahs. Thus, we may reason that our genus was adapted at an early stage to rather high daytime temperatures, cool, but above freezing, night-time temperatures, moderately low humidities, high solar radiation, and a relatively high atmospheric pressure of oxygen. This does not imply that there was a perfect adjustment to these climatic characteristics, but does mean that most of the individuals living in these conditions tolerated the thermal, water vapour, radiation, and atmospheric oxygen environment well enough to develop and reproduce. It is also logical to assume that some limitations on their adaptability must have kept them confined to this natural environment until the development of a *Homo erectus* form. Given these tropical origins for hominid forms it is enlightening to explore how well contemporary populations are adapted to this hot lowland environment.

450

Heat tolerance

A core body temperature of 37°C is the approximate neutral condition for our temperature regulation system. However, as in all homeostatic systems, numerous factors, including age, sex, and the time of day, cause small variations around this average. The environmental and individual metabolic conditions under which this neutral core temperature is found are, indeed, quite limited. For reclining young adults whose internal heat production is restricted to resting metabolism, an external temperature of about 28–30°C is neutral. However, even this is a neutral zone only if radiant heat loss and gain are nearly equal, and if air movement is minimal.

While many individuals remain functional with a core temperature 3–4°C above its mean, the strain is substantial and permanent damage or death can result when core temperatures rise to this extent. Homeothermic responses are, therefore, induced in healthy individuals when core temperature rises significantly.

In most of the world's climatic environments the primary source of the increased heat load is the heat produced by the metabolism of muscle action. Externally produced heat loads occur only when there is a net radiant energy gain from the external environment and/or the surrounding media (usually air) has a temperature higher than the skin temperature.

As the heat load on the individual rises there is first a peripheral vasodilation and a rise in heart rate which increases blood flow to the surfaces. The consequent rise in skin temperatures increases the rate of heat loss to surrounding air, so long as the air temperature remains below skin temperature. With a further increase in heat load, sweating by means of eccrine glands begins. Rises in core temperature through its effect on the hypothalamus is the primary mechanism for stimulating total body sweating. Localized sweating can also be stimulated by an areal application of heat.

Human sweating is a highly efficient cooling mechanism making us one of the few species well adapted to high energy expenditure during the day in hot and dry environments. Eccrine sweat glands, which cover the human body, discharge mostly water with small quantities of electrolytes. This contrasts with the apocrine gland, common in other sweating species, which along with water excretes significant amounts of fat and protein. As a method of cooling the eccrine gland is more efficient since not only does the discharge evaporate more easily, but the body also loses fewer nutrients in the process.

The number of active human sweat glands (shown in Table 21.1) is well in excess of a million. Some differences have been reported between groups living in tropical as compared to temperate climates. It is now believed that the morphological number may not vary significantly between populations. Instead, research by Kuno suggests that in individuals exposed to high heat loads during childhood a greater percentage of the glands are activated. Since the dispersion of water over the skin surface will depend on the number of active glands and affect evaporation rates in hot climates, the climate in which an infant develops may affect at least in minor ways its adult heat tolerance.

TABLE 21.1
Total number of glands of adult males (means in millions ± SD)†

Population	N	X̄	SD	Source
Hindu	(6)	1.51 ± 0.19		(Knip
Indian	(19)	1.69 ± 0.16		(Weiner)
Dutch	(9)	1.47 ± 0.29		(Knip)
British	(2)	1.66 ± 0.19		(Weiner)
European in West Africa	(21)	1.75		(Thompson)
West African	(26)	1.66		(Thompson)
Bush Negro, Surinam	(8)	1.69 ± 0.14		(Knip)

† From *Human biology* (2nd edn)

With this large number of sweat glands the amount of water which can be excreted is substantial. For very short periods in extreme temperatures excretion rates for young males have been reported as high as 4 litres per hour while over extended periods rates of 2 litres per hour are common. In various hot desert studies a water loss of 8 litres per day has been found to be the average for young men. Body size is a significant factor in the amount of sweat produced when the individual is active. As illustrated in Fig. 21.1 a young acclimatized man of 55 kg will sweat only 1 litre per hour while walking in moderately hot (40.5°C) desert conditions, while a 90 kg man will sweat over 2 litres per hour. At lower temperatures body weight has less effect on sweat rates. Body fatness probably also affects not only the sweat rate in active individuals but, as the correlations in Table 21.2 suggest, fatter individuals may also have higher core temperatures and heart rates when walking in a hot desert.

The effectiveness of sweating as a cooling mechanism in a hot dry climate is shown by the fact that the evaporation of only 1 litre of sweat can in theory produce a loss of 2500 kJ of heat. The limitation on sweat loss as a cooling mechanism is the ability of the ambient environment to induce water evaporation. Within the hot environments of the natural world the hot deserts and savannahs generally produce total evaporation whatever the sweat loss rate. In hot wet climates, however, this is seldom the case. At the extreme of hot wet conditions, such as those encountered in deep mines, even at 35–40°C, the atmosphere may be totally water-saturated.

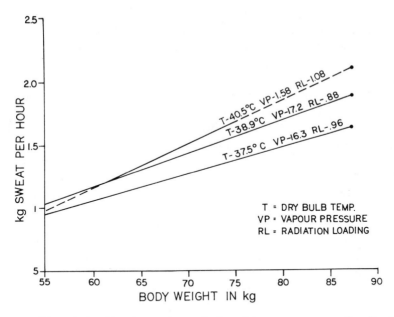

Fig. 21.1.   The relationship of sweat loss to body weight among groups of acclimatized young men, walking at 5 km/hr in hot desert conditions. (Adapted from Baker, Q.M., *R & D Tech. Rept. 7*, 1955.)

TABLE 21.2

The correlations (r) of estimated body fatness to physiological responses after walking for 2 hours (5 km/hr) in a hot desert *

Group	N	Environmental temperature (°C)	Rectal temperature	Pulse rate	Sweat loss
I	10	40.5	0.49	0.68	0.50
II	19	38.9	0.32	0.17	0.53
III	21	37.5	0.44	0.12	0.65

* From Baker, Q.M. *R & D Tech. Rept. 7*, 1955.

Under those conditions sweating is of little or no cooling value. Even in tropical forests or on the shores of the Persian Gulf where vapour pressures are high, evaporation rates are quite low.

In such environments the sweat glands for reasons related to skin hydration do not produce sweat loss at the extreme rates found during exposure in hot dry environments. Nevertheless, sweat loss frequently exceeds the evaporative rate and excess water drips from the body surface without producing evaporative cooling. When sweating has limited utility as a cooling mechanism, the rate of heat loss is highly dependent on the thermal gradient between the skin and surrounding air. The size of the skin surface over which the heat loss can occur then becomes a significant factor in heat loss.

Sweating does not become maximally effective in a specific hot environment until a physiological process termed *'heat acclimatization'* occurs. When an individual is newly exposed to a heat load, his or her sweat production over subsequent days of exposure gradually increases. At the same time core temperature during the exposure stabilizes at a lower temperature. Cardiovascular strain is also reduced as shown by lower heart rates during the stress period. Figure 21.2 shows the contrast in physiological responses between unacclimatized and acclimatized young men after 10 days of working in a hot environment for 4 hours. It has also been observed that salt excretion in the sweat declines over the period of acclimatization so that the danger of serious symptoms from electrolyte imbalance in the body also decreases. Heat acclimatization appears to be specific to the heat load since further acclimatization will occur if the heat load is increased and acclimatization is completely lost if the heat load is not maintained on a fairly continuous basis.

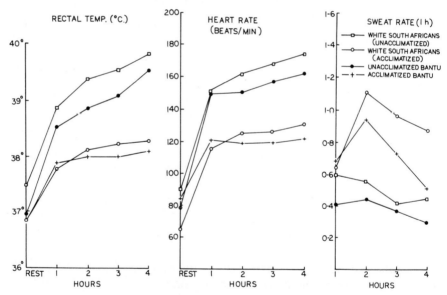

FIG. 21.2.   The effect of heat acclimatization on the physiological responses of whites and Bantu. (Modified from Wyndham, in Baker and Weiner 1966.)

Comparative heat tolerance in primates

The resting adult human being can, with an adequate water supply, physiologically tolerate the hottest of world desert conditions without inordinate stress. This includes a full solar radiation load with daytime temperatures well in excess of 50°C. Acclimatized individuals can also walk for several hours during the day at slightly lower air temperatures, c. 47 or 48°C. In very humid natural environments a sustained rapid walk over two hours produces near tolerance limits in temperatures below 40°C. In hot desert conditions water loss during the day substantially exceeds water intake even when water is readily available. This deficit is normally replaced by excessive intake during the evening and morning hours. Failure to replace the water loss within the 24-hour cycle is likely to produce serious dehydration and even death, if a second day of high heat load is experienced.

Most of our primate relatives seem to have less effective thermoregulatory responses to hot dry conditions. Even the chimpanzee appears to have less effective eccrine sweating and does not tolerate temperatures about 40°C well. Of the non-human primates studied only the rhesus monkey (*M. mulatta*) has been found to have an eccrine sweating response comparable to humans. Whether this produces a heat tolerance similar to that of people, remains untested.

While the high dry-heat tolerance universal to *Homo sapiens* suggests the possibility of a long period of evolutionary development under almost desert conditions, it must be remembered that our high heat tolerance required frequent access to major quantities of naturally occurring water. Even for a human population which has containers for carrying water the need to carry between 4 and 10 kg of water per day for every day away from a source would have significantly restricted how far adults could travel from natural and reliable water sources.

Individual differences

Age, sex, body size, shape, and composition all affect responses to heat stress. While most people have a high tolerance as adults, the infant because of small size and several morphological immaturities does not tolerate heat well. The small size of the infant results in a large surface in relation to the amount of metabolically active tissue. Thus, heat loss per unit of heat production will be high when the surrounding air is below skin temperature. However, both neurological and sweat gland immaturity makes the sweating response quite ineffective in most newborn infants. Without effective sweating, air temperatures above skin temperatures will cause rises in body temperature and the relatively large surface area to body mass will hasten the core temperature rise. A radiative heat load will similarly increase the infant's heat load problem. Little (1973) showed that infant mortality is probably high in very hot weather, even though he suggests they are much better adapted to heat than cold. During old age, some people show lower heat tolerance for reasons not fully understood.

Individual variation in body morphology has a substantial impact on responses to heat. As noted in Fig. 21.1, the weight of the individual shows a linear relationship to sweat loss for men walking in a hot desert. The data also show that, as the heat stress level rises, the amount of water needed per unit of body weight rises. While the water costs of hot dry conditions are greater for large individuals, it does not appear that the amount of muscle mass is important to final heat tolerance. This conclusion is based on the fact that in heat stress studies the core temperature of large men was no higher than that of small men, provided they had comparable levels of body fat. It does appear that the amount of subcutaneous fat may affect heat tolerance since core temperatures were higher among fatter men during heat tolerance tests. Furthermore, a study of heat stroke deaths among US soldiers during World War II showed that men 10 per cent above average weight for height had death rates nine times those of men 10 per cent below average weight for height.

In the hot wet environments of the world, the morphological characteristics of the individual may affect heat tolerance differently because of the reduced effectiveness of sweating. In theory, having a large surface area per unit of heat-producing body mass should effectively improve heat tolerance during exercise so long as the body is not exposed to a high radiant heat load. The volume of the body as measured by weight is a critical factor in the surface area to body mass ratio since, with increasing size, this ratio inevitably declines. A human infant has a surface area to weight ratio (SA/WT) of 0.6 (cm^2/g), while the adult has only a 0.027 ratio. Even a small adult compared to a large one has a significantly larger ratio. The distribution of body mass may also have a minor effect. For example, the SA/WT ratio is slightly greater in long-extremity peoples such as the Nilotics of Africa than it is in the short-extremity natives of Eastern Asia. These differences in form, which increase relative surface area, should increase the evaporative cooling surface and enhance heat loss relative to heat production.

Population comparisons

Research comparing the heat tolerance of human populations has taken two forms, laboratory and field testing. In the laboratory members of various world populations have been studied in identical heat stress situations. In field testing most studies have involved comparing the responses of members of two populations as they worked in the natural environment.

The controlled laboratory studies have shown that, while various groups such as Europeans, Australian Aboriginals, North Africans, and South African Bantu show major differences when first tested under hot working condititions, most of the differences may be attributed to group differences in the initial level of their heat acclimatization. Most of this difference disappears after heat acclimatization to a high level of stress. The residual differences in response can probably be totally accounted for by population differences in body size, physical fitness, and levels of subcutaneous fat.

The field tests tend to yield similar results, but point up some population-specific effects produced by solar radiation. Thus, in a desert environment nude dark-skinned American blacks experienced a greater heat stress than nude light-skinned American whites. A study comparing a pygmoid population in Zaire with a Bantu group showed that when walking in the open grasslands the pygmoids showed evidence of equal or greater heat stress than the Bantu. Because of the high humidity in this climate the results were rather unexpected, but the investigators believe that radiation heat input to the relatively larger surface area of the pygmoid group may explain the results.

Culture and heat

Despite our tropical origins and the evidence that we continue to have an unusually high heat tolerance, human populations have developed elaborate mechanisms to reduce heat exposure. In the recent past, houses in the hot deserts were built so as to keep inside temperatures cool during maximum daytime heat. Clothing usually covered almost all of the body so as to reduce radiative heating. In hot wet environments housing was open and clothing minimal to promote sweat evaporation. While these responses date from the development of agriculture, even hunters and gatherers tend to restrict physical activity during mid-day in hot climates. These cultural adaptations to heat may simply be responses to the discomfort people feel, but the performance of many physical and mental tasks tend to decline at air temperatures above 28°C, and heat stress imposes a measurable if slight metabolic cost on all human activity.

Cold tolerance

As a tropical species human cold tolerance is low in spite of the fact that as a relatively large adult animal we have a proportionally small surface area over which to lose heat when compared to our heat-producing mass. Without clothing our metabolic response to low air temperatures differs dramatically from native arctic mammals (see Fig. 21.3). The differences between our response and that of the arctic mammal is attributable to the fact that we do not have any significant surface insulation, although like aquatic mammals our subcutaneous fat provides some insulation for the body core. For the average adult male to remain in thermal neutrality while inactive at 21°C requires 1 clo. of insulation, about the equivalent of wearing a business suit. This compares to a rabbit whose fur provides up to 4 clo. or the arctic white fox whose fur provides 7 clo.

When air temperatures fall below 28°C the arterials below our subcutaneous fat vasoconstrict. The warmed blood from the core of the body is then shunted away from the skin and, as the skin temperature declines, the loss of heat from the core of the body is reduced. In totally inactive individuals with very heavy subcutaneous fat deposits, the vasoconstrictive response will keep the core body temperature in the neutral range in air temperatures as low as 15°C, but for thin individuals further homeostatic responses in the form of shivering may be triggered by 25°C. Given freedom of response most people will also react to lower skin temperature by huddling, which reduces the surface area for heat loss, or by voluntary muscular activity, which increases heat production.

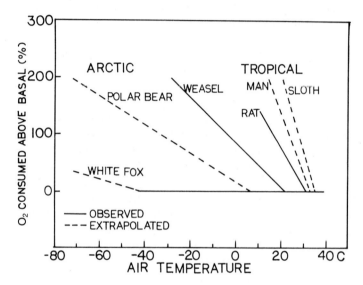

Fɪɢ. 21.3.   Increase in oxygen consumption for different species exposed to cold air temperatures. (Modified from LeBlanc 1975.)

If the individual remains inactive, involuntary muscle contraction, which eventually results in visible shivering, begins as core temperature begin to fall. Shivering can increase metabolic level 100 per cent above basal. Voluntary activity, such as running, can increase heat production much more, but the length of time over which voluntary activity at higher metabolic output can be maintained to prevent further declines in core temperature is limited.

Perhaps because of our tropical origins, human populations appear to have a limited acclimatization response to cold as compared to our substantial ability to heat acclimatize. Some other mammals of tropical origin, such as the rat, show, upon continuous exposure to cold, major increases in basal metabolism, reduced metabolic sensitivity to cold, and increases in the quantity of a metabolically active form of fat called brown fat.

Extensive searches for evidence of similar acclimatizational responses in people have shown that brown fat is present in infants and that adults may undergo minor seasonal changes in metabolic responses to cold. There is, also, rather clear evidence that continuing exposure to cold modifies sensory responses so that individuals experience less discomfort and are better able to judge the potential danger of significant body cooling or frostbite.

Nevertheless, these changes are relatively minor compared to those found in some other species and fail to confer upon the individual the greatly enhanced adaptation manifested by heat acclimatization.

The absolute limit of human cold tolerance is difficult to determine. Unlike heat tolerance which is generally limited by fatal heat stroke, detrimental responses to lowered core temperature are sequential and often reversible. Most individuals suffer a gradual loss of mental alertness as core temperatures decline, and most become unconscious or semi-conscious at core temperatures of 3 or 4°C below normal. Nevertheless, adult individuals whose core temperature have fallen nearly 10°C below normal have been known to recover fully if the cause of the low core temperature was entirely thermal. Conditions which produce unconsciousness in adults are approached when nude inactive thin individuals are exposed to still 5°C air temperatures for more than 2 hours; individuals with heavy fat layers remain fully functional under the same conditions. Most physically fit young adults can tolerate temperatures close to freezing without clothing for 8–10 hours providing wind speeds are low. This is possible because of their ability to generate adequate body heat through physical activity.

These experimentally derived measures of cold tolerance limits for individuals undoubtedly suggest a greater cold tolerance than is possible for the continuing survival of a whole population. In naturally occurring conditions, groups such as the central Australian aborigines have been described as living without shelter or clothing in winter conditions which varied from night-time temperatures just below freezing to daytime temperatures above 15°C. In Tierra del Fuego the American Indians (wearing only skin capes), lived in very wet and sometime snowy conditions. During the winter temperatures ranged from near freezing to only 7 or 8°C. Both groups used fire for warming at night and the Tierra del Fuegans constructed crude shelters of branches. Thus, it appears that with fire, human populations could survive without effective clothing or shelter in most environments which did not involve temperatures significantly below freezing. Without fire, the tolerable minimums must have been some 10–15°C higher. It should also be noted that since people originally depended on subcutaneous fat rather than hair or fur for insulation, cold tolerance, unlike heat tolerance, is virtually unaffected by the moisture content of the air. With the invention of clothing the effectiveness of the insulation provided by the clothing was inversely related to the water permeability of the material used.

Comparative cold tolerance in primates

Non-human primates have not been extensively tested for cold tolerance. The large-bodied gorilla prospers in the relatively cool mountains of Central Africa, but only monkey species are found in relatively cold environments. The rhesus species of North-west Africa and Gibralter live in winter conditions of occasionally below freezing temperatures while the langurs of South Asia range in the Himalayas up to quite cold and occasionally snowy environments. The macaques of Japan live continuously in what is probably the coldest climate encountered for non-human primates. Nevertheless, experiences with primates in captivity suggest that it may be the food supplies which limit the temperate climate penetration of some primates. Judged by their natural habitats, macaques probably have cold tolerances which physiologically exceed those of the unclothed human being.

Individual differences

Age, sex, size, and body composition all have an effect on cold tolerance, and these effects appear to be greater than the effect of the same variables on heat tolerance. Infants are particularly susceptible to the effects of total body cooling because of the combined effect of neurological immaturity, small subcutaneous fat thickness, and size. This is partially compensated for by higher non-shivering heat production produced in part by the deposits of brown fat. The disadvantage of small size for maintaining body temperature remains during childhood. As shown in Fig. 21.4, the measurement of core temperature in children and adults in the cold environments of highland Peru confirms that children suffer greater core temperature reduction during the cold night hours than adults. Older adults also tend to have a lower cold tolerance as a result of their reduced ability to increase metabolism to the levels possible during youth.

Body mass or weight variation in adults is a factor in cold tolerance since weight is a measure of heat-producing muscle mass, and the heat produced in such ordinary activities as walking and running will be proportional to weight. As has been explained earlier, the surface area over which this heat will be lost is not proportional to the mass; therefore, larger body size will result in better maintenance of core temperature during activity in a cold environment. For inactive young adults the thickness of the subcutaneous fat layer may be more important than body

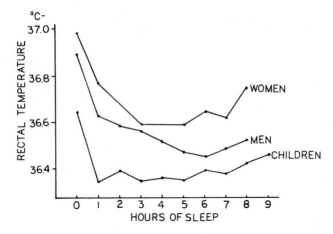

FIG. 21.4. The rectal temperature of Quechua Amerindians sleeping in unheated houses during temperatures below freezing. (Modified from Hanna, in Baker and Little 1976.)

size. As shown in Figs. 21.5 and 21.6 the metabolic responses and minimum rectal temperature elicited by resting inactive in a 15°C air environment are strongly affected by body fatness.

Theoretically, differences in average morphology between men and women should affect their modal responses to cold. Adult women are on average smaller, but have shorter extremities in relation to trunk size than men. As a consequence their surface area per unit mass tends to be

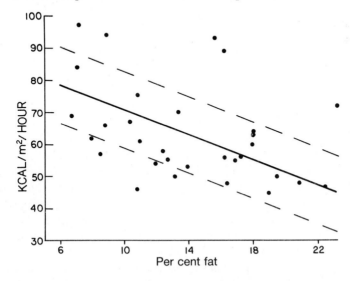

FIG. 21.5. Metabolic rate of young men at the end of 2-hour nude exposure to 15°C regressed on estimated body fatness. (Modified from Daniels and Baker, 1961, *J. Appl. Physiol.* **16**, 421.)

somewhat greater than that of men, but is less than that of men of the same weight. This should result in lower body temperature in the same environment. However, women have a greater average thickness of subcutaneous fat in all populations, which should result in lower skin/higher core temperatures.

When exposed to total body cold stress white inactive women have been found to maintain somewhat higher core temperatures with lower metabolic cost as expected. During studies of exercise in the cold, sex differences in response to cold tend to disappear.

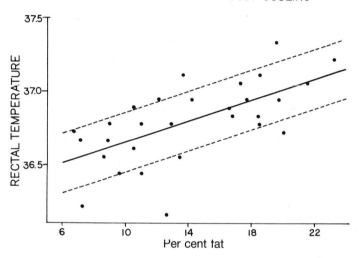

SKINFOLD THICKNESS AND BODY COOLING

FIG. 21.6. Minimum rectal temperatures in young men exposed to 15°C for 2 hours regressed on estimated body fatness. (Modified from Baker and Daniels, 1956, *J. Appl. Physiol.* **8**, 409.)

Extremity cooling

When only the hand or foot of an individual is exposed to severe cooling conditions such as water near the freezing point, there is a strong vasoconstrictive response which almost completely stops blood flow to the affected part and the skin temperatures drop rapidly. After a few minutes many individuals show a rise in skin temperature produced by a temporary vasodilation (see Fig. 21.7). Still more cycles of vasoconstriction and dilation follow in some individuals (often called a Lewis wave). A great deal of individual variation is found in this response with some individuals experiencing no rewarming response, while others maintain very high average skin temperatures with short cycles.

Individual variations in this response have several functional implications for the person who is cold stressed. Individuals who maintain complete vasoconstriction in the hand will lose less body heat, but will have less sensitivity in the fingers and lose more manipulative capability than the high-cycling responder. If the hand or foot of the lower responder is exposed to air temperatures below freezing, it is also more likely to develop severe frostbite. Because of the very limited circulation in the low-responder, chronic cold exposure above freezing can produce tissue damage. In World War I when soldiers spent long periods in the trenches with cold wet feet, a crippling deterioration of foot skin tissue called trench foot was a common occurrence.

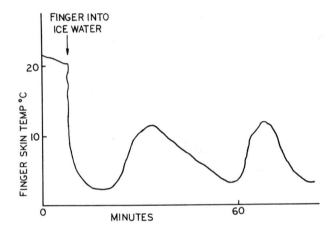

FIG. 21.7.   Representative adult Lewis waves of finger temperature during ice water immersion.

Studies on individuals with European backgrounds indicate that the intensity and duration of the vasoconstrictive response is modified by what many physiologists call habituation. In testing situations where the extremity is repeatedly exposed to very cold water or air, a number of responses change. First, the hand temperatures indicate that, over repeated exposures, the vasoconstriction response declines. Secondly, the rise in blood pressure that occurs in response to the pain and shock of the cold exposure also decreases with repeated testing. Thirdly, these short-term studies show that the reduced vasoconstrictive response tendency is restricted to the extremity being tested. Finally, retesting after periods of up to 30 days shows that the higher blood flow persists. This series of findings is generally interpreted to mean that the central nervous system gradually reinterprets the sensory input from the cold in such a way that return signals for vasoconstriction in the specific part of the body affected are modified. This interpretation has been further supported by studies on such groups of people as cold-water fishermen who maintain warmer hands in cold water than non-fishermen from the same villages.

Despite the strong evidence that habituation modifies the peripheral temperature-regulatory response of the individual, it does not appear to explain the bulk of individual variations in response. This conclusion is based on the fact that, even in a group of individuals who have not been exposed to significant extremity cooling during their lifetime, the response variability is much greater than any changes which can be produced by habituation.

Population differences in response to cold stress

A number of experimental studies have been conducted comparing the physiological responses of populations to various types of cold stress. The individuals tested were generally young men, and most often the studies involved a comparison of the responses of men from European genetic backgrounds with men from other ethnically defined populations. Considerable caution is called for in interpreting the results of these studies, since the European-derived individuals tested came from diverse cultural and natural environment backgrounds as did some of the comparative groups which were identified simply by racial or ethnic group.

Total body cooling

Table 21.3 briefly summarizes some of the larger sample studies in which the cold stress involved total body cooling. Most of the studies reported some differences between the groups, but investigators varied in interpretation of the causes. Some of the differences between the groups studied appear to have been a result of population differences in body size and composition. Often the European-derived sample had a larger body mass per unit of surface area and a thicker layer of subcutaneous fat. As a consequence, their skin temperatures were lower, their rectal temperature higher, and their metabolic response lower in comparison with the other group being studied.

A study comparing Australian Aborigines from the desert areas with white investigators showed quite a different pattern. In this study both groups were monitored over night, while they slept or attempted to sleep without significant insulation in conditions near freezing. Under these conditions the aborigines developed lower skin temperatures, lower core temperatures, and maintained lower metabolic rates than the investigators (Fig. 21.8). Hammel (1964), one of the investigators in the Australian studies, later combined these results with those from studies on the Bushmen of South Africa to suggest that populations whose evolutionary history was confined to the tropical savannahs might have a genetic adaptation to cold sleeping conditions, which had been lost or altered in the populations native to other climates. An alternative explanation for the

TABLE 21.3
Studies of total-body cooling

Groups	Sample	Format	Response Total metabolism*	Response Rectal T (°C)	Response Mean Skin T (°C)	References§
Inuit White	10 2	7 hr at 3°C Insulation = 2.9 clo	Inuit>White (30%)	Inuit>White (0.4°)	Inuit>White (1.5°)	Hart et al. 1962
Canadian Indians White	8 7	7 hr at 0°C Insulation = 3.1 clo	Indians>White (15%)	Indians>White (0.5°)	No Difference	Elsner et al. 1960
Lapp White	7, 5 5	7–8 hr at 0°C Insulation unavailable	Lapp>White	Lapp>White (1.0°)	Lapp>White (2.0°)	Lange Anderson et al. 1960
Quechua White	26 15	2 hr at 10°C Nude	Quechua> White (6%)	Quechua> White (1.0°)	Quechua> White (0.8°)	Baker et al. 1966
Alacaluf White‡	9	8 hr at 2–5°C Insulation = 2.5 clo	No Difference	No Difference	Alacaluf> White (1.5°)	Hammel 1964
Pitjandjara Australians White	6 4	6 hr at 0–5°C Insulation = 3.4 clo	Australians< White (20%)	Australians < White (0.5°)	Australians< White	Scholander et al. 1958
US Black US White	16 17	2 hr at 10°C Nude	No Difference	No Difference	No Difference	Iampietro et al. 1959

* Metabolism is measured in O_2 ml/minute.
‡ The control subjects cited here were from an earlier test.
§ For references see Steegman (1975).

absence of increased metabolism, or perhaps only a corollary characteristic, is the lack of conscious discomfort from the very cold skin temperatures the aboriginals experience.

Some of the other differences between the groups may be the result of a physiological acclimatization to continuous cold exposure. A debate continues over the extent to which cold acclimatization characterizes human beings. Some studies, particularly those involving the prolonged cold experienced by divers, suggest that under certain conditions resting metabolism and the metabolic response to cold stress may be increased.

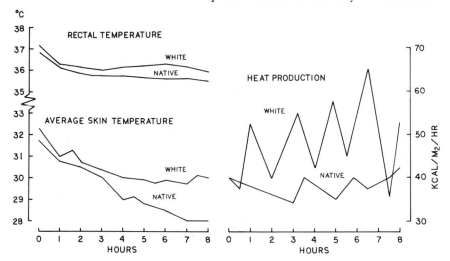

Fig. 21.8.   Comparisons of the thermal and metabolic responses of Australian aborigines and whites sleeping under cold conditions. (Modified from Hammel, in Dill *et al.* (eds), *Adaptation to the environment*, 1964, Am. Physiol. Soc., Wash. D.C.)

Extremity cold stress

Population comparisons of responses to localized cooling of the extremities are much more numerous than those involving total body cooling. Table 21.4 presents some of the study summary data developed by Steegman (1975). As the summaries suggest, a consistent difference has been found among the responses of people with African, European, and American native genetic heritages. Of these groups those with the most continuous vasoconstriction and lowest average re-warming trends were those with African heritage. It should be noted that no populations in Africa have been tested and the studies have been confined to US residents who classified themselves as blacks. From other data it must be assumed that many of these individuals also had European and American Indian ancestors as well as those from sub-Saharan Africa. It is, also, perhaps significant that an epidemiological study of frostbite casualties in US

TABLE 21.4

A summary of ethnic differences in response to hand or foot cooling*

Sample compared	Number	Cold testing procedures	Assessment of response	Results ($>$ = warmer; $\simeq$ = equivalent)	Citation
Adult M Eskimos, N. Alaskan	28	Fingers immersed in 0°C *water*, 30 min	Mean finger skin temp.; min. finger temp.	Native Alaskans > Europeans > Negroes; 4% of Alaskans failed to keep finger temp. >0°C; 63% of Negroes failed. Those with cold-injury history generally showed low response	Meehan 1955a
Adult M Indians, N. Alaskan	24				
Adult M Negroes, Am., South	38				
Adult M Europeans, Am., South	168				
Adult M Negroes, with histories of frostbite	9				
Adult M Europeans, with histories of frostbite	12				
Site: Alaska					
Adult M Negroes, moderately cold habituated	8	Hands exposed to −12°C *air*, 90 min (environmental room); rest of body exposed but clothed. One Summer, one Winter test	Finger temp.; other skin temp.; core temp.; CIVD; metabolic rate	Europeans > Negroes as to finger temp. Europeans showed more CIVD. This advantage was seen both Summer and Winter. Negro Summer ≈ Negro Winter; European Summer ≈ European Winter	Rennie and Adams 1957
Adult M Europeans, moderately cold habituated	8				
Site: Alaska					
Adult M Negroes, matched for physical characteristics	16	Fingers immersed in 0°C *water*, 45 min	Yoshimura's and Iida's criteria (see below)	Europeans > Negroes, all criteria	Iampietro *et al.* 1959
Adult M Europeans, matched for physical characteristics	17				

TABLE 21.4 (continued)

Sample compared	Number	Cold testing procedures	Assessment of response	Results (> = warmer; ≃ = equivalent)	Citation
Site: Massachusetts					
Adult M Negroes	17	Hands immersed in 5°C *water*, 30 min. Repeated after a 8-week cold acclimatization to 5°C air	Heat loss from hand; finger skin temp.	Europeans > Negroes, heat loss and skin temp. Results similar for post-cold acclimatization tests	Newman 1967
Adult M Europeans	22				
Site: Massachusetts					
Adult M Eskimos, hunters	22	Sub-studies with small sub-samples. Immersion of hand in 5, 10, and 20°C *water* for 2 h	Blood flow vol.; fluctuation of flow; skin temp	Under all cold-testing conditions Eskimos > Europeans relative to initial resistance to cooling, volume of extremity blood flow, CIVD, and skin temp.	Brown and Page 1952
Adult M Europeans, students	37				
Site: Central Canadian Arctic; Southern Ontario					
Young and adult F Eskimo	33	Immersion of hand in calorimeter starting with 0.5 to 1.5°C *water*, warmed by hand during 20-min exposure	Heat loss from hand to water; blood pressure rise ('cold pressor response'); CIVD as judged by finger temp.	Greenland Eskimo > European lumberjacks in temp. and flow values	Lund-Larson *et al.* 1970
Young and adult M Eskimo, hunters	39				
A group of European lumberjacks					

Site: Greenland, Norway

Highland Peruvian Indians, farmers	12	Foot immersed in calorimeter *water*; starting temp. 5, 10 and 15°C (each on a different day); water is warmed by foot heat loss during the 30-min tests. Subjects comfortably clothed	Skin temp.; heat loss from foot; difference between foot and water temp.	Indians>Europeans in initial resistance to temp. decrease but differences decreased with time. Indians lost more heat in 15°C and probably in 10°C baths, but not at 5°C. Overall Indian response was better impressionistically, not clearly statistically	Little 1969
Europeans, University people	12				

Sites: Peru, Pennsylvania

M Highland Peruvian Indians farmers	41	Hand exposed to 0°C, moving *air*, 60 min. Subjects comfortably clothed and tested twice each	Skin temp.	Lowland Indians≈Highland Indians. Indians>Europeans, especially as to finger temp., but may have become ≈ had tests been longer	Little *et al.* 1971
M Lowland Peruvian Indians, farmers	10				
M Europeans, scientists	8				

Site: Peru

Young adult Japanese (cool or cold climate)	98	Immersion of left index finger in 0°C *water*, 30 min	Index based upon timing and degree of CIVD by finger temp.; mean finger temp.	Cold climate Chinese>Cold climate Japanese. Warm climate Chinese≈Warm climate Japanese	Yoshimura and Iida 1952
Young adult Japanese (warm climate)	36				
Adult Chinese laborers (cold climate)	33				
Adult Chinese laborers (warm climate)	27				

Site: Local

* Modified from Steegman (1975).

soldiers during the Korean War showed that the incidence and severity of frostbite was much greater in blacks even when social factors, previous cold exposure, and conditions at the time of frostbite were statistically controlled.

The studies of native American populations indicate that individuals of this background tend to maintain a higher hand or foot temperature during cold exposure than individuals from a European background. This seems to occur even when their lifelong experience has not involved significant cold stress. Matched comparisons between Arctic American Indians and Inuits suggests that the Inuit may maintain the highest peripheral skin temperature of any group tested when challenged by localized cold stress.

The physiological findings from these studies have been interpreted by many investigators as evidence of a genetic adaptation to the severe extremity cooling which occurs in the colder world climates. The evidence is, however, incomplete. No genes which could produce the observed differences are known, and as Steegman's (1975) careful study of Algonquin American Indians living in the Boreal Forest indicates, it is unlikely that even historical studies of living populations will reveal how selection for such a response might have occurred.

Cultural adaptations to cold and biological adaptability

Given the very limited human ability to acclimatize to cold and the evidence for only minor genetic adaptations to cold, it is obvious that a combination of material culture inventions and behavioural adjustments were the principal adaptations which allowed our ancestors to spread and survive over most of the world's climates. Basically, the important material culture items used were fire, shelter, and clothing. While all contributed to reducing temperature stress, the variety of ways in which they were constructed and used variously affected the biology and health of populations. As an example, the use by Melanesian groups of smoky fires within shelters for night-time protection against both cold and mosquitoes resulted in exceedingly high rates of respiratory disease, reducing the work capacity of young adults and increasing respiratory death rates. It is also quite possible that the practice found among populations in North Eastern Siberia of families sleeping in insulated boxes may have subjected them to the stress of oxygen deficiency (hypoxia).

Of even greater interest to the problems of human adaptability are the ways in which population variation in physiological response to cold relates to cultural adaptations. The Australian aborigines' method of coping with the night-time cold of the central desert was to build a windbreak and then sleep nude between fires, which were maintained periodically during the night. Given the extreme cooling of arms and legs which characterize their response to sleep in the cold, it has also been observed that these limbs are very insensitive to pain. As a consequence, they frequently received quite serious burns.

The most suggestive evidence that the cultural adaptations to cold may have produced genetic adaptations is the Inuit and American Indian response to extreme hand and foot cooling. The Inuit had a broad array of material culture items which protected them from extreme cold, but the most impressive was their highly sophisticated clothing. This clothing can prevent a serious decline in core body temperature under almost all arctic winter conditions. However, no insulative clothing known can prevent hand and foot freezing without a constant high heat input from the body core. Even moderate exercise provides sufficient blood flow to transport the heat, but total inactivity, such as is required for seal hunting or resting, will not. As demonstrated in many studies, the Inuit have the greatest heat flow to the extremity during localized cooling with American Indians ranked second. Considering that the ancestors of all American Indians passed through the Arctic, it may well be that the invention of effective cold-weather clothing contributed to altering the genetic structure of American natives by providing a micro-environment which favoured individuals who maintained high peripheral blood flow during cold exposure. Clearly, this would have been a reversal of the warm savannah situation which favoured individuals with maximal vasoconstriction.

Altitude tolerance

Increasing terrestrial altitude results in a number of changes in the physical environment including decreasing water vapour pressure and increasing radiant energy penetration. Of particular importance to life in all forms are the nearly linear declines in air temperature and atmospheric pressure. Air temperatures in the tropics are reduced to an average near 0°C at 5000 m, and are proportionately reduced in relation to sea-level values at higher latitudes. Throughout all latitudes atmospheric pressure at 5000 m is only one-half of sea-level values. As shown by the ascent of Mount Everest without the use of bottled oxygen, some individuals can at least briefly tolerate the highest of terrestrial elevations. However, in a study of mountain climbers residing at about 5700 m for 6 months, it was found that health deteriorated so rapidly that the investigator in charge concluded that all of the climbers were rapidly approaching death. The principal difficulty is the reduced partial pressure of oxygen which probably restricts permanent human populations to residence under 5000 m.

Despite the apparent stresses of the high-altitude zones, the three major high-altitude plateaus of the world in Ethiopia, South America, and Tibet had by the fifteenth century all developed major civilizations with large populations. Today the same areas contain well in excess of 25 million people living above 2500 m.

Hypoxia

While both atmospheric and oxygen pressure decline in a near linear manner over terrestrial altitudes, the effects on people are not linear. Only very slight physiological effects from oxygen pressure decline have been found in healthy individuals at altitudes below 2500 m. This is directly related to the oxygen-binding characteristics of haemoglobin. As illustrated in Fig. 21.9, at sea level the oxygen pressure is high enough so that haemoglobin becomes 97 per cent saturated with oxygen. After circulating blood has given up its oxygen to the tissues, it returns to the lungs 40 per cent saturated. In this process each 100 ml of blood delivers about 5 ml of oxygen. As the oxygen pressure of the air in the lungs falls with increasing altitude, the oxygen saturation of the blood falls and provides less oxygen per unit of volume. As shown the saturation capabilities of haemoglobin decrease in a curvilinear relationship to oxygen pressure. Thus, the blood leaving the lungs is still near full saturation at 2000 m, but by 4000 m blood oxygenation is substantially lowered.

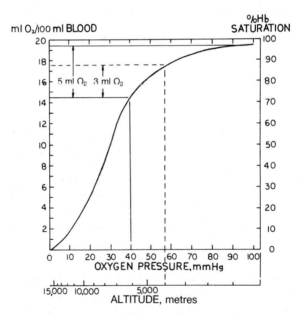

FIG. 21.9.   The oxygen dissociation curve of haemoglobin showing the amounts of oxygen which can be carried by blood at different oxygen pressures. (Redrawn from Little and Morren 1979.)

For low-altitude natives the immediate response to the hypoxia encountered at high altitude is an increase in breathing rate and a rise in heart rate. These responses partially compensate for the decreased pressure by increasing the partial pressure of oxygen in the lungs and by increasing the blood flow. Unfortunately for the comfort of the individual these changes also contribute to some of the symptoms reported by newcomers to high altitude. The increased heart rate is distracting and makes sleep difficult. More importantly, the increased lung ventilation decreases blood. CO_2 content triggering alternating bouts of hyper- and hypo-ventilation with headaches, possible tunnel vision, and even fainting. One of the more distressing respiration-related symptoms common after rapid ascent to altitude is the Cheyne–Stokes syndrome during sleep. The breathing rate oscillates with increasing severity often producing nightmares until the individual awakes with feelings of suffocation. These various symptoms have a frequency of 50 per cent or less at ascents up to 4000 m and tend to disappear within a few days as respiration cycling declines in response to a still poorly understood acclimatizational response.

A more persistent physiological response to the lower oxygen pressures at high altitude is the reduction in maximal oxygen consumption capacity (max VO_2/min). This measure is obtained by a work test which rapidly pushes individuals to the limits of their energy expenditure and measures the absolute amount of oxygen the body can obtain and utilize from the environment. It has proven the best measure of sustained work capacity at any altitude; when calculated per unit of body weight (max VO_2/min/kg), it is a good measure of individual ability to extract oxygen from the air and deliver it to body cells.

The summary of representative studies presented in Table 21.5 illustrates that most people lose between 20 and 30 per cent of their aerobic capacity when they go from low altitude to altitudes between 3500 and 4500 m. This loss is recovered if they return to low altitude. Whether they will typically recover their previous abilities after lengthy residence at high altitude remains open to debate. A stay of 1–2 months appears to have little effect and studies of residence up to 1 year have failed to show significant improvement in work capacity, but cross-sectional studies suggest that perhaps improvement does occur after some 5–20 years. One study (Frisancho et al. 1973) suggests that growing up at high altitude (developmental acclimatization) may be critical for developing aerobic capacity capabilities near sea-level values. Only longitudinal studies measuring this ability over many years are likely to resolve this question. The answer may be a very important one for explaining how so many people and their societies have prospered in the high-altitude zones of the earth.

TABLE 21.5

Change in oxygen consumption capacity upon upward migration

Population	N	Sex & age	Max VO$_2$ 1/min/kg STPD		% Decrease.	Reference
			Low altitude	High altitude		
US white Researchers	12	Male 27	50.4 (300 m)	38.1 (4000 m)	24.4	Baker 1969
English Mountain Climbers	4	Male 32	50.0 (sea level)	39.7 (4000 m)	20.6	Pugh 1958
US white Soldiers	24	Male ?	40.4 (sea level)	32.1 (4300 m)	20.5	Consolazio 1966
Peruvian Sailors	10	Male 18–21	39.0 (450 m)	31.4 (4500 m)	19.5	Velasquez 1964
US white Runners	6	Male 20	64.2 (300 m)	46.6 (4000 m)	27.2	Buskirk *et al.* 1967
US white Runners	5	Male 15–17	65.2 (300 m)	49.4 (3100 m)	25.1	Reeves *et al.* 1967
Peruvian Quechua (H.A. heritage)	10	Male 22	49.3 (100 m)	44.5 (4000 m)	9.7	Baker 1969

From Baker in Salzano (ed.), *The ongoing evolution of Latin American populations*, 1971, C.C. Thomas, Springfield.

High-altitude people

Most of what we know about the effects of the high-altitude environment on the health and biology of lifelong residents is derived from the many studies of people on the Andean high plateau. Recent studies in Ethiopia, Asia, and North America suggest, however, that either some traditional views of altitude effects are incorrect or human populations vary in their responses to low oxygen pressure.

To date, no studies have verified the frequent claim that some people native to the high mountains of Eurasia and South America live to unusually old ages. On the other hand, there is also no evidence that the physical environment shortens life expectancy. The possible effects of high altitude on fertility remain open to debate (Clegg 1978). Temporary changes in human testicular and ovarian function have been observed in new migrants to high altitude. Downward migrants may be more fertile since time between births is shorter than at high altitude. A number of studies both in the Andes and North America show that pregnancy at altitude results in newborns some 10 per cent lower in birth weight. Nevertheless, some high-altitude groups have quite high completed fertilities with women past the reproductive age reporting that they had borne an average of eight live children.

Studies of Andean natives show that high altitude reduces child growth rates in height particularly during the rapid-growth periods of infancy and adolescence. This reduction in growth rate does not appear to affect adult height because of a prolongation of the growth period. It also does not affect all parts of the body, since chest size seems to develop at the same rate in both low-and high-altitude children. In fact, some studies suggest that chest growth may be more rapid in high-altitude people because of the stress on respiration. In Asia child growth appears to be slowed by altitude in the Tien Shan Mountains, but not in Nepal. In Ethiopia high-altitude children grew faster than low-altitude ones. Given the major retarding effects that undernutrition and some kinds of deficiency diseases have on growth, it remains unclear whether the hypoxia of high altitude reduces the growth rates of all children or only those from specific populations. It appears that the altitude retarding effect on growth is much less than the potential effect of undernutrition. If so, the contrasting results obtained in the various studies may be a result of the overriding effects of nutritional inadequacy in such populations as the Ethiopians.

While the migrant to high altitude shows a definite loss in max VO_2/min/kg, high-altitude natives in the Andes and the Himalayas have quite high values even at high altitude. As the study summaries in Table 21.6 document, the work capacities of some high-altitude populations may actually average higher than those found in some modern populations and are comparable to the most physically fit populations at low altitudes.

TABLE 21.6

Physiological responses of young men to maximal short term exercise

	Test location (m above sea level)	N	Mean age	Aerobic capacity VO$_2$ (ml/kg/min)	VO$_2$/H.R. (L/beat)
Natives of high altitude					
Peruvian (American Indians)[1]	4000	25	25	49.1	15.7
Peruvians (Mestizos)[2]	3400	20	23	51.2	13.9
Chileans (American Indians)[3]	3500	15	22	47.2	
Sherpa (active)[4]	3400	11	25	50.4	14.8
Sherpa (inactive)[4]	3400	13	25	45.2	14.5
Ethiopians (not specified)[5]	3000	15	14	34.4	
Natives of low altitude					
US men (active)[6]	S.L.	15	22	52.8	
Inuit (active)[7]	S.L.	21	25	50.0	
Samoans (inactive)[8]	S.L.	34	24	38.9	
Yorubas of E Africa (active)[9]	S.L.	23	25	55.5	
Bozo of W Africa (active)[10]	S.L.	29	21	45.6	
Twa (Pygmy of Zaire)[11]	S.L.	23	—	47.5	

(1) Baker 1969; (2) Frisancho *et al.* 1973; (3) Donoso (quoted in Baker 1971); (4) Weitz 1973; (5) Weitz 1973; (6) Buskirk and Taylor 1957; (7) Shephard 1980; (8) Greksa *et al.* 1986; (9) Davies *et al.* 1972; (10) Huizinga 1972; (11) Ghesquiere 1972.

For sources see Baker and Little (1976), Baker (1978), Shephard, *Human physiological capacity* (1980), Cambridge University Press, and Baker *et al.* (1986).

While the max VO_2/min of relative newcomers rises when they return to low altitude, no increase has been noted in natives when they migrate to lower altitudes. This contrast may be interpreted either as evidence of a long-term acclimatizational process or evidence that natives have developed some form of genetic adaptation to the reduced oxygen pressure.

It has been assumed in the past that the high red blood cell counts and haemoglobin values found in high-altitude natives might help explain their high oxygen extractive ability. If this were the case then their ability would be acclimatizational since not only do upward migrants show rises in red cell counts after a few months, but so do people at low altitude who are hypoxic as a result of such diseases as congestive heart failure and some lung disorders.

Contrary to this view is the fact that at very high red cell counts blood circulates more slowly because of its high viscosity. Thus, the current view is that while a red cell count, modestly higher than normal sea level value, may slightly enhance oxygen extraction abilities at high altitude, it is primarily an indicator of the amount of hypoxic strain suffered by the individual.

In summary, the high-altitude populations of the world appear well adapted to their environment in spite of the fact that temporary sojourners suffer a variety of symptoms and perhaps a permanent reduction in work capacity. While a short-term acclimatizational process contributes to native abilities, it appears that either developmental change or genetic adaptation is necessary for people to become fully functional at the higher elevations. As the graph (Fig. 21.10) developed by Buskirk (1978)

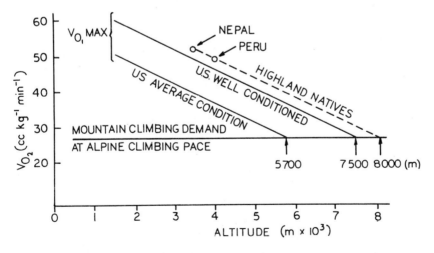

FIG. 21.10.   The ability of various population groups to extract the atmosphere oxygen required for climbing at various altitudes. (Modified from Buskirk in Baker (ed.) 1978.)

suggests, most of the people reaching the top of Mount Everest (Sagarmatha) are likely in the near future to be native residents of higher altitude regions of the world.

Other stresses of the physical environment

In this chapter only the stresses imposed by temperature and altitude have been critically examined. These represent only a small portion of the physical environmental stresses to which human population have in the past adapted. The stresses imposed by such forces as gravity and mild doses of short wave radiation are old ones. Contemporary phenomena, including high radiation and the concentration of gases, minerals, and new biochemical compounds, represent potential stress for which the past provides few guidelines. While it is clear that the adaptive capacity of human populations to stress contains a repertoire of adaptive mechanisms, our ability to cope with the removal of stresses such as gravity and to survive new toxicological threats is unclear.

22   Nutritional stress

Since the biological stresses on human populations are so complex, only two aspects of human adaptability will be explored. These are, first, our need to obtain our food from the biological environment (nutrition), and, secondly, the attempts by other organisms to use us as a living source of nutrition (infectious disease). In this chapter the responses of human populations to the problems of obtaining food energy and the structural components needed by the body will be examined.

Human energetics

In simplistic classification of eating behaviour, animal species are often termed herbivorous, carnivorous, or omnivorous according to whether they obtain their basic nutritional needs from plants, animals, or both. Since only plants have the ability to convert radiant energy into a form which can be used by living organisms, it is obvious that the energy available to the herbivore for the production of biological mass and energy expenditure is greater than it is for the carnivore. These energy transfers ecologists describe as energetic trophic levels. The long sequence of who eats whom among some carnivores results in a top carnivore species (end of the sequence), which has a minute total biomass compared to the plant base on which it ultimately depends. This loss of mass is a result of the fact that each trophic transition results in a major degradation of energy from a stored chemical form to a radiant form.

Within this schema, the omnivore would presumably have an advantage, since it could tap the energy at all levels of the trophic chain and produce a biomass exceeding any species restricted to a given trophic level.

For the omnivore *Homo sapiens* the classification system and concepts are too simplistic. In spite of our technological adaptations to the energy problem, many species, including the herbivore termite, vastly exceed us in biomass. This is the case because the human digestive system cannot extract food energy from most of the cellular structure in which plants store their carbon-bonded energy. What the human digestive system can utilize are only those plant parts which concentrate energy in the forms of carbohydrates, fats, and proteins. These include seeds, fruits, tuberous roots, and some plant leaves containing a few extractable calories and other essential nutrients.

Energy requirements

Fundamental to the assessment of human energy needs is basal or resting metabolism. This measure of energy use is determined while the individual is totally inactive and thermally neutral. It is considered an assessment of the energy conversion required for vital functions. It includes the energy used by such continuous functions as the neural, cardiovascular, respiratory, and liver functions, plus energy consumed for the survival of other body cells. The amount of energy expended in basal metabolism is higher during growth because of high cell multiplication rates and is high in women during pregnancy and lactation. Basal metabolism varies with other factors held constant according to body size. The basal metabolism of an average-size young adult male is in the range of 300 kJ/hr, but may be as high as 400 kJ/hr in pregnant and lactating women, and as low as 150 kJ/hr in young children. This results in individual energy requirements, which vary from 3600 to 9600 kJ per day in basal requirements.

Basal energy requirements are only a foundation on which the other requirements of living lay additional energy needs. Living in a hot or cold environment adds energetic requirements. In hot conditions the amount of extra energy required will depend in great part on the activity of the person, since muscular activity increases the heat load which must be dissipated through increased blood flow rates and sweating. It might be assumed that human energy needs in cold climates would occur because of heat loss. This may be so for some populations but recent studies suggest that in most populations the major cause for increased energy expenditure in cold climates is the increased energetic cost of wearing heavy and bulky clothing in the cold.

While these variables result in a significant variation in individual energy needs, the level of physical activity during waking hours is likely to produce the greatest source of variation among adults. The absolute energetic cost of most activities is determined to some extent by body weight so that the cost of a given task, such as walking, varies between individuals. As can be calculated from Table 22.1, a fit 70 kg young adult can expend energy at the rate of 2930 kJ/hr which is about 10 times his resting metabolic rate. Most adults, can expend energy the level of four times the basal rate for 8 hours or longer. As a result, total daily energy expenditure may vary between individuals of the same basal rate by more than 100 per cent.

Population requirements

Since the populations of the world vary substantially in average body size, child to adult ratios in numbers, and in the work requirements for adults, it is clear that they also have substantial differences in their food energy needs. Many attempts have been made to develop reliable estimates of

TABLE 22.1

*Rates of energy expenditure during various activities**

	Level	Cost above RMR (kcal/kg/hr)†	Example of activity
1.	Resting	0.00	Lying still
2.	Sitting 1	0.25	Sitting quietly
3.	Standing 1	0.50	Standing 'at ease'
4.	Sitting 2	1.00	Sewing, weaving
5.	Standing 2	1.50	Sweeping
6.	Walking 1	2.50	4.8 km/hr, no load
7.	Walking 2	3.50	4.8 km/hr, 15 kg load
8.	Heavy	4.50	Hoeing, ploughing
9.	Very heavy	7.00	Tree felling
10.	Extreme	10.00	Near maximum exertion

* From Leslie *et al.*, *Human ecology* (1984).

† 4.186 kJ = 1 kcal.

energetic needs for the populations of both small communities and whole countries. These models were often generated as part of attempts to assess the adequacy of dietary intake among groups or as a method for estimating the food requirements of regions or countries. Many of the models utilized recommended intakes developed in countries such as the United Kingdom or the United States. Since these recommendations were based on populations with a large body size and a low child to adult ratio, they often substantially over-estimated energy needs for populations in other world areas. A more detailed model, based on the variables described, suggests that, among many populations of concern in human biology, energy needs are lower than those estimated using models developed by the UN/Food and Agricultural Organization (Table 22.2). While the work requirements prescribed by cultures and climates create significant differences in a population's energy need, adult body size is also an important factor. This indicates that genetically determined population differences in average adult body size have a significant effect on the energy needs of populations.

Energy sources

The energy sources in human food are fat, carbohydrate, and protein. Fat provides about 38 kJ per g in energy while carbohydrate provides 17 kJ per g. The energetic content of protein is slightly higher than carbohydrate, but, because some of the energy is degraded to heat in the biochemical conversions necessary for human utilization, the available joules of effective energy yield are also about 17 kJ per g. The groups selected for illustrative intakes of these nutrients (Table 22.3) demonstrate that human populations have in the past survived on diverse sources for their energy input. The high Inuit energy input from fat is derived from

TABLE 22.2

*Energy needs in kcal for four populations estimated by two models and by food intake surveys**

(Values in kcal/person/day)

Population	Age and sex	Intake estimates	Leslie Model	FAO
Nuñoa (Thomas 1973)	All male	1632	1512	2211
	All female	1420	1435	1834
Kaul (Norgan et al. 1974)	Adult male	2347	2603	2970
	Adult female	1830	2146	2350
Lufa (Norgan et al. 1974)	Adult male	2570	2576	2970
	Adult female	2245	2070	2350
!Kung (Lee 1979)	Juvenile male	2000	1218	1953
	Juvenile female	2000	1178	1795
	Adult male	2250	2135	2477
	Adult female	1750	1818	1926

* From Leslie et al., *Human ecology*, 1984.

TABLE 22.3

Average daily intake per head of fat, carbohydrate, and protein

	British (1950's)		Kikuyu (1950's)		Eskimos (1855)		Tokelauans* (1972)	
	(g)	(kJ)	(g)	(kJ)	(g)	(kJ)	(g)	(kJ)
Fat (38kJ/g)	110	4180	22	836	162	6156	139	5282
Carbohydrate (17kJ/g)	400	6800	390	6630	59	1003	202	3434
Protein (17kJ/g)	100	1700	100	1700	377	6409	68	1156
Total	—	12 680	—	9166	—	13 569	—	9872

Data from previous editions of this book and (1) Prior et al., *Am. J. Clinical Nutr.* (1981).

animal fats, while the fats which provided a major energy source for the Polynesians on Puka Puka came from coconuts. The relatively low protein content of edible plant products means that only meat-eating populations are likely to derive a significant amount of their energy needs from protein, and, even in the extreme case of the traditional Inuit where meat provided most of the calories, fat provided almost as much energy as protein.

While carbohydrates provide the major energy source for almost all populations, some fat intake and more certainly a significant protein intake is essential. The exact needs for fat intake have not been determined, but it has been suggested that at least the intake of some unsaturated fats may be necessary for normal growth and development. The requirement for protein is a complex problem, since it involves the need for nitrogen and a variety of amino acids, which cannot be manufactured by the body. Plant proteins can provide the needed nitrogen, but natural food sources have incomplete complements of amino acids. Thus, it is only through an appropriate combination of plant protein intakes that an individual can obtain the necessary protein over the life cycle. On the other hand, animal sources of protein provide the needed amino acid complex, and it appears that small intakes are adequate when complemented by larger quantities of plant protein.

Although the need for significant nitrogen intakes has been physiologically documented along with the need for a balanced intake of amino acids, the minimum level needed for survival and reproduction remains uncertain. Evidence for malfunction in human beings as a result of low protein or inappropriate amino acid intake remains limited. The most prominent disease state which responds to improved protein intake is a relatively rare disorder named kwashiorkor, which appears in early childhood. This condition first described in South Africa occurs in post-weaning young children who have been nourished on almost pure carbohydrate. The symptoms include muscle wasting, oedema, lassitude, and hair and skin depigmentation, which are rapidly reversed by feeding with balanced protein. The disorder can be fatal without treatment.

A less serious, but more common disorder titled marasmus has somewhat similar symptoms. It appears to be caused by a combined protein-energy deficiency. Adults who have extremely low protein intakes are reported to show physical lassitude and mental slowness, but how this specifically relates to nitrogen and amino acid requirements remains poorly defined. The results of studies on New Guinea groups, who use the starch of the Sago palm as a source of energy, suggest that at least some human populations may have been able to survive on diets in which protein constituted as low as 4 per cent of total energy intake.

Energy balance in individuals

The balance between energy intake and expenditure in most people is controlled by one of the most efficient human homeostatic systems. It has recently been emphasized that this balance can be consciously modified to make the individual thinner or fatter, but it should be noted that the natural balance is maintained within remarkably close limits without conscious control. For example, one may consider the not uncommon instance of the adult male who without intent gains some 15 kg between the ages of 20 and 50. This appears to be a rather inadequate homeostatic response, since the adult is presumably not growing or significantly increasing in muscle mass.

Such an individual is increasing in weight by an average of 500 g per year over this time period, and it is possible to calculate roughly the energy imbalance involved in the change. Assuming the weight gained is all in body fat, then about 67 per cent of each kilogram of weight gained is fat, while the remainder is cell membrane and extracellular water. The amount of energy stored, then, for the weight-gaining individual is per year 500 g × 38 kJ × 67% = 12 730 kJ per year. Given that the average intake per day for an adult male in a modern society is about 12 680 kJ or 4 628 200 kJ per year, the difference between intake and expenditure per day would be only about 5 kJ or 0.27 per cent. Quite obviously this homeostatic error was of a rather minor nature even though the individual may have changed from a slender youth to a stout middle-aged individual.

The narrowness of this homeostatic control raises the question of how consciously controllable the adult weight gain or loss process can be. In a study of a West African group which valued large body weights, deGarine found that young men could gain up to 25 kg in 3 months, but to do so they subjected themselves to total inactivity and a compulsory energy intake of 28 000–50 000 kJ per day. Weight loss clinics in modern society have also experienced variable results with a low frequency of long-term retention of weight loss. It thus appears that the average weight of most human populations bears only a slight relationship to conscious desires, but is controlled by a still poorly understood interaction of genetic characteristics with environmental controls on energy expenditure and intake.

Energy flow studies

In recent decades ecologists, such as H. Odum (1971), have argued that an understanding of the energy flow through human societies would provide us with major insights on the origin and future of such societies. Even earlier some anthropologists argued that the increasing ability of our species to control and utilize energy was a key to understanding cultural development over the history and prehistory of human cultural change. More recently, several anthropologists have used the expenditure and intake of energy within selected human groups to explain certain aspects of their behaviour. Figures 22.1 and 22.2 show the results of two such studies which explored the energy exchange in contemporary communities.

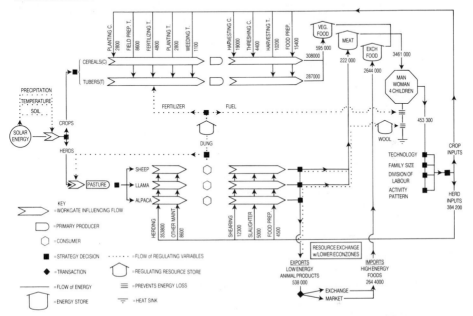

FIG. 22.1. The energetic flow of a typical Nuñoa American Indian family in 1969. (From Thomas, in Baker and Little 1976.)

Thomas (Fig. 22.1) expanded his study by noting that the community had, in fact, maximized its energetic yield per person by using individuals such as adolescents and women for such tasks as herding, which required constant attention, but minimal high energetic expenditure capacities. His results show that, while these highland pastoralists must trade their meat products for some food energy, they use some of the products produced to obtain other goods than food energy. Other modellers suggest that if the population he studied increased by about 30 per cent then all cash production would have to be spent on food energy sources to meet the population need.

Bedoian's study suggested quite a different situation for a community in southern Tunisia. In this population the possibility of earning cash income outside of the local system had a major influence. He concluded that many of the residents so valued their traditional lifestyle that they would forego the purchase of food energy and other items in order to invest in the lower

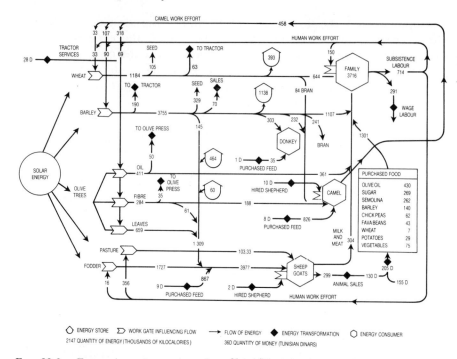

FIG. 22.2.   Energetic and monetary flows for a family in a rural village of Southern Tunisia. (Modified from Bedoian, *Energetic, ecological and cultural factors in economic strategies in Southern Tunisia*, 1979, PhD Diss., Penn. State University.)

energy production capabilities of their traditional pastoral and agricultural lifestyle.

Both of these studies have been criticized as not being representative of the broader societies from which they were sampled, and as not necessarily reflecting the behaviour in populations where food energy was more easily available. Nevertheless, the studies do show that two very different pastoral groups could not survive without a cash economy or the trading of food products with another society. The survival of human pastoral populations may for a long time have depended on an interchange with agricultural groups.

Studies of food energy production in other societies demonstrate a range of ratios between the amount of energy a group puts into food energy production and yields. Perhaps the highest yield found has been for maize in parts of the United States. For this specific food item the human energy input is only 1 kJ for every 1700 kJ available from the harvested products. Such high ratios are achieved by the use of stored energy in such forms as oil and coal to mechanically accomplish much of the work needed. In terms of the combined human and mechanical energy required to produce human food energy, traditional methods provide significantly higher yields per unit input for most crops.

Evolutionary ecologists suggest that such studies of behaviour at a population or societal level cannot provide insight into the causes for our food seeking and processing behaviour. They suggest that since behaviour has been determined by evolution, its function must be studied at the individual rather than the group level. Within this format the testing of optimal foraging models has concentrated on hunting and gathering populations. These studies indicate that individuals tend to apply some of the same strategies other animals use to obtain the most favourable energy yield from their behaviour. For example, the percentage of time an individual spends in hunting a given animal species may be predicted from a mathematical formula based on food yield in relation to the time required to capture and process the animal for eating. Nevertheless, as Bedoian found, some aspects of subsistence behaviour cannot be shown to be the most energetically efficient. Future research using optimal foraging models is made difficult by the fact that almost all traditional societies now obtain some of their food through the worldwide economic system and all are incorporating some aspects of modern food production technology into food production behaviour.

Other nutrients

In addition to meeting energy needs, the diet of an animal species must provide the basic material its genetic information requires to accomplish physical growth. It must also provide for continuous replacement of the inorganic and organic compounds lost in the processes of physiological functions. The nutrient intake required for healthy growth and function in individuals of our species depends upon age, sex, physical activity, temperature, and various disease states. Specific nutrient needs may also vary as a result of individual genetic differences. Selected hereditary anaemias, for example, result in iron intake needs many times greater than normal in order to maintain optimal haematocrit levels. For most species, the natural diet fulfils nutrient needs, and deficiencies occur only as a result of general undernutrition or an altered environment. This probably also applied to early forms of the genus *Homo*, but the geographical spread and dietary specialization of *Homo sapiens* has created not only individual problems, but also high frequencies of deficiency symptoms in some populations.

Water and minerals

Along with water, the list of minerals which must be ingested in some amount for normal growth and physiological function is quite lengthy.

The list of elements and minerals for which at least a small needed intake is discovered continues to expand with continuing research. Only in the early 1970s was it well established that small amounts of zinc are required for normal child growth. This was established by studies on children in regions of North Africa and the Near East where the soils used for agricultural production were unusually low in zinc content. Although instances of deficiency malnutrition for all such elements and minerals can be found in many populations, only water, salt, calcium, iron, and iodine will be examined in more detail to illustrate problems of adaptability.

Water. Water is not only the major structural component of the body, but is critical as a transport medium, including waste elimination. This use of water along with sweating as a cooling mechanism results in the human need for large water intakes. A loss of as little as 5 per cent of normal body hydration can produce death, and even less dehydration will impair many functions, including the brain's capability to process information. Such states of dehydration are found in infants as a result of severe diarrhoea and can occur, even in adults, as a consequence of restricted water intakes in a hot desert.

Within normal water intake variation, the kidney regulates body water content by dilution or concentration of urine. However, the physiological adaptation to water intake shortages may have a functional price. In hot desert populations, the frequency of kidney stones (calculi) is much higher than in populations from cooler climates where urine volumes tend to be higher.

Salt. Our evolution has provided us, along with most other land animals, with body water near the salinity level of the ocean at the time our predecessors came ashore. The sodium and chloride are both critical to many physiological functions. As a consequence, salt loss through urine and sweating must be restored. Salt loss can be quite high in individuals when they are first exposed to hot climates, but, as previously described, heat acclimatization substantially reduces salt loss through sweat. Even during the initial heat exposure period, incrased salt needs are modest. Studies of isolated tropical forest populations show that at least some populations in hot areas obtain adequate salt for normal functioning from food sources such as meat.

Despite the relatively low need for salt in its pure form, large intakes appear to have relatively few health risks for young adults. Curiously, salt has been the most sought after food supplement known. In areas where salt was not available, the desire for it appears to have been one of the strongest incentives for the development of trade, and even an excuse for war. The strong taste and desire for salt does not appear to be based on a shortage in our earlier diet as hunters and gatherers, but may be related to some aspect of our earlier heritage as a tropical savannah animal exposed to high heat stress and/or a low meat diet.

Calcium. With a skeletal support based on calcium hydroxyapatite, people need substantial calcium intake for growth. Because there is also a rather steady turnover of calcium in the bone even during adulthood, a continuing intake is necessary. A major source of calcium in food is milk and cheese. The eating of animals, especially when it includes the periosteal tissue and some bone, provides considerable calcium. Vegetable foods tend to be low in calcium, and, in some, the calcium is biochemically bonded in such a way that the body cannot use it. The diets of many populations that rely heavily on vegetable products were once thought to be calcium deficient, but more recent discoveries make it apparent that such deficiencies must be rare. The traditional diets of highland South American and Mesoamerican Indians were considered calcium deficient, since the basic foodstuffs showed a singular lack of calcium. Later research showed that in the South American region burnt limestone was used as a spice in children's food, and the same substance or plant ash was used with the coca leaves chewed by most adults. These practices provided significant amounts of usable calcium. In Mesoamerica the staple food, maize, was often prepared by soaking in lime water, which greatly enhanced the available calcium in the ingested food. How such practices arose is unclear, but it is doubtful that it was a conscious effort by the populations to fulfil a potential nutrient requirement.

There also appears to be a homeostatic or perhaps reversible adaptive response to low calcium intakes. Studies conducted in Europe and the United States found that calcium absorption in the intestine was relatively poor. Thus, intake standards were established at several times growth and replacement needs. More recent research on low-intake individuals and groups show a much higher absorption rate. The conditions, time, and age at which such changes in absorption will occur have not been fully documented.

Based upon the findings that there may be many non-food sources of calcium in traditional diets and the increased absorption ability found in many groups, it is doubtful that calcium shortage has been common. The previously recorded high frequency of rickets and adult osteomalacia now appears, in most populations, to be better explained by shortages in the vitamin D necessary for calcium utilization, and possibly other factors, such as hereditary or environmentally caused anemias which affected bone structure.

Iron. Red blood cells are among the shorter-lived cells in the body, and the recycling of the essential iron in the cells can involve some loss. The loss is particularly obvious for women through the menstrual cycle, and the loss to the placenta and foetus during pregnancy. As with many other nutrients, a significant meat intake provides an adequate iron intake from the stores contained in the animal, particularly when the blood itself is ingested or when such blood-rich organs as the liver and heart are eaten. While some leafy vegetable foods contain iron, such food staples as grains are very low in iron content.

A lack of sufficient iron stores produces anaemia which is defined as an abnormally low haemoglobin or haematocrit level. While medical practice uses a demarcation level in these variables to indicate a deficiency state, the symptomology attending anaemia does not show a similarly sharp demarcation. The effect of low blood iron is a reduction in oxygen transport capabilities so that the effects include reduced work capacity, and in the extreme states, noticeable lassitude. At least a moderate level of anaemia is found in many populations who derive most of their energy from grain crops. In addition, many of the hereditary abnormal haemoglobins, which are common in warm climate areas, result in an extreme form of anaemia when iron intake is low.

While an excess intake of many nutrients is either not absorbed or harmlessly excreted, excess iron intake can result in health problems. The soft iron cooking pots traditionally used in parts of Africa imparted a high iron content to the foods cooked in them. The continual use of these pots led to a very high iron intake in adults, resulting in frequent liver damage and to early death. The very high meat content in the traditional Inuit diet may, through its iron richness, explain the finding that many early middle-aged Inuit suffered from enlarged livers and cirrhosis of the liver.

Iodine. Among the most puzzling nutritional deficiency diseases found in human populations is goitre, and its more severe concomitant myxoedema. Goitre, which is often manifest as a massive growth in the lower throat, is a result of the enlargement of the thyroid gland. The accompanying symptomatology, depending on the amount of thyroxin produced by the gland, may range from an extremely high basal metabolism, body thinness, and extreme anxiety to subnormal basal metabolism lassitude, and even subnormal mental alertness. The children of mothers afflicted with thyroid malfunction often suffer from myxoedema which results in mental deficiencies and growth anomalies, including reduced adult size.

As shown in Fig. 22.3 goitre was widespread from the early 1900s. After it was discovered that low iodine intake was correlated with the disease, an iodine supplement to the diet was found to reduce symptoms. Later results showed that iodine supplements could eliminate the disease when administered throughout life so that many countries supplemented salt with iodine in order to eliminate goitre, and its consequences. The oceans of the world contain substantial iodine; thus, people who used sea-

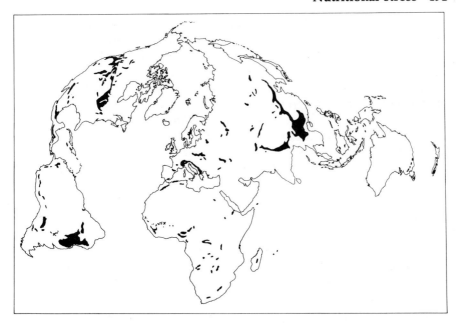

FIG. 22.3.   The worldwide distribution of goitre in the early 1900s.

produced food or agricultural foods from near the sea did not develop goitre and cases were concentrated in selected inland areas.

Iodine supplements in the diet have proved an effective preventative so that there is no doubt that a sufficient intake of iodine will prevent this form of thyroid gland malfunction in all individuals. Nevertheless, the study of the populations not reached by iodine supplementation during the first half of this century, leaves many questions unanswered. First, the presence of abnormal thyroid function was not predictable so that high frequencies could appear among the people of one mountain valley but be totally absent in the next. Secondly, within affected populations whole families were often unaffected while other families showed high frequencies of the disease. Several explanations were proposed including genetic differences in iodine need and the presence of goitrogenic factors in the diet which interfere with iodine utilization. While the exact reason for individual and locational variation in symptoms remains unknown, the worldwide distribution of this deficiency disease suggests that all of the major world population are susceptible to the effects of inadequate iodine intake.

Vitamins

The discovery of some vitamins and their essential role in human nutrition during the first half of the twentieth century was stimulated by the appearance of symptoms which could be cured by a diet supplement with specific foods. Since that time, a number of additional vitamins have been identified, but correlated symptoms traceable to deficiency states have been rare. This lack of symptoms is a result of the facts that either the supply is adequate in almost any diet or that the body is able to manufacture its needs from other biochemical compounds, which are readily available in the diet. The vitamins which have produced significant frequencies of serious ill health and death are the most pertinent to the problems of human adaptability. Table 22.4 lists these vitamins and some of the symptoms associated with inadequate intakes. Because of the differences in the role these vitamins have in biochemical action, they are separated according to whether they are water or fat soluble.

TABLE 22.4

Vitamin deficiency symptoms in people

	Deficiency symptoms
Water soluble vitamins	
Thiamine	Beriberi: chiefly neural and cardiovascular symptoms including loss of ankle movement and knee jerks; mental confusion, peripheral paralysis and heart enlargement. Oedema in wet beriberi and muscle wasting in dry beriberi.
Niacin	Pellagra: dermatitis, especially in areas exposed to sunlight; diarrhoea, mental confusion and delirium, with psychotic behaviour in severe cases.
Ascorbic acid	Scurvy: swollen bleeding gums with tooth loss, poor wound healing, subcutaneous haemorrhages, and joint swelling.
Fat soluble vitamins	
Vitamin A	Night blindness: hyperkeratinization of epithelial tissues, poor growth in children.
Vitamin D	Rickets: decalcified bones in children and permanent bone deformation such as bowed legs, osteomalacia in adults.

Thiamin. The thiamin-deficient diseases of wet and dry beriberi had been recognized for centuries, but it was only in the late nineteenth century that a surgeon in the Japanese Navy demonstrated these diseases to be nutritional in origin. He demonstrated that the addition of whole grains to the customary ration of polished rice and dried fish was a preventative. It was later found that the rice husks contained adequate thiamin, but that rice polishing which is common in rice-based diets removes the husk from the eaten form. The problem of thiamin deficiency may have been particularly acute in populations such as the coastal Japanese where fish is eaten raw. It has been found that both fresh and salt water fish contain an antithiamine compound called thiamase, which tends to render the vitamin inactive. Thiamase is, however, thermolabile so that cooking fish destroys the antithiamine action.

Niacin. Altough pellagra had reached epidemic proportions in several populations whose basic energy source was maize, it was not until the mid-1930s that the specific nutrient deficiency was identified. As with beriberi, the disorder caused serious disability resulting in psychotic behaviour and death. The exact reasons why a maize-based diet is associated with pellagra remain unclear. It is believed that maize may contain an anti-niacin compound which substantially increases the amount of niacin which must be ingested to meet needs, but such a compound has not been identified. The lower incidence of pellagra in some parts of Mexico led to the hypothesis that the treatment of maize with lime in this region might free the niacin, but experiments have failed to support this hypothesis. A genetic adaptation within the human population or the presence of an undetected critical nutrient in the diet both remain viable explanations for these pellagra groups.

Ascorbic acid. A high frequency of minor scurvy symptoms was probably common in the colder regions of the world where fruits and vegetables were seasonally unavailable. The severe symptoms and deaths arising from this disorder were not described until long-distance sailing became common. Lacking refrigeration, any fresh source of ascorbic acid rapidly deteriorated on long voyages in the tropics. The body store of ascorbic acid, which averages about 1500 mg, is catabolized at the rate of about 4.5 mg per day. As the symptoms of scurvy do not generally appear until the pool is down to about 300 mg, it appears that these should be enough to last nearly 9 months. In fact, some individuals have developed symptoms in as short a time as 30 days, while historical records show that voyages longer even than 9 months left some few individuals symptom free.

Vitamin A. Beyond meat, the sources of vitamin A are limited to plants containing coloured carotene pigments. As a result, many populations in the world suffer high frequencies of mild Vitamin A deficiency. Since more severe deficiencies are known to affect growth and reproduction in some mammals, it may have similar effects in human population. To date, no populations have been identified as suffering from these serious deficiency symptoms. In contrast to the water soluble vitamins where excessive intake is usually not harmful, excesses of both vitamin A and D can produce serious symptoms. The effects of excess vitamin A intake were first identified in arctic explorers who feasted on Polar bear liver, which contains nearly 600 mg of retinol per 100 g. The results can be fatal as a result of greatly increased intercranial fluid pressure.

A less acute complex of symptoms, including anoxia, vomiting, and irritability, occur with chronically high intakes. Among the many other symptoms of chronically elevated intakes is liver enlargement. The high frequency of enlarged and poorly functioning livers, which has been reported for some Inuit groups, may, therefore, have been related to high vitamin A intake in their traditional meat diet as well as the high iron intake previously described.

Vitamin D. The importance of vitamin D in preventing rickets has been noted elsewhere in this section, but it should be added that there are several forms of vitamin D. The plant form D_2 and the synthesized animal form D_3 both possess equal biochemical activity for people, but the animal form is much more potent in curing rickets.

An excess of Vitamin D can also be toxic. Mild excesses produce hypercalcaemia, while severe toxicity can create supravalvular aortic stenosis and mental retardation. It has been suggested, on this basis, that the heavy skin pigmentation found in most tropical populations might have been genetic adaptation to prevent excessive vitamin D_3 production, but convincing evidence that less skin pigmentation would result in toxic symptoms is still lacking.

Comparative dietary and nutritional adequacy

The diet of our early precursors cannot be directly determined; instead, we can only speculate on its composition, based on the diet of our nearest relatives, the apes, and the preserved evidence of the earlier hominid forms. The arboreal specialization of the orangutans and Hylobates make their diets unlikely sources of dietary information about our primarily ground-dwelling ancestors. The diet of the gorilla also seems an unlikely candidate, since gorillas are tropical forest rather than tropical savannah dwellers, and have both masticatory and digestive systems which allow them to obtain more nutrients from plant leaves and stalk than we can. The chimpanzees, who are morphologically closest to people, also have the diet closest to our own. As with many human hunters and gatherers, the chimpanzee's basic food sources are seeds and fruits. They also dine on a variety of insects, and apparently enjoy a limited amount of meat. Indeed, observation of their behaviour in natural conditions suggests they hunt for small animals, and may have a rather ritualistic manner of sharing the larger catches.

The teeth and jaw structures of the earlier hominids offer the best clues to their diet. In general, the teeth became smaller in the more recent forms, but did not show high degrees of specialization. Some of the earlier forms developed teeth which are presumed to indicate a more vegetarian diet. These finds are currently assigned by most palaeontologists to evolutionary lines which became extinct.

It was commonly thought, until quite recently, that during the evolutionary process hunting became an increasingly important food source so that, by the time *Homo sapiens* appeared, meat was the major food in the diet. The reasons for this belief were multiple. First, most of the artifacts found were utilized for hunting and animal slaughtering and, secondly, early art in caves and even the evidence for religion focused on hunting. Even in the contemporary world, most populations prefer a high meat diet when possible. The fact that our morphology did not particularly suit us for killing animals and digesting them has presumably been solved by tools, technology, and the mastery of cooking.

While hunting and meat eating were undoubtedly important in the early human diet, recent research makes it appear that meat was never the major energy source at any stage of human evolution. Aside from the morphological continuity of the digestive system, our current physiology requires certain nutrients, which cannot be obtained from a total meat diet. In addition, the energetic studies of hunting show that, even with the technology and techniques of the recent past, hunting is a very inefficient source of obtaining food. Finally, studies of recent hunting and gathering societies show that most of their food comes from non-meat sources. As to why earlier investigators focused on hunting as a source of food, it has been suggested that as men they failed to observe the women who brought home most of the food.

Hunters and gatherers

One of the first technologies to be modified by hunters and gatherers when they come into contact with technologically complex societies is that associated with food production. In addition, such groups quickly add imported food items to their diet when regular trade is established. For these reasons, it is doubtful that quantitative nutritional information on the diets of hunters and gatherers before the development of agriculture will ever be precisely reconstructed. Table 22.5 shows an attempted reconstruction of the nutrients in an Inuit summer diet based on an historical record of food intake. This diet, if correct, is adequate in some regards, but perhaps excessive in fat soluble vitamins. It lacks fibre and must have been even lower in carotene during the winters. Inuits are notable for their low life expectancies prior to European contact, and particularly the lack of individuals over age 50. Aside from the dietary factors noted this may have been partially a result of the extremely rapid tooth wear resulting in meat mastication problems by middle age.

The diet of some San in southern Africa remained relatively unaffected by external factors throughout the 1960s. The diet of the !Kung San (see Table 22.6) has been partially analysed for nutrients. For this group, meat contributed less than a third of the daily energy intake, and the foods were diversified in spite of a heavy dependence on the mongongo nut during the season of the dietary survey. Lee (1979) noted that, in the food category of other vegetables, over 20 species of roots, melons, bulbs, and dried fruits

TABLE 22.5
Nutritional value of the adult Eskimo daily diet in 1855*

Foodstuffs	Weight of edible portion (g)	Energy intake (kJ)	Total protein (g)	Animal protein (g)	Carbohydrate (g)	Fat (g)	Iron (mg)	Calcium (mg)	Phosphorus (mg)	Vitamin A (i.u.)	Carotenoids (µg)	Vitamin D (i.u.)	Thiamine (mg)	Nicotinic acid (mg)	Riboflavin (mg)	Ascorbic acid (mg)
Sea flesh	860	7100	163	163	26	103	23.2	95	1686	7740	0	0	0.95	42.1	1.2	69
Other flesh	225	1855	43	43	7	27	6.1	25	441	2025	0	0	0.25	11.0	0.32	18
Capelin (salmon)	620	2710	105	105	0	19	6.2	155	1500	508	0	6144	1.30	4.61	0.87	56
Other fish	370	1862	61	61	0	11	3.3	67	699	56	30	0	0.15	8.5	0.18	7
Eggs	5	34	1	1	0	1	0.1	3	10	35	27	3	0.01	0	0.02	0
Berries	50	60	0	0	3	0	0.6	30	22	0	0	0	0.02	0.2	0.02	45
Bread	27	269	2	0	13	0	0.3	6	20	0	0	0	0.01	0.2	0.01	0
Barley and peas	6	60	1	0	2	0	0.4	11	19	0	0	0	0.03	0.1	0.02	0
Sugar	6	100	0	0	6	0	0.	0	0	0	0	0	0	0	0	0
Coffee	6.5	80	1	0	2	1	0.3	9	10	0	0	0	0.06	0.6	0	0
Total	—	14 130	377	373	59	162	40.5	401	4407	10 364	57	6147	2.78	108.8	2.64	195
Oxford Nutrition Survey Standard	—	12 600	72	36	432	102	10	750	1000	833	3000	200	1.2	12	1.8	30
Proportion of standard met (per cent)	—	112	524	1036	14	159	405	54	441	1244	2	3074	232	907	147	650

* After Sinclair (1953).

were included. A similar diversity of foods has been reported for most other hunters and gatherers. It thus appears that the diets of hunters and gatherers must have been nutritionally balanced and the common deficiency diseases found in more recent populations were absent or quite rare. Such a conclusion does not preclude the frequent occurrence of undernutrition and starvation.

TABLE 22.6

Estimated calorific and protein levels in !Kung diet: July–August 1964*

Class of food	Percentage contribution to diet by weight	Per capita consumption		Calories/ person/ day
		Weight (g)	Protein (g)	
Meat	31	230	34.5	690
Mongongo nuts	28	210	58.8	1365
Other vegetables	41	300	3.0	300
Total	100	740	96.3	2355

* After Lee (1979).

Agriculturalists and pastoralists

As suggested by !Kung San dependence on mongongo nuts, many hunting and gathering groups used single naturally occurring plant or animal species as a primary source of food. Often these species were so concentrated that dense human populations with rather complex technology developed. Examples in North America included the acorn-based groups of California, the wild rice harvesters of Minnesota and Wisconsin, and the salmon-eating societies of the North-west Coast. All of these populations continued to rely on supplemental hunting and gathering to provide a nutritionally balanced diet. It is probable that nutritional deficiency diseases first became common in some of these populations when the population depleted the supplemental foods.

Early agricultural societies with a limited repertoire of domesticated plants also continued to depend on wild plants and animals for certain essential nutrients. Only a few populations in the world achieved an adequate repertoire of domestic food sources so that foraging was unnecessary. As might be surmised, only these populations developed densities large enough to support large villages, craft specialization, and the beginnings of urbanization and civilization.

The nutritional adequacies of the plant and animal complexes were not uniform, and the types of nutritional deficiencies, which were likely to be manifested in times of shortage, appear to have varied according to the basic crops. Based upon the probable origins of domesticated crops and animals and the archaelogically recovered remains from various sites, the basic traditional food sources in some regions of the world are shown in Table 22.7.

TABLE 22.7
Some food complexes in the world as of AD 1500

Region	Domestic plants	Domestic animals
North China	Millet, oats, soyabeans, apricot, peach, radish	Pigs, chickens, cattle
South China and South-east Asia	Rice, soyabeans, bananas, coconut	Chickens, cattle
South-west Asia	Wheat, barley, peas, carrots, dates, olives	Goats, sheep, cattle, camels
Mesoamerica	Maize, kidney beans, squash, tomatoes	Turkeys, dogs
Andean Area: Highlands Lowlands	Potatoes (white), chenipodia, maize, sweet potatoes, tomatoes, lima beans	Llama, alpaca
Africa — South of the Sahara	Millet, oil palm	Chickens
Melanesia and Polynesia	Taro, sweet potatoes, breadfruit, coconut	Chickens, pigs

As suggested in the lists, the densely populated centres of Eurasia had a diversity of crops and domestic animals. The basic carbohydrate sources of wheat and millet contained substantial plant protein, and a variety of fruits and vegetables potentially provided adequate minerals and vitamins. The rice-based food complex, extending from Japan and Korea, south to Indonesia, and east to parts of India, supported a substantial population, but had the potential to produce a vitamin B1 deficiency. The deficiency may have been common because polished rice is quite deficient in B1, and other sources are in short supply without a meat supplement. European and American prisoners of war during World War II and the Korean War developed high frequencies of beriberi on the diets provided, although it appears that they were generally fed on the same diet as their guards. These findings suggest that in times of relative food shortage a deficiency of B1 must have been a common problem. It has not been determined whether these populations developed a special tolerance for the low intake of B complex vitamins.

In the Americas, the potato-based diet of highland South America and the maize-based diet of Mesoamerica supported over 90 per cent of the hemisphere's population. Of the two, the South American complex was nutritionally more secure for several reasons. First, although the potato is a high-carbohydrate, low-protein food, it was supplemented by two domesticated seed producing chenopodia, which contain more protein than the wheats. Secondly, the highland areas had the only domesticated herbivores in the hemisphere, so that reasonable supplies of meat were available. Finally, the steep Andean mountain chain provided the opportunity for a fast exchange of foods which require different types of climatic conditions. Clear evidence of the inherent nutritional balance in this food complex is shown by the fact the deficiency symptoms are still quite low in this food complex region, in spite of the fact that undernutrition is relatively common as a consequence of its currently poor economic status.

The maize-based food complex depended upon a much more delicate balance. Maize does contain a substantially better percentage of protein than root crops, such as potatoes, but falls considerably short compared to chenopodia or wheat. Domestic varieties of beans were a major crop and critical as a protein supplement. A significant problem was the lack of appropriate domesticated animals. In central Mexico, dogs and turkeys were raised as food. In fact, special hairless breeds of dogs were developed for cooking. The problem with both species as food sources is that their diets significantly overlap that of people. Production on a significant scale, therefore, required that a population sacrifice edible plant food to obtain the much lower energy yield the animals provided in the form of meat.

It has been suggested that the practice of mass cannibalism of prisoners captured during Aztec wars was related to meat scarcity. This seems doubtful, since the energy costs of obtaining meat in this fashion is even higher than raising dogs and turkeys. Cannibalism would also provide only a very limited food source no matter how constant the warfare. Even so, it is curious that this is one of few examples where cannibalism provided a measurable nutrient input for a population.

Another maize-based population which may have been affected by nutritional problems was the Maya. This civilization which developed cities before AD 1000 was reduced to a few large towns by the time the Spanish arrived in the region during the early sixteenth century. The analysis of historical Spanish records and archaeological survey information shows that this decline not only involved cultural change, but also included a reduction of population size from about 3 to 5 million in AD 900 to about 800 thousand in 1528. Archaeologists have not found evidence that this loss was a result of warfare or climatic change.

Whether the population decline was the result of rapidly deteriorating soil quality, devastating maize disease, or even an unidentifiable human disease is unknown. The small amount of skeletal material available suggests serious nutritional diseases, very slow statural growth, and multiple skeletal defects. It, therefore, appears that, whatever the cause for the population decline, it broke the delicate nutritional balance and resulted in widespread nutritional disorders.

The most concrete evidence that the maize-based diet can lead to serious tests of a population's nutritional adaptability relates to niacin deficiency. The basic food stuffs are very low in niacin, and in times of economic disruption, the symptoms of pellagra become common. As late as the early 1900s the lower classes of the southern United States who had a maize-based diet had a high incidence of this serious nutritional deficiency disorder. During the low points of economic cycles, death rates were substantial.

Protein and population density. While by 1500 most population centres had, to various degrees, developed adequate diets based on domesticated species, the vast majority of agricultural populations continued to depend on foraging to provide a significant part of the diet. For reasons that are not nutritionally obvious, the availability of fish, edible mammals, or milk and its products are closely associated with the density of population supported by a particular system.

Many authors have assumed that protein with the required complement of amino acids was the key element. However, as discussed earlier, only young children are susceptible to seriously debilitating symptoms as a consequence of very low protein intakes, and, in recent studies, viable populations growing in numbers have been found whose protein intake is extremely low. Meat also provides a substantial array of required minerals and vitamins, so that, without it, diets based on only a few plant sources may result in a number of serious deficiency syndromes which could be more serious than low protein intake for population survival.

Whatever the cause for the meat and population density association, the evidence for correspondence is strong. Among the maize-growing American Indians of the eastern United States the early explorers found the largest densities around such fish- and shellfish-rich waters as Narraganset Bay in Rhode Island and the brackish Chesapeake Bay.

The Polynesian-inhabited islands of the Pacific provide an even better documented example. The island populations brought with them in their initial migrations crops which included coconut, breadfruit, and taro. These provided a very high energy yield for very little work input. Between the fat soluble vitamins in coconut and the vitamins and minerals in the edible taro leaves, the crops also provided most nutrients. The pigs and chickens they also brought with them, on the other hand, found very little wild food they could use and depended upon a portion of the human food.

A recent analysis of pre-contact population size on the inhabited volcanic islands show that population numbers were closely correlated to the size of the coastline, and relatively unrelated to the land areas of the island. Furthermore, the highest densities occurred on such islands as Oahu in Hawaii where the Polynesians had developed special fish ponds to increase the fish supply. How such a close association between the fish supply and population density may have developed is not at all clear. Early explorers report that the Polynesians on all these islands appeared very healthy and both the skeletal measurements and studies of living groups show that they have not even shown signs of a secular increase in height with the advent of modern nutrition and disease control.

While examinations of the fish supply and population density in the Amazon basin and of the game supply and population in sub-Saharan Africa provide equally interesting examples of the protein and population density association, the nutritional relationship between herding and agriculture provides a different perspective on the problem. As noted earlier in this chapter, the study of energetics in two recent herding societies suggests that the human energetic cost of animal herding may be so high that populations could not be supported nutritionally by the meat provided by their herds. This finding conforms to the fact that most known herding societies also practise a limited form of agriculture or are involved in a trade system in which the energetic yield of their animals is traded for food with a much higher energetic yield. Some ecologists have referred to this as a multiplying effect where usable nitrogen is traded for a greater quantity of usable carbon. The trade in recent times has also involved the significant value placed on hides and wool, but still involves a substantial value placed on meat. Whether any pastoral society could have survived without the agricultural supplement is doubtful, but natural foraging could in some environments have significantly supplemented the energy supply. Studies of the Turkana herding population in Kenya suggest that the use of cattle primarily as sources of blood and milk may provide a more efficient use of energy input than the eating of meat.

An unusual example of the meat–carbohydrate trade is provided by the relationship between the Pygmoid hunters of central Africa and the Bantu agriculturalists who moved into this area in relatively recent times. Several variants of the relationship between these populations have been described, but usually the Pygmoid group hunts and provides meat for a particular village and, in return, the village agriculturalist provides the carbohydrates necessary for adequate energy intake. This system appears to be breaking down at present with the Pygmoids increasingly assuming a role as the sedentary lower class in the villages.

As an alternative to a nutritional cause for the human desire for meat, it might be argued that people simply find the taste desirable. Protein, as such, has no taste for human beings as anyone who has eaten pure protein gel can attest. On the other hand, fat does and until recently the quality of meats such as beef were rated on their desirability by fat content. Even among fish, some societies rate the value of a fish in relation to its relative fatness. It might be further deduced that, in the evolutionary process, the genetic development of a taste for fat has been genetically selected, since it tends to guarantee that the hominid would seek to capture animal food, which would provide nutrients essential for its survival and well-being. (Such an argument still leaves unanswered why some human populations still pursue meat when such fat-rich sources as mongongo nuts, peanuts (ground nuts), and avocados are available.)

Lactose deficiency

Finally, let us consider an example where the food source appears to have shaped the adaptive capacity of the species in terms of its ability to utilize a major energy and nutrient source. In the period following World War II, a number of the more prosperous countries of the north temperate region, particularly the United States, decided that the food shortages of many tropical countries could be aided by a broad-based programme which included food supplementation. Among the readily available excess food supplies in the temperate countries was milk, which was dehydrated and provided to many of the countries short of food. By the 1960s, it was observed that in many populations these food supplements often resulted in widespread diarrhoea, and further research showed that many adults developed a diarrhoea when they ingested significant quantities of milk.

Subsequent research showed that a major proportion of the world's population cannot digest the lactose contained in fresh milk products, because of the lack of an enzyme called lactase. This gut enzyme converts lactose to a sugar which may be converted to energy in the usual biochemical chains. Lack of the enzyme results in a loss of food energy and an irritation of the intestine causing diarrhoea and other undesirable effects. The causes for lactase deficiency in adults remain uncertain. Infants and young children in all populations have this enzyme but, in many groups, the enzyme disappears in older children producing lactose-intolerant adults.

It has been suggested that the lack of lactose in the diet after weaning may inhibit the production of lactase in older individuals, but the predominance of evidence now suggests that the age-related stoppage of lactase production in the intestine is more probably related to genetic control.

Whether lactase deficiency is purely an individual genetically controlled phenomenon or a process partially controlled by the genetic plasticity in our species, it is associated with the traditional uses of milk by human populations. The processes of fermentation involved in the production of yoghurt and most cheeses break down lactose into a digestible form. Therefore, the problem arises only when populations use milk or its products in fresh form, such as milk, cream, or very fresh cheeses. From the data available (Fig. 22.4), it appears that the very high frequencies of lactose intolerance occur among the traditional non-milk users while the lowest levels of intolerance occur in the European populations with a long tradition of using unfermented milk products.

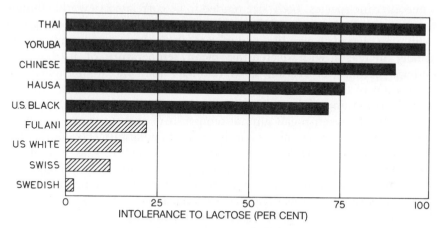

Fɪɢ. 22.4.   The percentage of lactose-intolerant individuals in select human populations.
(Modified from Kretchmer 1972.)

Although there remains some doubt about a specific genetic adaptation to lactose in the diet, the findings reinforce the idea that either some adaptational process which involves permanent change during growth or a specific genetic adaptation has occurred. In either case the studies of lactose intolerance show that a population is best adapted to its traditional nutrient environment and changes in that environment are likely to be stressful.

Adult body-size and shape variation

Nutritional inadequacies often have a strong impact on the growth of children. While other environmental characteristics such as disease and the hypoxia of high altitude may also slow growth in some aspects or at particular times in the growth processes, it is quite obvious that in many recent populations nutritional inadequacy, resulting in either undernutrition or malnutrition in the form of nutrient deficiencies, is a primary cause for slowed growth rates. Less well understood are the causes for the major inter-population variations in size and body proportions. As Tanner notes, this variation is probably, in part, a consequence of population differences in genetic structure and, in part, a consequence of the environmental factors which govern the rates of growth. In this section the variations in adult body sizes and shapes in various populations will be examined in relation to the possible causes.

Body size

As early as the mid-nineteenth century, it was documented that, for homeothermic species, body size tended to be correlated to world climatic variation. C. Bergman in 1847 stated this association as a rule noting that 'within a polytypic warm-blooded species, the body size of the sub-species usually increases with decreasing mean temperature of its habitat'. Perhaps because of resistance to the idea that we are animals, the application of this rule to the human species was not carefully explored until the beginning of the 1950s. At this time, D. F. Roberts analysed the relationship between mean annual temperature and the mean weight of adult males in 116 samples of indigenous populations. He found that simple correlation yielded an r value of -0.600. Using a multiple correlation technique for simultaneously examining how weight and stature were related to mean annual temperature, he obtained a value of $r = 0.820$. A smaller series suggested a similar relationship in women since the weight to mean annual temperature r value was -0.704.

Roberts further subdivided the sample of male populations according to the geographical regions in which they were indigenous and obtained the regression lines plotted in Fig. 22.5. It is probably significant that in Africa and Europe where population mobility was limited during the known past there is a much steeper regression than among the Amerindians and South-east Asiatic Mongoloids, both of whom are known to have settled in their immediate environments from 2000 to 20 000 years ago. Within the

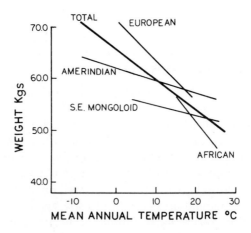

FIG. 22.5.   The regression of mean body weight on mean annual temperature for human populations. (Modified from Roberts 1978.)

Americas, an analysis by M. Newman (1953) using a greater number of populations than those available to Roberts confirmed his findings.

The human adaptability question raised by these proofs that Bergman's rule applies to *Homo sapiens* is what are the causes for the correlations. Roberts (1978) suggested that most of the variation between populations is genetic and is related to climatic adaptation as a result of natural selection. M. Newman (1960) argued that the primary cause may have been nutritional, but did not state whether the nutritional stress produced the variation as a result of genetic selection or through its effects on the genetic plasticity of the American Indian population.

Information exists to support both of these possible explanations. R. Newman and E. Munroe (1955) examined the weights of Army inductees in the United States during World War II and found that, based on the state of residence, body weights between those inducted in the hottest and coldest states varied by slightly over 4 kg. They also showed that the resulting regression of weight to mean annual temperature was mostly a consequence of the association between the average temperature of the coldest month and weight, but only slightly associated with the temperature during the warmest month. On this basis, they suggested that the association found might have occurred because cold weather stimulated a positive energy balance. M. Newman's nutritional argument was broader, based on the demonstration that the diets of the American Indians in hot climates tended to contain less meat and fat and was less nutritionally balanced than the diets of populations in colder regions. Roberts' arguments for climatic selection are based on the theoretical postulates and experimental studies summarized in the previous chapter. As these show, small body size could provide signicantly improved heat tolerance in hot wet environments, and large body size, particularly when accompanied by thick subcutaneous fat layers, could improve the tolerance of the cold found in cold temperature climates.

While Roberts' regressions clearly establish that Bergman's rule applies as validly to *Homo sapiens* as to any other polytypic species, it is probably inappropriate to choose a single cause for the climate–weight association. The increasing body size, which has occurred in many countries over the past century or two, makes questionable the relative importance of genetic information to average body size. At the present time some tropical populations have developed heavy body weights. It appears that Samoans who live in a hot wet environment have the heaviest average weight of any population in the world. Nevertheless, nothing discovered so far suggests that Pygmies in Africa or elsewhere, even with appropriate diet, will attain the adult average weight of Northern Europeans. As with many other basic problems in human adaptability, it appears likely that future research will show a complex interaction which regulates both the genetic information and the genetically plastic development of the adult morphological and functional organism.

A compilation of data by Allen (1877) led him to formulate still another rule in relation to polytypic warm-blooded species. In this rule, he stated 'in warm-blood species, the relative size of exposed portions of the body decreases with the decrease of mean temperature'. The operation of this rule is particularly obvious in such animals as rabbits, where ear size in cold climate rabbits is dramatically smaller relative to body size than it is in warm climate varieties. To examine the applicability of this rule to *Homo sapiens*, Roberts used, among others, two measurements commonly recorded for human populations: relative sitting height and relative span. The first, sitting height, is an index of the height of a seated individual to his standing height, while the second, relative span, is an index of the measure from extended finger tip to finger tip divided by height. The regressions of these indices to mean annual temperature are illustrated in Figs 22.6 and 22.7. The correlation of the relative sitting height to mean annual temperature is similar among men to that found for body weight with an r value of -0.619. The relative span value has a somewhat lower $r = +0.470$. Both are based on population samples nearly twice as large as those available for body weight.

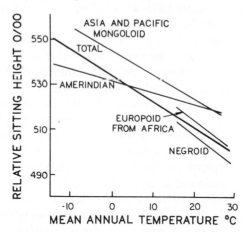

FIG. 22.6. The regression of mean relative sitting height on mean annual temperature for human populations. (Modified from Roberts 1978.)

For this larger series of populations, Roberts used a different set of subdivisions in order to examine regional population variation. He included all of the East Asian groups often classified as Mongoloid into a single group, and separated the North African groups that resembled Europeans from the other Africans. He included the indigenous groups from Australia and New Guinea in a sample he identified as Negroid.

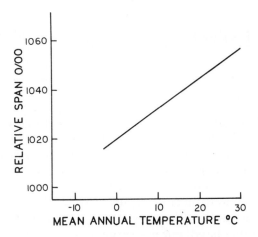

F<small>IG</small>. 22.7.   Mean relative span regressed on mean annual temperature for human populations. (Modified from Roberts 1978.)

While significant correlations were found in the identified samples, the regression lines differed. The Mongoloid, Europoid from Africa, and Negroid samples all had similar slope changes in relation to mean annual temperature, but the Mongoloids had the shortest legs in relation to height, while the Negroids had the longest for a given annual temperature. Among the American Indian groups, the regression slope was less steep, a result similar to that found for weight.

An examination of extremity proportions suggested that among human populations, the Allen rule also applied to a proximal-distal trend in arms and legs. Thus, the lower leg-to-thigh length ratio is greater in hotter climate populations, and even the foot length relative to breadth is greatest in the population native to hot climate regions.

While body weight shows great phenotypic lability in relation to diet and lifestyle, the body proportions appear to be very little affected by these aspects of the environment. Certainly, a major iodine deficiency can result through myxoedema with many individuals in a population having short limbs in relation to body size. It has also been found in secular trends studies that the increase in height, which has been occurring in many populations, results in slightly greater leg length than trunk length growth. Nevertheless, most studies suggest that body proportion differences between populations are strongly affected by genetically controlled growth factors and at this time some form of genetic selection, perhaps through the physiological responses to temperature, seems indicated.

23 Infectious disease

The stresses of climate and the need for food were clearly perceived by people throughout their evolution as a conscious species even though some specific stresses such as high-altitude hypoxia and the subtleties of nutrient requirements were unknown. Certainly, the dangers of the carnivores and poisonous life forms were also appreciated. In response to these known stressors human populations gradually developed a series of culturally transmitted responses which allowed them to cope with many of the strains. On the other hand, our major predators, the viruses, bacteria, and multicelled animals which use us as hosts, remained unknown. Until recently, the symptoms and deaths caused by these stressors were attributed to the supernatural or sometimes blamed on such recognized environmental characteristics as temperature or humidity. Since a multitude of infectious diseases left no record except the number of dead, it is doubtful that we will ever be able to fully reconstruct their role in our evolution or the adaptive responses of human populations to them. Even if records were more complete, an understanding of the history of infectious disease would be difficult since it is a problem of co-evolution in which selection processes favour the invading organism which can reproduce in the largest numbers and the host which can do the same. Within this constant process of co-evolution the invading organism has the advantage over people since all of the infectious disease organisms go through several, if not thousands, of generations for each human generation. Nevertheless, an endemic equilibrium such as that which occurred with malaria may be stressful enough for the human population to produce a genetic adaptation.

The nature of infectious disease

Taxonomic classification

All species of living organisms are affected to some extent by the parasitic predators and, as just another species affected by this mostly invisible world, it has proven difficult to classify those organisms particularly affecting our species. One may, of course, classify them in morphological taxonomic categories which tell us something about their detectability and the level of organization at which they attack our complex organisms. At the organizationally simplest level are the viruses which are no more than encased DNA. These diseases flourish within the cell and often within the cell nucleus itself. Indeed, some become attached to our DNA and may become generationally transmissible. Our antibody systems attack these alien DNA systems and in most forms finally control their reproduction. Nevertheless, a virus such as chicken-pox has mechanisms which allow it to survive in the DNA of some cells. It may later proliferate in these cells causing shingles after many years of quiescence. Among the many viral diseases which affect people are colds, influenza, measles, mumps, chicken-pox, and herpes simplex.

Another aggregate of species which infect us are single-cell animals such as bacteria and fungi. This group of parasites is known to have such variable effects upon us that some cannot even be called predators since they must be present in the gut for us to obtain some of the necessary nutrients from our food. In these instances the relationship is more properly termed symbiosis, since the life form in providing food for itself also provides the necessary by-products for our nutrition. Such organisms are certainly not all benign since this taxonomic group also includes the staphylococcus, streptococcus, and other often lethal forms of infection which are now partially controlled by antibiotics.

Finally, there exist a variety of multicelled parasites which range in size and complexity from the spirochetes which produce syphilis to some gut-dwelling helminths, forty or more feet long, which feast on the food we ingest.

Transmission

Another classification which aids in understanding the infectious disease process is the mode of transmission. Because of the high frequency of rhinoviruses (colds) and childhood diseases, the transmission of viruses in water droplets exhaled or sneezed is widely recognized as a transmission form. However, the majority of the infectious diseases to which our species is prone have other, often complex, transmission modes.

Many of the bacterial intestinal disorders and the multicellular forms are transmitted by ingestion in the gut. Among the most familiar are the diarrhoeas caused by *E. coli*, typhoid, salmonella, and amoebas. Trichinosis which bores from the intestine to the blood stream is another familiar disease spread through the gut.

A route which tends to spread many serious diseases is that which involves an intermediate biting insect vector. Of diseases spread by this route the most serious include the malarias, yellow fever, and encephalitis all of which depend on a variety of mosquito species. An equally complex transmission route through rodents and fleas provides for the transmission of plague and some forms of typhus. Varieties of biting flies which spread veruga in the Andes and both onchocerciasis (river blindness) and trypanosomiasis in Africa have all contributed to our species' share of diseases.

While these are some of the common methods of contagion many diseases have very low tolerances for exotic environmental conditions and require unusual situations for transmission. Among these are the spirochetes causing yaws, syphilis, and pinta, that require sexual contact or lesion to lesion contact for transmission. Even more unusual is rhesus B, a viral sub-lethal disease in rhesus monkeys, which is lethal to people, but must be transmitted by direct contact between the blood of a live, infected rhesus monkey and an individual. Perhaps the most exotic if not complex of all transmission modes is evident in the newly evolved disease called Acquired Immune Deficiency Syndrome (AIDS). In this disorder transmission appears to be through blood to blood and possibly through semen to the large intestine.

Mode of action

A more traditional way of classifying diseases is based on symptoms produced. This folk system tells us very little about the actual cause of the disease or how it is spread but, since almost all historical records of disease and mortality are based on symptomatology, it is a classificatory system which must be recognized. Within such a system major classifications may include such categories as fevers, neurological pathologies, diarrhoeas, and skin eruptions. Even today, death certificates include such categories as dehydration, heart failure, myocardial infarction, cerebrovascular accident, kidney failure, and pneumonia as immediate causes of death. The underlying causes of death often cannot be determined from the certificates unless fuller documentation of underlying cause is presented.

Determinants of disease presence and severity

Since each infectious disease has unique survival and reproduction needs, attempts to draw generalizations about the conditions which lead to the development and spread of human diseases have been rare. Traditionally, epidemiologists have restricted their analysis to the environmental requirements for the survival of a specific disease. As the body of information on the growth and reproductive cycle of specific diseases has improved, attempts to define, in broader terms, the climatic, biotic, and human behavioural conditions needed for the spread of disease have improved. Even so, the mutagenic capabilities of the disease organism and the adaptive capabilities of such intermediate hosts as insects have made predictive models subject to considerable error.

In general, the simpler the transmission mode the more predictable the disease spread will be. When a disease, such as the influenza virus, is easily transmitted through the water droplets of breathing or sneezing, the spread is rather predictable based on the frequency of personal contact and the situation in which contacts occur. We can predict that rhinoviruses will affect a higher percentage of the population in a northern town in winter than in the summer and that children at school are more likely to be affected than those not in school.

Diseases with a faeces to oral route of human transmission such as typhoid are also reasonably predictable. While modern chlorinated water supplies inhibit the spread of such diseases, the more traditional towns and cities offered optimal conditions for endemic disease. Small permanent settlements also provide the opportunity for endemic disease, but if they are in cold temperate or seasonally rainy environments then at least the shorter-lived parasitic diseases will tend to be seasonally epidemic. For migratory small bands such as pastoralists and hunters, the frequencies and severity of these infections are likely to be low, particularly in dry climates.

For the great variety of diseases which require animal reservoirs or vectors, general principles about the spread and infection rates are very limited and strongly influenced by culturally prescribed human behaviours. Nevertheless, in some regions the patterns of infectious diseases may be related to characteristics of the natural environment.

Within the relatively short distance from the eastern escarpment of the Andes to the western slope the physical environment shifts from a relatively warm and wet climate to a cold dry one down to a cool desert. As the data presented in Table 23.1 shows, the pattern of infectious disease also shifts. Part of this shift may be explained by the impact of aridity on mosquitoes and flies while another part is related to the fact that high solar radiation and extreme aridity inhibit the spread of parasites such as the helminths. While climatic variables in this instance strongly affect disease patterns, it should also be noted that the culturally prescribed pattern of keeping of guinea-pigs in the house throughout the Andean plateau is the primary reason why this region is subject to periodic sweeps of flea-borne typhus.

Human adaptability and disease

The co-evolution of a disease, its vectors, and its hosts must produce an interacting complex which allows all three to survive as populations. The effective worldwide vaccination programme for small-pox disrupted one such complex. Theoretically, such a disease could reappear as a result of a mutation from a related form, but for the moment the species is extinct. From the disease perspective an evolutionarily fit disease is one that produces a minimum disruption in the functioning of host and vector. Such a state of what might be called co-fitness probably represents the situation for most diseases in the past and appears to have characterized the infectious disease status of early hominids including early *Homo sapiens*.

TABLE 23.1

Infectious diseases and parasites in four Peruvian villages (frequency in per cents)

	San Antonio (152 m)	Cachicoto (730 m)	Yacango (1870 m)	Pusi (3840 m)
History of†				
Malaria	25.3	13.3	12.6	0.7
Yellow fever	1.1	1.8	0.0	0.0
Typhus	0.0	0.0	1.0	8.4
Parasitic frequency‡				
Nematodes				
Ascaris	79.7	76.6	1.6	32.5
Trichuris	76.6	89.7	1.6	65.8
Hookworm	49.4	68.3	0.0	0.4
Strongyloides	1.0	12.9	0.0	1.2
Cestodes				
Taenia	0.0	0.5	0.2	1.6
H. diminuta	0.0	0.7	0.0	0.0
H. nana	0.0	1.4	8.9	3.8
Trematodes				
S. mansoni	0	0	0	0
Fasciolidae	0.6	0.2	1.4	0.0
Protozoa				
Giardia	0.8	0.5	5.8	3.6
Balantidium	1.8	3.1	0.0	1.9
Chilomastix	3.6	1.3	12.4	8.4
Endolimax	12.6	7.9	16.9	12.8
Iodameba	11.3	3.5	18.5	13.7
E. coli	75.1	54.1	23.6	76.1
E. histolytica				
Small cysts	17.5	15.1	26.5	28.0
Large cysts	12.2	10.2	2.5	5.3
Trophozoites	0.6	0.0	0.0	0.0

* Modified from Baker (1978).
† Percentages listed are age-adjusted.
‡ Based on single stool examination for ova or parasite.

From the disease perspective this does not mean that all possible vectors and hosts must be unaffected since the disease may survive quite well if it has multiple potential hosts and most are fit while some others die. Certainly, a disease may also survive quite well even on a single host with a high mortality as long as there are adequate replacement host animals available and susceptible to the infection.

Human infections display the diversity of the potential co-evolution solutions. For a disease such as tetanus it is irrelevant to the disease that it results in the death of people since it does not need the human host for its growth and reproduction. For people, it is also an evolutionarily insignificant disease since only a very few individuals under unusual environmental and behavioural circumstances are infected. At the opposite extreme are some of the bacteria which in the co-evolutionary processes have actually become beneficial to human survival through their action in the digestive process. It must be assumed that such a circumstance also occurs among some viruses. Between these extremes are the mild viruses like herpes simplex which produce minor if annoying symptoms in some people and none in others.

For the majority of infectious diseases it is impossible to determine the extent to which the disease has evolved so as to minimize damage to the human host versus the extent to which we have evolved a series of defences which limit our susceptibility to infection and its potential stress. Certainly, all of us have a complex system of defence mechanisms which current research is rapidly illuminating. These range from the immunoglobins provided to infants in mother's milk to the complex RNA-coded antibody formation system which can provide protection against foreign proteins even when they have never been encountered previously. These systems in part pre-date our development as a hominid or even a primate. It may not be an exaggeration to state that at the present time this highly adaptable system represents the most complex explicable example of how genetic plasticity allows the human organism to respond successfully to a set of environmental stresses.

Even though human populations have a variety of disease defences available which fit into the categories which have previously been termed acclimatizational or adaptive in the non-genetic sense, it is demonstrable that at least in one instance, if not many, *Homo sapiens* populations have responded to specific infectious diseases by specific genetic adaptations. The best documented example of such an adaptation to a disease is falciparum malaria and haemoglobin S which has been discussed in previous sections of this volume. Many other examples have been suggested. However, a broader perspective on how infectious disease and human adaptability interact may be obtained by reviewing our co-evolution with disease.

The co-evolution of people and disease

In the relatively short time span since we emerged as *Homo sapiens* our cultural evolution produced a major disruption of the world's life forms. The development of a food supply based on human effort was particularly disruptive since it changed not only the relative balance of plant species in a given ecosystem, but also the relative abundance of various invertebrate and vertebrate species. It even changed the distributions and densities of human populations. The environmental changes produced by people placed many diseases in new environmental situations leading sometimes to shifts from animal to human hosts, and often to the evolution of genetically modified parasites which could utilize the new environment. The new diseases were often poorly adapted to their hosts, resulting in high mortalities with maximal opportunities for new selection opportunities in both the invading organism and the host.

The diseases had the evolutionary advantage of rapid generation time, but the human host had the advantage of behavioural plasticity. If the process had been one which the human population could understand, the results might have been different, but as it was the human response was only occasionally significant in reducing the health and mortality impacts of a particular disease. Given the limited possibilities for people to use cultural adaptive responses the probabilities of genetic adaptation were high in spite of our long generation time. How often changes in gene frequency occurred or new mutations were selected for remains unknown. However, the relative isolation of the Western Hemisphere from the evolving diseases of the Eastern Hemisphere provides some clues concerning the limits of genetic adaptations which may have occurred in Eurasian and African populations.

The rapid expansion during the sixteenth to nineteenth centuries of people from Europe into North America, South America, and the islands of the Pacific Ocean resulted in massive depopulations in the aboriginal populations. These depopulations were caused by both high death rates and low fertility. It is now believed that the depopulations were almost totally a result of introduced diseases. While there can be no doubt that in some instances high mortalities were the consequence of warfare and social disorganization, it is clear that most of the population loss was disease-related. Among the prime killers and causes of infertility were small-pox, measles, influenza, and venereal diseases.

Some of the mortality was the result of cultural responses which enchanced the probability of mortality from new diseases. As Neel showed for a previously uncontacted tribe in the Amazon Basin, the response to measles was population dispersal which left measles sufferers to die without the care which would have kept many alive. Despite demonstrations that European cultural systems included adaptive behaviour to the diseases which were not present in the Western Hemisphere, most epidemiologists doubt that the extremely high mortalities such as that from measles (about 14 per cent among the Fijians) can be explained by the lack of proper health care systems. It seems more likely that the low mortality in the Europeans reflected a genetic adaptation. This hypothesis is further supported by the fact that a number of generations later in the late nineteenth to early twentieth century the indigenous populations began a sharp increase in population size as a result of increased fertility and lower mortality.

Cultural evolution and disease

As information has accumulated on the transmission modes of infections, the severity of the symptoms, and the size of the human populations required for the survival of various diseases, it has become possible to partially reconstruct the kinds of diseases which stressed human populations at the various stages of cultural development and diversification. For many contemporary groups, factors controlling parasitic disease prevalence continue to operate. In these groups, characteristics including methods of food production, population density, natural environment, and contact with other populations are closely associated with the types of infectious disease manifested. A time scale of change and its relation to cultural form is provided by Table 23.2.

TABLE 23.2

Cultural characteristics in relation to the number of human generations and population aggregation *

Years before 1985	Generations	Cultural state	Size of human communities
1 000 000	50 000	Hunter and food gatherer	Scattered nomadic bands of < 100 persons
10 000	500	Development of agriculture	Relatively settled villages of < 300 persons
5500	220	Development of irrigated agriculture	Few cities of 100 000; mostly villages of < 300 persons
250	10	Introduction of steam power	Some cities of 500 000; many cities of 100 000; many villages of 1000 persons
130	6	Introduction of sanitary reforms	—
0	—	—	Some cities of 5 000 000; many cities of 500 000; fewer villages of 1000

* Modified from Fenner (in Boyden (ed.), 1970).

Infectious disease in hunters and gatherers

Throughout our long history as hunters and gatherers, average life expectancy was never more than 30–35 years and very few people lived beyond the fifth decade. The causes for this low life expectancy are not clear, but do not appear to be primarily the result of infectious disease. Accidental and traumatic death rates appear to have contributed significantly to early death as did such exotic predators as reptiles and carnivores. Social mortality in the form of infanticide, geronticide, cannibalism, and warfare may have also contributed.

There are several reasons for believing that the infectious diseases were less implicated in these mortalities. Of these reasons, the large population sizes and rapid transmission time required for many known fatal infectious diseases are of prime importance. The relatively low incidence of childhood deaths compared to early adult deaths in recent foraging groups is another reason for believing that infectious disease was not a major cause of death. This mortality pattern contrasts with recent agriculturalists where the many bacterial and viral infections produce high mortality rates in infants and young children, but result in the immunity of most young adults to high-fatality infections.

A comparison of the disease prevalence pattern in two hunting and gathering groups in central Africa with some groups of settled agriculturalists and village populations was summarized by Desowitz in Table 23.3. Such a comparison shows that the hunters and gatherers have lower frequencies of the malarias, yellow fever, and schistosomiasis even though the near equal rates of the sickle-cell trait suggest comparable rates of falciparum malaria. On the other hand, the rates of most parasitic infections are quite high in the hunting and gathering groups. Since these groups of hunters and gatherers have had a significant interaction with the agriculturalists in recent times, this data fits reasonably well with the reconstructions of the development of infectious disease.

A reconstruction of disease patterns based on the factors previously described suggests that parasites which have rather benign effects on people are probably quite old while diseases such as malaria which depend on the dense agricultural human populations who promote the growth of specific mosquito vector populations are more recent. Yellow fever, although endemic in African primate populations, also requires large human populations to become epidemic. Fenner (1970) suggests that, since yellow fever produces very mild symptoms in African populations compared to American Indians and European descendents in the Americas, there was probably a genetic change which improved adaptation in African groups.

Schistosomiasis (bilharzia) which produces rather severe symptoms appears to be an exception to the rule of co-evolution because it is a very ancient human disease. The trematodes which cause the urinary form *Schistosoma haematobium* and the intestinal form *Schistosoma mansoni* are dependent on snails as intermediate hosts. The incidence of the diseases depends on the individual wading or swimming in snail-infested waters.

TABLE 23.3

Results of health surveys of populations (per cent infected) of the tropical African forest zone*

Habitat	Deep forest (hunter-gatherers)		Forest and forest settled agriculturalists		Village in forest
Population group:	Ituri forest pygmies (Zaire)	Cameroon: Babinga and other groups	Bantus of Congo	Banaka langa (Uganda)	Akugo, Nigeria
Entamoeba histolytica	36				12
Giardia lamblia	7				10
Malaria	Children 12 *falciparum* 17 *malariae* 40 Adults 22	51	Infant spleen rate 28	24	51 parasite rate all *falciparum* with 14 mixed *malariae* 70 spleen rate 2–5 yr group 11 spleen rate 20-yr group
Trypanosomiasis	none reported	3.4 and 0.2			none
Hookworm	86, 40		80		71
Ascaris	58, 22		73		70
Trichuris	70, 27		11	100	45
Onchocerca	Adults 43–82	4	5		
Loa loa	2	24			4.1
Dipetalonema perstans	60–100	78	endemic	35	1
Schistosomiasis	11		28		
Leprosy	7, 6–9	rare	endemic		rare
Syphilis, yaws and treponemal antibody	high prevalence syphilis, >50		yaws 20 syphilis 60		
Rickettsiosis	typhus present in rainforest				
Yellow fever			rare 0.1		75 immune by 5 yrs
Sickle-cell trait	26		26	24	25

* UNESCO (1978).

The free-swimming larvae released by the snail then penetrate through the skin and migrate to the liver where they mature into adult worms which settle inside blood vessels and lay eggs. The snail hosts which carry *S. mansoni* require permanent water, but those which are hosts for the urinary form of bilharzia, *S. haematobium*, can survive for long periods in the mud of periodically dry ponds and streams.

As these trematodes survive for long periods in their various hosts, it seems probable they they were a significant human parasite even when all human populations were hunters and gatherers. Although these infections cause diminished vigour and a progressive loss of work capacity, they are only occasionally a direct cause of death.

Schistosomiasis was probably initially limited to Africa and, as shown by the data on the Ituri forest pygmies, probably affected only a small percentage of the population. Irrigation agriculture greatly increased the number of environments where the snail hosts for *S. haematobium* could survive. Increased population mobility also aided in the spread of both snail and disease. The construction of dams and irrigation systems in Africa during recent times has promoted the rapid spread of the disease with the number of infected individuals now undoubtedly well in excess of 100 million. The intestinal bilharzia now found in tropical Central and South America was introduced from Africa, but it is not certain whether the variant *Schistosoma japonicum* found in East and South-east Asia is indigenous or a mutant form derived initially from Africa.

People may have some form of active protective immunity to the diseases since children in highly infected communities have higher frequencies of ova in urine and faeces than adults. What is more certain is that, as with many other diseases, the human adaptations which improved the food supply and population numbers have also, to date, continued to increase the frequencies of these parasites.

Agriculture and the development of infectious disease

The growth and permanent settlement of human populations following the development of agriculture is believed to have promoted the development of most of today's familiar infectious diseases. Among the viral diseases, only a few such as chicken-pox, and slow virus diseases such as kuru could have been present. Even such a disease as kuru, if present in the environment, must have had little or no impact on populations before rather high densities were reached.

Kuru. The discovery of kuru, which revealed the existence of a delayed lethal virus, was based on a long series of studies on an exotic disorder apparently confined to the agricultural Fore people in Eastern New Guinea. Sufferers of the disease developed neuromuscular symptoms which progressed slowly. A fatal degeneration of the nervous system then followed over a period of up to a year before death. The virus which caused the disease appears to be a human limited one, since even among our related primate species only the chimpanzees proved susceptible. Subsequent research showed that the virus remains resident in significant quantities only in the brain and neural tissue. It also appears to be a delicate virus which can be transmitted only by brain and neural tissue to blood contact or possibly by the ingestion of relatively uncooked brain tissue. While other transmission modes theoretically are possible, it was closely associated in the Fore with frequent cannibalism and is disappearing following the suppression of this practice.

While such slow viruses would have had a better survival chance than ones which stimulate immunity responses, certainly a fatal one such as kuru which required a dense enough population for extensive cannibalism could not have become established in the sparse pre-agricultural populations.

Tapeworms. Among the many diseases which are favoured by a settled agricultural life, tapeworms *Cistoda* are aided by the use of domestic animals as intermediate hosts. A very common form of tapeworm which affects human populations is *Taenia solium*, which is normally acquired by eating poorly cooked pork. In its normal life cycle, the eggs of the tapeworm, which are produced by the mature worm in the human digestive system, are consumed by pigs. Once in the pig's digestive tract the eggs hatch, producing microscopic embryos, that penetrate the intestine, circulate in the blood, and finally develop somewhere in the pig into a bladder-like form called a cysticercus. Finally, when the cysticercus is eaten by people in the undercooked pork, it attaches to the intestine and develops into its mature egg-producing form. While these worms may consume some of the food passing through the gut, they produce almost no symptoms in the individual and probably have little effect on general health or longevity.

As Desowitz (1981) has described, this rather harmless parasitism can be altered to a deadly one when the human ingests the eggs and is attacked by the cysticercus. He describes the situation which occurred when pigs from Bali, which were infected with the tapeworm, were given to the agriculturists of Western New Guinea. In Bali, where the population often ate poorly cooked pork but were very careful in personal cleanliness, the consequence of the disease for people was almost completely limited to tapeworm infections. In the New Guinea groups which lived in close contact with both their pigs and their own faeces, the eggs were often directly ingested by people. As a consequence, the people were infested by not only the tapeworms, but also the embryos. These embryos formed cysticercus in all parts of the body, including the brain. Within the brain many dysfunctions occurred, including epileptic-type seizures. To complete the deadly chain, the people in the highlands slept near their fires and often acquired massive burns or death from falling in the fire during epileptic seizures.

This rather exotic example of how culturally prescribed behaviour influenced a particular disease illustrates that, for whatever reasons, the rigid hygienic proscriptions found in many societies may be among the few successful cultural adaptations populations have made to disease.

Towns, cities, and disease. While many of the bacterial diseases which cause dysentery, including such fatal ones as cholera, probably became well established in early agricultural populations, most of the viral infections, especially those which stimulate long-term or lifetime immunity, are more recent. This list contains most of the familiar, so-called, childhood diseases, such as measles, mumps, whooping cough, and poliomyelitis. Based on the size of the population in which the virus must circulate to stay alive, these diseases could not have existed until some areas of the world became quite densely populated and organized into social systems such as towns or cities where human contact was frequent. Data such as that compiled by Black for measles exemplify the effects of population size and interpersonal interaction on the prevalence of one of these viral diseases (Table 23.4).

TABLE 23.4

Endemicity of measles in islands with populations of 500 000 or less, all of which had at least four exposures to measles during 1949–64*

Islands	Population	Annual population input†	% months with measles (1949–64)
Hawaii	550 000	16 700	100
Fiji	346 000	13 400	64
Samoa	118 000	4440	28
Solomon	110 000	4060	32
French Polynesia	75 000	2690	8
Guam	63 000	2200	80
Tonga	57 000	2040	12
Bermuda	41 000	1130	51
Gilbert and Ellice	40 000	1260	15
Cook	16 000	678	6
Falkland	2500	43	0

* Modified from Fenner (in Boyden (ed.), 1970).

† 1956 births less infant mortality.

Among the semi-isolated island populations shown in this table, measles was constantly present only in Hawaii where the population exceeded 500 000. Indeed, it is believed that within a self-contained human population measles would disappear in a population below 500 000 persons. However, this is an approximation since a smaller number would suffice if both birth and death rates were high. A larger population number is required in groups with low birth rates and greater longevity.

On the other islands shown in Table 23.4, the percentage of months in which measles was present was governed by the size of the population and the frequency of outside contact which could reintroduce the disease. The population size, of course, affects the rate of spread; measles prevalence was affected by high reintroduction rates in Guam that, at the time, had a high turnover of US military personnel, and Bermuda, which had many tourists. On the other hand, even relatively large populations, like those in French Polynesia and Tonga, had few visitors and were measles free much of the time. These data along with what has been summarized about prehistoric population sizes suggest that the human forms of the childhood diseases probably did not exist until six or seven thousand years ago when some agricultural centres developed dense population clusters with high levels of personal interaction.

All of the childhood diseases have less serious effects on children than on adults. Thus, the rubella form of measles in a pregnant woman can cause serious damage to the embryo and foetus; mumps can cause sterility in adult men. All are more likely to result in fatality among adults. Whether this difference in age-related severity is the product of the virus' genetic adaptation to a young host, a genetically flexible human adaptive response to their attack during youth, or an accidental accommodation is unknown. However, it appears that the most successful example of this probable co-evolution occurred with three antigenically different varieties of polio viruses. Normally, these infections produce a symptomless infection of the superficial gut cells. Development of poliomyelitis symptoms occurs in less than 1 per cent of infections. As with the measles virus whether symtomless or not the polio viruses produce immunity. In populations where the faecal–oral transmission required for these viruses was easy, most people had been exposed during infancy.

For reasons that are not clear the viruses rarely produce the neurologically paralytic form of poliomyelitis in infants. They are more likely to produce poliomyelitis as the age of the infected individual increases. Thus, in areas where the viruses were easily communicated to the infant population, the neurological form of the disease and its frequent fatal or crippling effects were generally low. As shown in Fig. 23.1, the population of a place like Malta was unlikely to have many individuals suffering serious residual effects or death from the disease, while in previously uninfected groups, such as the people on the isolated island of St Helena, the number so affected was high.

The development of effective vaccines has currently controlled the polio viruses in many parts of the world. However, unlike small-pox,

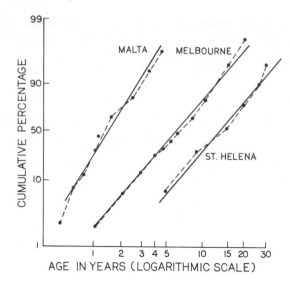

Fɪɢ. 23.1.   Cumulative cases of poliomyelitis by age in epidemics in three locations. (Modified from Fenner, in Boyden (ed.) 1970.)

whose ravages were apparent to all groups, this disease did not noticeably affect many groups. It, therefore, remains alive and ready to reinfect non-immunized populations.

Human mobility and disease. While most of the major infectious diseases had probably appeared and become established in some of the denser human populations by five or six thousand years ago, the epidemics they caused were limited. It appears that the early cities were not capable of population self-maintenance. The nearby rural populations retained a strong growth potential so that the cities and total areal populations could increase. A more complex and productive food production technology also diffused so that local populations protected by their small numbers and virtual isolation from serious infectious diseases could grow at a rapid pace. As the techniques of food production improved and frequencies of contact increased, at least two major changes in the co-evolution of disease and people occurred.

    First some of the diseases which had survived on animal reservoirs became able to depend and prosper on the basis of human hosts alone. The malarias described by Harrison earlier in this volume provide a good example. As these endemic malarial diseases continued to exert selective pressures on their human hosts, some highly specific mutations in the genetic structure of haemoglobins and its production became established in human populations because they enhanced reproduction and/or survival in their carriers. It is logical to assume that these endemic diseases must have also resulted in selection for improved immunological responses, which were less specific in action and acted in a manner similar to the acclimatizational response described for heat exposure. No such changes in the immune systems have been identified, but current research on the HLAs suggests that such generalized adaptive responses may soon be identified.

A second consequence of the rapidly elaborating technology and growing human populations of Eurasia and North Africa was an increasing rate of human contact across this broad expanse of land, which by 2500 BP contained about 90 per cent of the world's population. By this time and long before the famed trip of Marco Polo, the Chinese documented the arrival of Mediterranean traders. They arrived by sea and had presumably sailed around India, the Malay Peninsula, and up the eastern Pacific coast to China. The appearance of trade items from China, India, and elsewhere in the East rapidly rose after this time in the Mediterranean basin. As written history records, by 2300 BP Alexander controlled territory all the way to the Indus river.

Historians such as McNeill (1976) believe that the epidemic diseases which passed through the Mediterranean region following the opening of these trade routes were major contributors to the decline and fall of the Roman Empire, as well as an explanation for the perturbations in political structure and population size of China. Certainly, by the late Middle Ages when population size and deaths were more fully recorded, epidemic diseases (introduced from South Asia) were rapidly spreading through Europe.

Most notable among these invading diseases were the bubonic plagues, which were usually spread from rat hosts via fleas to people. These highly lethal diseases were initially slow to spread since the carrier rat species first had to arrive and disperse prior to the disease. Later, the epidemics, while spreading faster, were restricted to people living in urban areas which were infested by the host rat species. Finally, with the development of new plague foci in central Asia and a form which could be spread by human contact, the massive Black Death of the mid-fourteenth century hit Europe. Figure 23.2 shows how rapidly the Black Death spread through this now densely populated region. Even so the distribution was not uniform, and both selected cities and rural areas were spared. Among those who did contact the disease, between 30 and 90 per cent died. Even with the best care available, before antibiotics mortality from plague was between 60 and 70 per cent. While the total mortality rates for Europe are not exactly known, estimates of English mortality range from 20 to 45 per cent of the population, and it is believed that about one-third of Europe's population died in this epidemic.

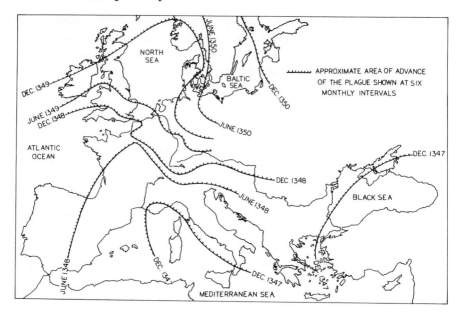

FIG. 23.2. The spread of the Black Plague in fourteenth-century Europe. (Modified from McNeill 1976.)

Voyages of discovery. While Eurasian populations appear to have suffered high mortalities from the disease epidemics which swept through them from 5000 BP, until recently the most massive human mortality rate known occurred in the Western Hemisphere and Pacific populations, when they were initially contacted by Eurasian people following the voyages of discovery.

It appears that these regions had been spared the massively infectious viral diseases which evolved in Eurasia, even though population concentrations in such agricultural centres as Mesoamerica and the Andean region of South America were sufficient to support such diseases. Why these centres had not developed such a broad array of highly lethal forms is not certain, but the recency of the dense human population and the lack of variety in domestic animals probably reduced the sources and time for the evolution of disease. In addition, such important endemic diseases as malaria and yellow fever were absent. This favoured the rapid growth of populations in the sub-tropical environments. Attesting to the lack of red cell attacking diseases is the absence of any established abnormal haemoglobins in the Western Hemisphere.

The arrival of the Europeans created the possibility for a major exchange of diseases. The only disease which may have been transferred from the Western Hemisphere to Europe is syphilis. Certainly, an epidemic of this venereal disease began in the Mediterranean region shortly after contact with the Western Hemisphere began. Furthermore, it acted like a new disease since the severity of its symptoms among Europeans slowly diminished over subsequent generations. Belief that the disease was introduced from the New World was questioned when it was discovered ·that the spirochetes which produced the disease were morphologically and physiologically undistinguishable from the spirochetes which cause yaws. Yaws, a tropical disease which produces open skin sores, is transmitted through contact with the sores. This led to the hypothesis that the spirochete simply adopted a human sexual mode of transmission since wearing of clothing in colder climates inhibited skin contact.

Nevertheless, a New World origin seems highly probable since a third form of symptomatology called pinta occurs in the Western Hemisphere as a result of another morphologically identical spirochete. This disease, again presumably spread by skin contact, causes first a depigmentation of large skin areas followed over a number of years by the spread of super-pigmented areas which begin in the centre of depigmented areas. This disorder causes no other significant symptoms and, prior to antibiotic treatment, was distributed from the Amazon basin groups to the southern United States. This wide distribution and low level of symptoms suggests a long history of the spirochete in the Americas.

For the populations of the Americans the invasion of the European-borne disease was devastating. In the now famous first instance of germ warfare, Cortez entered the Aztec capitol of Tenochtitlan to find a decimated population and disintegrated political structure caused by the advanced sweep of small-pox brought to Mexico by his army. Pizarro's 'conquest' of the Inca Empire followed the small-pox epidemic which had killed the Inca, his successor, and a large percentage of the population by several years. Even the European settlers in the more temperate zones believed that God had ordained their takeover of the lands, because the epidemics which destroyed large percentages of the resident American Indians left them large tracts of deserted lands to cultivate and use.

The extent of population decline in the Western Hemisphere related to the arrival of the Europeans is unknown, but based on the selected mortalities recorded in the epidemics which swept the American Indians various authorities estimate that the total population of native Americans was reduced by between 50 and 90 per cent. While the social disorganization loss of food production, lack of appropriate behavioural responses, and loss of social control systems must have all contributed to the population loss, it does appear that the European invaders may have arrived with some form of genetically-based resistance to the various diseases which the American natives lacked.

For the populations contacted by the Europeans in the late eighteenth and early nineteenth century, survival was sometimes better. For the

Samoans in the South Pacific, who were not contacted by Europeans until the 1830s, mortality was modest. On the other hand, as shown in Table 23.5, some of the Pacific island populations were finally as decimated as those in the Americas. The decline of the population and sociocultural system on Easter Island exemplifies what must have been a common experience. This Polynesian group had apparently survived initial European contact with a reasonably intact society. However, in 1863 a Peruvian-sponsored ship conscripted many Easter Islanders and other Polynesians for labour in Peru (Maude 1981). Many died in Peru, but more devastating was the small-pox they and other repatriated Polynesians introduced to Easter Island (Table 23.6). When explorers arrived on the island a few years later they found only a few surviving natives.

TABLE 23.5

Population declines on Polynesian Islands 1790–1902 (numbers in thousands)*

	Initial population estimate (year)	Population estimate (year)	Population estimate (year)
Samoa	42 (1840–1850)	35 (1850)	35 (1900)
Tonga	20 (c. 1830)	—	20 (1900)
Fiji	140–150 (c. 1850)	115 (1881)	94 (1901)
Cooks			
South	9.8 (1845)	7.3 (1871)	6.0 (1902)
North	?	—	1.8 (1902)
French Polynesia			
Tahiti and Moorea	35 (1790?)	11 (1863)	13 (1902)
Marquesas	20 (1840)	5.0 (1880)	3.6 (1902)

* Population estimates based on McArthur, *Island populations of the Pacific* (1967), ANU Press, Canberra.

Even in the later days of European exploration, introduced diseases often had devastating effects. The cause for the rapidly progressing depopulation of Yap an island in Micronesia was unknown until treatment with antibiotics after World War II showed that gonorrhoea was a prime cause of population sterility. Among the Inuit, the introduction of measles

TABLE 23.6

Population decrease on Polynesian islands[1] due to labour recruitment[2] and disease introduced upon repatriation[3]

Area	Total pop.	Recruited died in Peru	Deaths from small-pox introduced	Pop. loss	Remaining	% Loss
Total	11 449	2942	2800	5742	5707	50
Easter Island	4126	1386	1000	2386	1740	58

1. The total population represents only those Polynesian islands on which recruitment was practiced.
2. The term 'recruitment' (Maude 1978) included kidnapping as well as deception.
3. Fifteen Polynesians were eventually repatriated.
   Data from Maude, *Slavers in Paradise* (1981), Stanford Univ. Press.

in the early 1900s resulted in a mortality up to 7 per cent in some communities. Again social responses to these disease may have contributed significantly to the resulting sterility and morbidity, but it is difficult not to conclude that the Europeans who introduced the disease did not have some form of genetically induced or developmental response which reduced the health impact of the diseases in them.

Overview

Infectious disease has played a major role in the evolution of *Homo sapiens*, and during the last 6000 years may have led to a variety of genetically plastic and population-specific genetic adaptations. The factors which governed the development and spread of the immunity-producing viral diseases seem to have been in many ways a force that helped shape the cultural development of the species during recent history.

Within the last 200 years an increasing understanding of the causes for infectious disease and the discovery of how to control some of them has probably led to the most major revolution in life expectancy which has occurred in human evolution.

Despite these discoveries and their implications for the numerical growth of human populations, epidemiologists and parasitologists warn that we should not be complacent. Infectious diseases remain in continuous co-evolutionary progression with *Homo sapiens*, and the worldwide rapid transmission possibilities for infectious diseases open the possibility that at any time a new disease could rapidly encompass our species. At the same time the bacteria and other life forms which prey upon us are already developing genetic adaptations which survive in the media of our repertoire of antibiotic and vector-controlling chemicals.

24 Modernization and human biological responses

During their very long history as hunters and gatherers human populations spread over much of the world, developing specialized technologies, social organizations, beliefs, and behaviours, which at minimum allowed them to survive in their new environments. These adaptive systems were supplemented by the genetic plasticity in morphological and physiological systems, that improved their functional capacity in the various environments. Finally, many became resident long enough so that the stresses present in the social and natural environment acted by evolutionary mechnisms to provide some genetic adaptations. The development of agriculture and the dense population aggregates which followed led to nutritional specializations and infectious disease forms which provided stresses so strong that genetic selection operated rapidly on some diseases. Nevertheless, the increasing rate of interaction between populations grew and by the sixteenth century the populations of the world were in close enough contact so that the technology, beliefs, food sources, and infectious diseases were, to varying degrees, affecting a vast majority of the world's previously self-contained populations.

This does not mean that all of the many small populations had been in continuous contact with the rest of the world. Some remained uninformed about the existence of other peoples as late as the 1960s, even though most of them through diffusion and trade had been influenced by selected aspects of the technology and infectious diseases present in other distant populations. Major geographical regions, such as South-east Asia and Indonesia, had been massively changed by contacts with India and China. The Western Hemisphere had undergone major population changes as a result of European contact, and only the most isolated of Pacific regions and remote tropical areas in South America remained relatively isolated.

This last isolation broke down rapidly as European populations and their expanding areas of influence underwent major technological and population changes during the industrial revolution. By this time the transport of domesticated crops from one agricultural system to another had been widespread and many areas such as northern Europe had seen major population expansion as a consequence. In northern Europe the important crop introduction was the white potato from highland Peru. This crop was a high energy producer with the added advantage of frost resistance. It greatly enhanced the population support capacity of northern European countries and regions from Ireland to Russia. Unfortunately, the exotic environment undoubtedly led to the potato blight of the early 1800s which caused widespread starvation in Ireland and other parts of Europe.

The development of science and technology in Eurasia and America led to an increasing ability to produce food with decreasing human energy requirements and an expanding knowledge of how to control and limit the major infectious diseases. With the concomitant improvement in transportation capabilities and production of consumer goods the impact of European society on the other populations of the world grew at an accelerated pace. The process of change effected on other populations has been generally called Westernization, but in most recent times, the flow of change has not necessarily been from West to East and, indeed, most change has been the result of the influence of urban centres on rural populations. The most acceptable term for this process appears to be modernization, although this influence could more properly be considered the impact of worldwide technologies and regional culture on traditional populations.

The effect of these common environmental changes on the populations has not resulted in uniform biological changes. Indeed, one would not expect the responses to be comparable, since the groups affected manifested a diversity of biological and cultural adaptations based on their indigenous environments. Thus, in spite of the common genetic bond implied by our being a single species, the environmental changes produced by modernization should result in differential responses to what is a common new environment.

Demographic change

Population size

Perhaps the most uniform change which occurred as a result of modernization was the growth of population numbers. This does not mean that all of the world's populations began to increase at the same time or rate, but as shown in Fig. 24.1 accelerated population growth started somewhere between 1800 and the 1920s for most groups. The numerical growth rates in the more industrialized areas increased at an earlier date, but the growth of most of the populations, which had been previously decimated by contact with the diseases of Eurasia, was extremely rapid from 1900 or so onwards. The control of transmission of many infectious diseases which had been achieved by this colonial and trading period must have reduced death rates in these societies. Of course, in some of the Western Hemisphere populations the time span from first contact was also

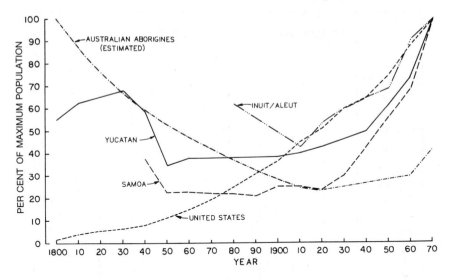

FIG. 24.1. The relative size of selected populations between 1800 and 1970. The 100 per cent level of each population is based on its maximum size during the time period.

long enough so that a genetic resistance to highly lethal diseases could have developed. However, the proximity of the times at which groups such as the Polynesians and Inuit were contacted, and the time when their numbers began to increase suggests that a principal explanation must be the control of infectious disease introduced from external sources.

An example of how this might have happened is provided by the Samoan Islands. Beginning in 1900 political control of this island archipelago was divided between Germany and the United States. The Western group of islands were German controlled, while the smaller Eastern islands (100 km away) were controlled by the United States. In 1918 when the influenza epidemic which was sweeping much of the world attacked the Western Samoan islands, the US commander governing the eastern section proclaimed a quarantine of the American-controlled islands. The subsequent changes in population size (see Fig. 24.2) show that, as a consequence of this act, population growth was more rapid in the Eastern Samoan islands than in the Western ones.

Fertility and mortality

The process of modernization in Europe and other slowly modernizing countries resulted in a combination of a decline in mortality followed by a decline in fertility that is often described as the *demographic transition*. As presented in Table 24.1, low fertility among hunters and gatherers changed to higher fertility among agriculturalists followed by very low fertility in recent modern societies. While this pattern was apparent on a national scale, it was not equally valid for all of the sub-groups, even

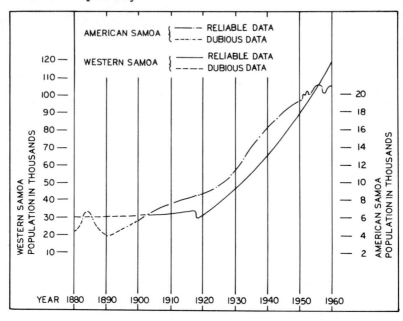

Fig. 24.2. Population sizes in Western and American Samoa between 1880 and 1960. (For data sources see Harbison, in Baker *et al.* 1986.)

within modern countries. In Canada and the United States, for example, there are some religious groups, such as the Hutterites, who have fertility rates as high as any group in the world, although mortality has declined at a rate similar to the pattern found for the respective countries.

Traditional populations who have been significantly affected by modernization during the past 50 years or less have generally not followed the transition pattern. In many groups fertility rose sharply almost immediately and, in many, it remains at a high level. It has been argued that this occurs because in such societies the social and economic benefits of a large number of children outweigh their cost to parents.

Mortality changes with the modernization of traditional populations have followed a more predictable pattern. As shown in Fig. 24.3 the hunters and gatherers had a rather uniform probability of dying at any age with almost no one living to what we now consider early middle age. The settled agricultural populations through to the modern era had very high infant and child mortalities, but adult mortalities were generally lower than among hunters and gatherers with a significant percentage living until old age. With modernization many traditional populations rapidly approximate the mortality pattern illustrated for a country such as the United States. One of the unusual features of the modern mortality pattern is the lower age-specific mortality of females. This is particularly

TABLE 24.1
Population fertility for various groups

Study group	Data source	Type of measure	More traditional	More modern
!Kung San	Census of 1968 Women aged 45 +	Completed fertility	4.6	—
Karen (Thailand)	Census of 1968 Rural & urban groups	Total completed fertility	6.82	3.61
Samoa	Census of 1980 Village & urban groups	Total completed fertility	5.9	4.9
US whites	Census of 1983 Women aged 45 +	Completed fertility	—	2.9

For data sources see Lee and Devore (1976), UNESCO (1978), and Baker *et al.* (1986).

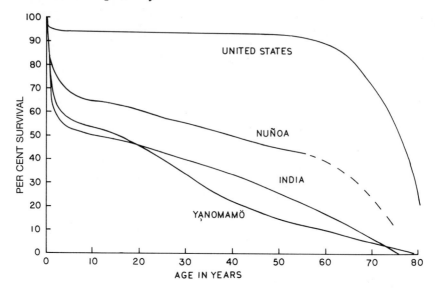

FIG. 24.3.   Cohort survivorship estimates for four populations: United States in the 1960s; a rural district in highland Peru (Nuñoa) 1960s; India in the 1910s, and the Yanomamo, a recently contacted tribe in the Amazon, in the 1960s.

evident in middle and old age. While the greater importance of women to a population's reproductive capacity has been suggested as an evolutionary cause for their better survival capabilities, the specific biological characteristics which enhance their survival compared to that of men remain unexplained.

As the demographic data emphasize, the changes in knowledge and behaviour which have occurred in the last two hundred years resulted in the most effective short-term cultural adaptations ever made by *Homo sapiens* to environmental stressors. In the next sections, the specifics of these adaptations and their potential future effects are considered.

Nutritional change and response

The worldwide dispersion of new domestic crops and animals followed the voyages of discovery. Maize and cassava from the Americas were rapidly adopted in Africa. Rice cultivation was brought to the Americas for feeding the increasing white and black populations. The potatoes also spread making increasing supplies of food energy available in most parts of the world. The introduction of horses, donkeys, cattle, and sheep into the Americas provided essential nutrients in the previously delicately balanced diets of many native populations. The net effect was expanding populations, even though the mixing of the world's plants and animals in new environments also brought periodic catastrophic food shortages, such as the European experience during the potato blight.

A more common problem must have been the malnourishment diseases, including mineral and vitamin deficiencies, which occurred when traditional diets were disrupted by the introduction of new foodstuffs. How serious these deficiencies were in earlier centuries is unknown, but by the twentieth century epidemics of goitre, pellagra, and beriberi were common. The movement of masses of people into the cash economies of urban areas made the poor among them subject to such unusual diseases as kwashiorkor. As recently as the 1970s the migration of Indians into such United Kingdom cities as Glasgow with its low solar radiation resulted in a high frequency of rickets and osteomalacia.

The development and effective dissemination of scientific knowledge about nutrition appears to be gradually reducing the worldwide prevalence of deficiency diseases, but in its place a rather unusual set of changes are happening to the rates of physical growth, sexual maturation, adult height, and the body composition of adults. Children in modern societies are now growing faster, maturing sexually at an earlier age, and developing into taller adults. These changes are generally attributed to better nutrition, although it has been suggested that the control of infectious disease and increased genetic heterosis are also possible contributors to this change. Rapid growth and larger size are generally viewed as signs of improved health. Such an interpretation is, however, only partially supported. In populations which have until recently contained significant number of undernourished or malnourished children, these changes are associated with better levels of physical fitness as measured by such physiological capabilities as aerobic capacity. However, as reported by Pariskova, Tunisian children were physiologically fitter than Czechoslovakian children even though they were smaller for age and slower growing. It may be that more rapid growth and larger size is harmful to some members of a population, while it is a sign of improved health in others.

Among adults the process of modernization has certainly resulted in substantially greater body weights. In virtually every study which has examined the changes in populations affected by modernization, body weights increase. As can be noted in Table 24.2, the weights of those segments of populations which are still modernizing show some degree of weight increase compared to their traditional modern society counterparts. In some, the weight change is relatively modest, while, in others, massive weight changes occur in a very short time. Most of this change is produced by an increase in quantities of fat rather than muscle.

The causes for these weight changes are not entirely clear. Most certainly the input–output ratio must change, but, as explained in the

TABLE 24.2

Average weights (kg) for adults living in traditional societies compared to modernized societies

Population	Life style	N	Men	Difference	N	Women	Difference
South African Venda[1]	Trad.	241	56.7				
	Mod.	240	64.1	+ 7.4			
South African Pedi[1]	Trad.	202	56.2				
	Mod.	223	60.6	+ 4.4			
New Guinea Papuans[2]	Trad.	40	60.6		65	57.5	
	Mod.	139	70.2	+ 9.6	46	66.3	+ 8.8
Tokelau Polynesians[3]	Trad.	284	74.8		377	73.5	
	Mod.	464	80.0	+ 5.2	363	77.0	+ 3.5
Samoan Polynesians[3]	Trad.	128	76.6		190	71.6	
	Mod.	225	91.8	+ 15.1	304	86.0	+ 14.4

Data from (1) Vorster in Harrison (1977); (2) Coyne *et al.* (1984); (3) Baker, in A. Boyce (ed.), *Migration and mobility* (1984), Taylor & Francis Ltd., London.

nutrition section, only very slight changes in this ratio are required to result in the kind of weight changes observed in even the heaviest of modern populations. The most common explanation for the weight gain is that the calorifically dense foods which were developed in the processes of modernization trick our appetite controls into ingesting excess calories. Thus, the preferred foods often contain a large amount of fat which yields nearly twice as many units of energy per unit of bulk as carbohydrates. Refined carbohydrates such as sugar are also popular and concentrate energy in relation to bulk.

Another view is that modern societal systems reduce the required physical work of a population so that fat accumulation occurs because the population maintains its food intake pattern, but gains fat because it immediately reduces its energy expenditure. Of course, there is also the middle view that both contribute to weight gain.

The available studies do not favour any one of these hypotheses. Historic descriptions suggest that members of the upper class may have been subject to obesity long before the energy-dense foods of modern societies were invented. The study of Samoans (Baker *et al.* 1986), the heaviest populations in the modern world, suggests that Samoans may become heavy without major dietary change, but with significant activity reductions. Yet, other studies show that in modern societies even sub-populations with high energetic outputs often have high average body weights.

The accumulating evidence on Polynesian and American Indians indicate that an explanation based on diet and/or energy expenditure may be too simple. In fact, it appears that in the past certain groups may have lived in environmental situations which resulted in a genetic potential to increase body energy stores (fat) quite rapidly when food was available as a protection against frequent food shortages. While such a selective process would have been of great advantage to populations subject to periodic food shortages, it would not have been helpful for those with continuous and adequate food supply. In these groups, obesity would have become common and work capacity reduced as it now is among modernized Polynesians and certain other populations.

The development of future nutritional stresses and their impact on populations is not obvious. In spite of continuing areas with malnutrition and the periodic reappearances of starvation, the technologies and educational approaches to solve these problems exist. Of course, these solutions depend on the use of mechanical energy in food production and a large human energy input into education, food redistribution, and the maintenance of vast trade networks. The problems of obesity and low physical fitness may turn out to be population-specific and certainly could be solved, but the solution may be difficult to effect because the presently known way of controlling weight gain requires that individuals choose foods and quantities other than those which are preferred, and increase exercise levels above those desired. Efforts to produce such changes in behaviour have so far had very limited success.

Infectious disease

While an adequate supply of nutrients fuelled population growth in many populations through the nineteenth century, it was undoubtedly the understanding and effective control of infectious diseases which permitted the explosive population growth of our species during the twentieth century. Even before the antibiotics of the mid-century, the cow-pox vaccinations for small-pox and the understanding of disease transmission through drinking water and a faeces to mouth passage had dramatically altered disease patterning. In fact, calculations on the effect of public health measures show that by these measures alone life expectancy would have been raised in a population to nearly 65 years. This theoretical calculation appears to be partially supported by the analysis of countries such as Western Samoa where food supplies are adequate, but financial constraints result in a health care system concentrated on the control of infectious disease. Life expectancy in Western Samoa is now about 65 and population growth is rapid. These public health measures for the first time reversed an age-old trend in which urban mortalities were so high in relation to fertility that only constant migration of people from rural areas maintained population numbers.

The problems of reducing infectious disease mortality in the cities was more complex than in rural areas because of the higher rates of personal contact, the difficulties of effective waste disposal, and greater contact with other parts of the world. In many tropical cities, these problems still persist and infectious disease remains a serious problem. In the early 1900s, cities such as New York with migrant populations from very diverse physical and cultural environments had particular difficulties because the types of disease problems varied greatly according to sub-population. Living in comparable slum conditions, Russian Jewish groups had very low infant mortalities while Italian migrants had the highest rates. The Jewish group also had the lowest death rate from tuberculosis of all migrants. The Irish migrants and even second generation natives maintained the highest overall death rate. Whether these differences were entirely related to initial cultural practices or related partially to differences in pre-existing biological adaptations, remains unknown but, even at present in the Americas, populations from American Indian backgrounds continue to manifest high rates of tuberculosis compared to other groups living in comparable conditions.

The increasing control over infectious disease and the ability to cure many infections had in recent decades created a rather optimistic view that eventually infectious diseases could be brought under full control. The development of DDT and other insecticides led to the view that even the malarial and other insect vector diseases could be eliminated. This optimism has been greatly diminished by the demonstrably rapid evolutionary capability of both the infecting organisms and their vectors to adapt to new environments. Many epidemiologists believe that the evolutionary capacity of our invading organisms may outstrip the speed with which modern culture can develop cultural adaptations to the diseases. Particular concern has been aroused by the fact that the large size of the world's population and the rapidity with which both biotic materials and people move from one region of the world to another, may open the possibility for a lethal virus to spread through almost all of the human species in a few months giving insufficient time for either effective human adaptive responses or selection to work on the virus.

The new virus, human T-cell lymphotropic virus (HTLV-IV/LAV), which produces AIDS, exemplifies the reason for these concerns. The disease, once manifest, appears to be uniformly fatal because of the breakdown of critical immune systems. A report from an investigative committee of the US National Academy of Sciencies reported that among the individuals in the United States who were diagnosed as having AIDS in 1983, 85 per cent had died by the time of the report in 1986. The long-term survival potential of the virus is also shown by the fact that the virus which results in AIDS may be present in the individual for years before symptoms appear, and it is possible that some carriers of the virus may never develop symptoms. Examples of such virus carriers without symptoms are common including carriers of polio and chicken-pox.

The origins of the AIDS virus remain unknown, although it may have started in Central Africa. We are more certain that the form which now causes the AIDS symptoms must be of recent origin and is spreading very rapidly throughout the world's population. The transmission pattern of the virus currently appears to require either a blood to blood or a semen to lower intestine contact. If these viral transmission modes continue to be the only ones possible for the virus, it will probably remain a highly lethal but manageable virus. Even so, it is a sharp reminder that the co-evolutionary process which produced an increasingly tenuous balance between human populations and their primary predators, the infectious diseases, continues.

Degenerative disease

With the decline in infectious diseases as causes of death and disability, the degenerative diseases have emerged as the leading causes of death. Among most traditional populations changing to a modern lifestyle, a clear indication that the major transition has occurred is that the two leading causes of death become cardiovascular disease and cancer. Until the 1950s, it was assumed that changes in causes of death were entirely related to the fact that in traditional societies almost everyone died before late middle age and the degenerative disease of old age simply didn't have the time to develop in most individuals.

Subsequent research on the remaining traditional populations suggests a much more complex situation in which some of the common degenerative disorders in modern populations are absent or rare in many isolated and self-sufficient populations. While these disorders cannot be properly called diseases of modernization, since they were present in some of the traditional groups, they clearly are not inevitable disease states built into our ageing process. Among the non-fatal forms are such functional disorders as myopia, tooth decay, and the middle-aged loss of high-frequency hearing, all of which have been found totally absent in some populations.

Among the more deadly diseases, the development of cancer has not been demonstrated to bear any simple relationship to modernization. Malignancies attack many life forms, including all vertebrates. Traditional human populations, as well as modern ones, have high overall death rates from malignancies among individuals who live long enough. Whether there are any age-specific or age-standardized differences in overall malignancy rates remains unknown, even though specific rates vary. Certainly, the use of cigarettes in modern societies has resulted in an increase in fatal lung malignancies. It has also been suggested that traditional populations as they modernize show differential frequencies in the types and locations of the cancers they develop. Whether these are related to population differences in genetic structure or particulars of culturally prescribed behavioural differences will require future population by population studies; but findings, such as the highly elevated gall bladder cancer rates among American Indians irrespective of their previous traditional culture, are highly suggestive of a heritable predisposition.

Cardiovascular disease

Many of the cardiovascular diseases (CVD) of modern society and the precursor physiological states which contribute to them appear to be absent in most traditional human populations. Certainly, congenital defects in the cardiovascular system were probably always present in human populations and heart defects were often common in adults as a result of childhood 'rheumatic' heart disease caused by bacterial infection. On the other hand, deaths during middle age from myocardial infarction and cerebrovascular accidents appear to have been exceedingly rare. It is difficult to know exactly how rare these events were since very few well kept records are available for such traditional groups.

The possibility of surveying traditional populations for risk factors which contribute to the probability of death from CVD provides the prime evidence for believing that CVD was quite rare. Aside from cigarette smoking, elevated blood pressures and elevated levels of blood serum cholesterol have been the two prime physiological characteristics identified as individual risk factors from studies of modern populations. High body weights relative to height have also been found significant as a risk factor in some studies. It appears the risks are elevated at resting blood pressures of 140 mm for systolic and/or 90 mm for diastolic pressure, and the risk probably increases in a logarithmic function at higher pressures. Risk standards in relation to circulating cholesterol are not as well defined but increased risk occurs above levels of 180 mg/dl. Relative body weight risks are not well defined either. Average adult values for these risk factors in a few traditional and modern populations are given in Table 24.3. Within a number of traditional populations there appears to be no increase in average blood pressures with age and, in some surveys on these groups, no individual has been found to have a blood pressure high enough to reach the risk category. A similar lack of high cholesterol levels and relative body weights occurs for many populations.

Based on present knowledge, it would probably not be correct to assume that CVD, which forms a prime stress on modern human populations, is entirely caused by some unique and new aspect of modern lifestyles, but it is clear that the lifestyle is causing a CVD epidemic of large proportions.

Why modern lifestyles cause all human populations to develop high rates of CVD remains unexplained. It has been suggested that the relatively low salt intake, high levels of physical activity, low body weights, and presumably low psychological stress levels found in many traditional societies might all contribute to low blood pressures. While such characteristics tend to be associated with many traditional societies, the correlation of these traits with individual blood pressure remain, with the exception of relative body weight, weak or contradictory. Even the positive correlations of blood pressure to relative body weight which have been found in most populations explain very little of the differences between the blood pressures of traditional and modern populations of the same biological and cultural origin.

The differences in serum cholesterol levels, as the differences in relative body weights, are generally attributed to dietary and activity level differences. The results of inter-population comparisons, as well as selected studies of individuals, suggests that the higher levels of serum cholesterol found in modern populations might be related to the amount of saturated fat ingested. While these comparisons of traditional and modern populations are suggestive, it is important to note that the studies of East African populations, whose milk-based diet contains large amounts of saturated fats, have reported very low serum cholesterol levels in the

TABLE 24.3

Average blood pressures and serum cholesterol values for 40–60-year-olds in selected human populations

Group	Systolic blood pressure (mmHg)		Diastolic blood pressure (mmHg)		Serum cholesterol (mg/dl)	
	Men	Women	Men	Women	Men	Women
Traditional						
Baegu (Solomon Islands)[1]	116	120	80	75	115	119
Kwaio (Solomon Islands)[1]	114	112	72	69	114	125
Puka Puka (Cook Islands)[2]	117	124	78	79	178	190
Samoan villages (W. Samoa)[3]	128	130	82	82	148	—
Quechua (Highland Peru)[4]	110	120	75	76	150	—
Modernized						
American Samoan[3]	138	140	95	90	192	188
Raritonga (Cook Islands)[2]	143	164	96	101	212	239
US whites[5]	135	131	88	83	228	240

1. Page *et al.* in Harrison (ed.) (1977).
2. Prior and Evans, *Israel J. Med. Sci.* (1969); Prior *et al.*, *Am. J. Clin. Nutr.* (1981).
3. Baker *et al.* (eds.) (1986).
4. Baker, *Science* (1969).
5. Robert & Maura, *Vital and health statistics DHEW* (1978).

native populations. The Polynesian groups on some of the isolated atolls also have diets containing high levels of saturated fat from coconuts, but show average cholesterol values below the risk levels.

In sum, the author of this section and most investigators of the topic have shared the experience of observing traditional populations with the frustrating feeling of not understanding why they would be unlikely to experience significant frequencies of CVD, while their migrant relatives to the city would be most likely to die from these diseases.

Type II diabetes mellitus

Non-insulin-dependent diabetes mellitus (NIDDM) or Type II diabetes mellitus also appears to be a disease which has become epidemic as a result of the modernization of life-styles. All of the studies on modern populations have found it to be a middle-aged onset disorder, which has a strong hereditary component. Its onset is also strongly correlated to high relative weight in middle age. It has been hypothesized that, in societies where food supplies were intermittent or where occasional food shortages produced starvation, the individual with large fat deposits would have had a significant energy buffer allowing improved individual survival and reproduction. Thus, the gene or genes responsible for the ability to gain weight rapidly in a young adult would have been evolutionarily selected for, even if they were likely to increase the possibility of a serious degenerative disease in late middle age.

While Type II diabetes occurs in all modernized populations, its frequency is significantly greater in some groups than others. At our current state of knowledge, it appears that the modernized groups which show unusually high rates of Type II diabetes include Micronesians, Polynesians, and American Indians. It has been suggested, therefore, that the voyaging of the Micronesian and Polynesian settlers would have greatly enhanced the survival possibilities for fatter individuals, while the original migrations of American Indians through the Arctic could have produced a similar advantage to fatter individuals. The principal evidence against such an evolutionary explanation is the lack of a close correlation between the relative weight within such populations and the presence or absence of Type II diabetes. Nevertheless, modern populations from these populations with high frequencies of the disease tend to also have the highest relative body weights.

Environment and the future

While the continuing stresses of disease form the major contemporary challenge to the adaptability of human populations, the impact of human behaviour on the physical and natural environment has become a strong concern in recent decades. It appears that unified cultural efforts may be able to control such stressors as those caused by localized atmospheric pollution, fresh water pollution, and even the multiple localized stressors caused by our tendency to concentrate potentially lethal chemical compounds in locations which lead to the destruction of life. Of more global concern is the rapid destruction of the genetic diversity represented by the millions of species of living organisms. While sometimes preyed upon by many of these species, they also represent the sources of our nutrients and the structure of our perceptual systems.

Perhaps even more significantly in the co-evolution systems which involve us with the biosphere, we are modifying the gases of the global atmosphere. From the limited data available it appears that the ozone layer is being depleted and the percentage of CO_2 in the total atmosphere is being increased. The direct effects of these changes may not be a significant stress on human populations. Light-skinned individuals can culturally adapt to the increased ultraviolet penetration caused by ozone destruction by more skin covering and less sun bathing. The decreased oxygen percentage of the atmosphere would be respiratorily unimportant for all but those at the highest elevations. However, the atmospheric changes also open the possibility of significant increases in worldwide temperature and changes in rainfall patterns. Such a change would stress not only our heat tolerance but also our ultimate survival need — food.

Suggestions for further reading (Part IV)

General human ecology and human adaptability

BAKER, P. T. and WEINER, J. S. (eds). (1966). *The biology of human adaptability*. Clarendon Press, Oxford.

BAYLISS-SMITH, T. P. (1982). *The ecology of agricultural systems*. Cambridge University Press.

DYSON-HUDSON, R. and LITTLE, M. A. (eds). (1983). *Rethinking human adaptation: biological and cultural models*. Westview Press, Boulder, Colorado.

FRISANCHO, A. R. (1981). *Human adaptation*. The University of Michigan Press, Ann Arbor.

HARDESTY, D. L. (1977). *Ecological anthropology*. Wiley, New York.

LITTLE, M. and MORREN, G. E. G., JR. (1976). *Ecology, energetics and human variability*. Brown, Dubuque, Iowa.

MORAN, E. F. (1979). *Human adaptability: an introduction to ecological anthropology*. Wadsworth, Belmont, California.

—— (ed). (1984). *The ecosystem concept in anthropology*. Westview Press, Boulder, Colorado.

NETTING, R. M. (1977). *Cultural ecology*. Cummings, Menlo Park, California.

SARGENT, F., II (ed.) (1974). *Human ecology.* North Holland Publishing, Amsterdam.

WINTERHALDER, B. (1980). Environmental analysis in human evolution and adaptation research. *Human Ecol.* **8**, 135–70.

—— and SMITH, E. A. (eds). (1981). *Hunter-gatherer foraging strategies: ethnographic and archeological analyses.* The University of Chicago Press.

Research design and methods

BAKER, P. T. (1976). Research strategies in population biology and environment stress. In *The measures of man* (eds E. Giles and J. S. Friedlaender). Peabody Museum Press, Cambridge, Massachusetts.

—— (ed.). (1977). *Human population problems in the biosphere: some research strategies and designs. MAB Technical Notes 3.* UNESCO, Paris, France.

BEALL, C. M. (1982). An historical perspective on studies of human growth and development in extreme environments. In *A history of American physical anthropology* (ed. F. Spencer). Academic Press, New York.

HAAS, J. D. (1982). The development of research strategies for studies of biological variation in living human populations. In *A history of American physical anthropology* (ed. F. Spencer). Academic Press, New York.

HARRISON, G. A. (1966). Human adaptability with reference to the IBP proposals for high altitude research. In *The biology of human adaptability* (eds P. T. Baker and J. S. Weiner). Clarendon Press, Oxford.

LITTLE, M. A. (1982). The development of ideas about human ecology and adaptation. In *A history of American physical anthropology* (ed. F. Spencer). Academic Press, New York.

WEINER, J. S. and LOURIE, J. A. (EDS). (1981). *Practical human biology.* Academic Press, London.

Temperature regulation

HANNA, J. M. and BROWN, D. A. (1979). Human heat tolerance: biological and cultural adaptations. *Yb. Phys. Anthropol.* **22**, 164–81.

LEBLANC, J. (1975). *Man in the cold.* Thomas, Springfield, Illinois.

NEWMAN, R. W. (1975) Human adaptation to heat. In *Physiological anthropology* (ed. A. Damon). Oxford University Press, New York.

ROBERTS, D. F. (1978). *Climate and human variability.* Cumming, Menlo Park, California.

STEEGMAN, A. T. (1975). Human adaptation to cold. In *Physiological anthropology* (ed. A. Damon). Oxford University Press, New York.

Altitude

BAKER, P. T. (1978). *The biology of high-altitude peoples.* Cambridge University Press.

FRISANCHO, A. R. (1975). Functional adaptation to high hypoxia. *Science,* **187**, 313–9.

HEATH, D. and WILLIAMS, D. R. (1977). *Man at high altitude.* Longman, Edinburgh.

Human energetics

HARRISON, G. A. (ed.). (1982). *Energy and effort.* Taylor & Francis, London.

ODUM, H. T. (1971). *Environment, power and society.* Wiley-Interscience, New York.

SHEPHARD, R. J. (1978). *Human physiological work capacity.* Cambridge University Press.

Culture and nutrition

KATZ, S. H. HEDIGER, M. L., and VALLEROY, L. A. (1974). Traditional maize processing techniques in the New World. *Science.* **184**, 765–73.

LEE, R. B. (1979). *The !Kung San.* Cambridge University Press.

NEWMAN, M. T. (1975). Nutritional adaptation in man. In *Physiological anthropology* (ed. A. Damon). Oxford University Press, New York.

SIMOONS, F. J. (1969). Primary adult lactose intolerance and the milking habit: a problem in biological and cultural interrelations. Part I. *Am. J. Digest. Dis.* **14**, 819–36.

—— (1970) Primary adult lactose intolerance and the milking habit: a problem in biologic and cultural interrelations. Part II. *Am. J. Digest. Dis.* **15**, 695–710.

Infectious disease

ANDERSON, R. M. and MAY, R. M. (eds). (1982). *Population biology of infectious diseases.* Springer-Verlag, Berlin.

COCKBURN, T. A. (1971). Infectious diseases in ancient populations. *Curr. Anthropol.* **12**, 45–54.

DESOWITZ, R. S. (1981). *New Guinea tapeworms and Jewish grandmothers.* Norton, New York.

GAJDUSEK, D. C. (1977). Unconventional viruses and the origin and disappearance of Kuru. *Science,* **197**, 943–60.

MCNEILL, W. H. (1976). *Plagues and peoples.* Doubleday, Garden City, New York.

Modernization

BAKER, P. T., HANNA, J. M. and BAKER, T. S. (eds). (1986). *The changing Samoans: behaviour and health in transition.* Oxford University Press, New York.

BOYDEN, S. V. (ed.) (1970). *The impact of civilisation on the biology of man.* University of Toronto Press.

COYNE, T. BADCOCK, J., and TAYLOR, R. (eds). (1984). *The effect of urbanisation and Western diet on the health of Pacific Island populations.* Technical Paper No. 186. South Pacific Commission, Noumea, New Caledonia.

HARRISON, G. A. and GIBSON, J. B. (eds). (1976). *Man in urban environments.* Oxford University Press, Oxford.

HIERNAUX, J. (1982). *Man in the heat, high altitude and society,* Chapter 5. Thomas, Springfield, Illinois.

WOLANSKI, N. and SZEMIK, M. (eds). (1984). *Industrialization impact on man.* Polish Scientific Polish Scientific Publishers, Warsaw.

Population adaptability studies

BAKER, P. T. and LITTLE, M. A. (eds). (1976). *Man in the Andes: a multidisciplinary study of high altitude Quecha.* Dowden, Hutchinson & Ross, Stroudsburg, Pennsylvania.

BAYLISS-SMITH, T. P. and FEACHEM, R. G. (eds) (1977). *Subsistence and survival.* Academic Press, London.

HARRIS, D. R. (ed.) (1980). *Human ecology in savanna environments.* Academic Press, London.

HARRISON, G. A. (ED.) (1977). *Population structure and human variation.* Cambridge University Press.

JAMISON, P. L., ZEGURA, S., and MILAN, F. A. (eds). (1978). *Eskimos of Northwestern Alaska.* Dowden, Hutchinson, & Ross, Stroudsburg, Pennyslvania.

LEE, R. B. and DEVORE, I. (eds). (1976). *Kalahari huntergatherers.* Harvard University Press, Cambridge, Massachusetts.

MILAN, F. A. (ed.). (1980). *The human biology of circumpolar populations.* Cambridge University Press.

STEEGMAN, A. T. (ed.). (1983). *Boreal forest adaptations.* Plenum, New York.

UNESCO. (1978). *Tropical forest ecosystems,* chapters 15–18. UNESCO, Paris, France.

—— (1979). *Tropical grazing land ecosystems,* chapter 7. UNESCO, Paris, France.

Author index

Subject index